Semiconductor Processing and Characterization with Lasers
- Applications in Photovoltaics -

Semiconductor Processing and Characterization with Lasers

- Applications in Photovoltaics -

Proceedings of the First International Symposium
Stuttgart, Germany, April 18-20, 1994

Editors:

M. Brieger
H. Dittrich
M. Klose
H.W. Schock
J. Werner

TRANS TECH PUBLICATIONS
Switzerland - Germany - UK - USA

Seplae
Chem.

Volumes 173-174 of
Materials Science Forum
ISSN 0255-5476

Distributed *in the Americas by*

Trans Tech Publications Ltd
LPS Distribution Center
52 LaBombard Rd. North
Lebanon, NH 03766
USA

Phone: (603) 448 0037
Fax: (603) 448 2576

and worldwide by

Trans Tech Publications Ltd
Hardstrasse 13
CH-4714 Aedermannsdorf
Switzerland

Fax: (++41) 62 74 10 58
E-Mail: ttp@transtech.ch

*Ser. Div
12/12/ 95
MZ*

Preface

This book contains the papers presented at the *First International Symposium on Semiconductor Processing and Characterization with Lasers - Applications in Photovoltaics*. The symposium was held at the Max-Planck-Institut in Stuttgart, Germany, from April 18 to 20, 1994, following the 12th European Photovoltaic Solar Energy Conference in Amsterdam, and was jointly organized by:

- DLR, Institut für Technische Physik
- Universität Stuttgart, Institut für Physikalische Elektronik und Institut für Strahlwerkzeuge
- Max-Planck-Institut für Festkörperforschung

There were a number of incentives for holding this conference. Firstly, lasers are playing an increasingly important role in various fields of semiconductor and device technology. Secondly, they contribute to the advanced technologies that are needed for economic solutions in photovoltaics. Thirdly, lasers are used in processing and characterization of photovoltaic materials, solar cells and module technology.

The specific aim of this symposium was to bring together specialists in laser applications, semiconductor processing, diagnostics and photovoltaics.

In order to realize this conception, eight scientists recognized in their field were invited to give reviews of their respective work. These expert representations were complemented by 25 oral and 26 poster contributions.

The combination of these different types of presentations led to very lively and stimulating interdisciplinary discussions, supported by an atmosphere of informal scientific exchange at the Max-Planck-Institut. Visiting tours to laboratories of the organizing institutes and an exhibition by 19 sponsoring firms completed the conference program.

The written versions of the invited papers, oral and poster contributions contained in this volume cover a wide range of laser applications, including deposition and film treatment, surface treatment and structuring, patterning and interconnecting, diagnostics of stuctural, optical and electronic semiconductor properties and mapping.

We would like to thank all conference delegates from 19 countries for their active participation, and particularly all invited speakers for their excellent presentations.

In addition, we are very grateful for the financial support provided by the Deutsche Forschungsgemeinschaft, Bonn, and the Wirtschaftsministerium Baden-Württemberg, Stuttgart.

Further, we gratefully acknowlege the generous help provided by all staff members who supported us during the conference.

Stuttgart, July 1994

M. Brieger
H. Dittrich
M. Klose
H. W. Schock
J. Werner

Organizing Committee

W. Bloss[1]

M. Brieger[2]

H. Dittrich[5]

H. Hügel[3]

M. Klose[2]

G. Krutina[2]

H. Opower[2]

H. W. Schock[1]

J. Werner[4]

[1] Institut für Physikalische Elektronik, Universität Stuttgart

[2] Institut für Technische Physik, Deutsche Forschungsanstalt für Luft- und Raumfahrt e.V.
Stuttgart

[3] Institut für Strahlwerkzeuge, Universität Stuttgart

[4] Max-Planck-Institut für Festkörperforschung Stuttgart

[5] Zentrum für Sonnenenergie- und Wasserstoff-Forschung Stuttgart

Exhibiting and Sponsoring Firms

B.E.S.T

Coherent GmbH

Dilor GmbH

Goodfellow GmbH

High-Speed Photo-Systems

Instruments S.A., Div. Riber

Messer Griesheim

Micos

Microplan

Owis

Pink

Profile

Radiant Dye Laser Accessories GmbH

Soliton

Spectra Physics GmbH

Spindler & Hoyer GmbH u. Co

Steeg & Reuter

VAB

Wittwer

TABLE OF CONTENTS

LASER INDUCED PROCESSES

LASER PATTERNING AND FABRICATION TECHNOLOGIES

SOLAR CELLS

SOLAR CELLS

Materials Science Forum Vols. 173-174 (1995) pp. 1-6
© 1995 Trans Tech Publications, Switzerland

FEMTOSECOND TIME-RESOLVED MICROSCOPY OF LASER-INDUCED MORPHOLOGY CHANGES

D. von der Linde

Universität Essen, Fachbereich Physik, Institut für Laser- und Plasmaphysik,
D-45177 Essen, Germany

Keywords: Femtosecond Laser Pulses, Laser-Induced Morphology

Abstract: The treatment of semiconductor surfaces by intense femtosecond laser pulses is studied by means of a variant of optical microscopy in which a femtosecond optical probe pulse is used as illumination. Picture frames corresponding to an exposure time of 10^{-13} s are recorded over a time span of 75 ns following exposure of the semiconductor material to an intense femtosecond excitation pulse.

Introduction

During the last decade impressive progress has been made in the generation and application of laser pulses with durations ranging down to about 10^{-14} s. These ultrashort laser pulses are nowadays used in many different scientific and technical areas to perform optical measurements with a time resolution corresponding roughly to the pulse duration. In many applications time resolved measurements are carried out by using a simple pump-probe scheme in which two laser pulses are used: a first laser pulse to excite the phenomenon under study, and a second, delayed pulse to interrogate the evolution of the process in time. This basic scheme can be readily combined with optical microscopy to provide both temporal *and* spatial resolution. With the electronic techniques for image recording and processing a powerful tool is available today for measuring the temporal and spatial evolution of many processes of scientific or technical interest on a spatial scale of about 1 µm and a temporal scale of several femtoseconds. Application of this technique to study femtosecond laser-excited semiconductor surfaces has been pioneered by Downer et al. [1]. Here we demonstrate recording of the complete sequence of events from the initial electronic excitation all the way to the final changes of the surface morphology following intense femtosecond laser surfaces excitation of semiconductors.

Experimental

The laser pulses used in this work were obtained from an amplified dye laser system operating at 620 nm and putting out pulses of approximately 100 fs duration and a total energy of about 1 mJ. As shown in Fig. 1 the pump beam was incident on the sample surface at an angle of approximately 45 degrees (polarization in the plane of incidence). The probe was produced by splitting off a small fraction from the pump pulse. A microscope objective lens was used to produce an optical image of the sample surface on a CCD camera. The surface area under observation was illuminated by feeding the probe pulse through the microscope. To suppress scattered light from the strong pump pulse the polarization of the probe was chosen to be normal to the plane of incidence, and an optical polarizer was placed in front of the CCD camera to block the pump light.

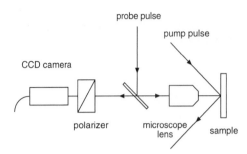

Fig. 1:   Schematic of the experimental setup

The time delay between pump and probe could be accurately controlled by means of an optical delay line (not shown). By varying the optical delay between the pump pulse and the probe pulse snap shot pictures of the surface with an exposure time of about 100 fs could be obtained over a time span of 75 ns (limited by the length of the optical delay line).

The CCD camera was controlled by a properly synchronized frame grabber to insure that for each laser pulse an optical micrograph of the surface could be caught and stored in a computer for further processing. The spatial Fourier transforms of the pictures were analysed to make sure that all the details of the laser-induced changes of the surface morphology were well resolved.

The samples used in our experiments were nominally undoped standard wafers of (111) silicon and (100) gallium arsenide which were raster scanned using a x-y-translation stage to provide a fresh sample surface for each laser pulse.

Femtosecond snap shot pictures

Figure 2 shows a series of pictures of (111) silicon for a pump pulse energy of 0.9 J/cm^2 which is five times greater than the threshold energy for melting. With the exception of the last frame the pictures shown in Fig. 2 represent a surface area of 300 μm × 220 μm.

Zero delay time is defined by the time when the leading edge of the pump pulse (oblique incidence from the left) meets the maximum of the probe pulse (normal incidence). The instantaneous pump area on the surface -determined by the longitudinal spatial extension of the wave packet- has a width of approximately 50 μm which sweeps across the surface from left to right.

At zero delay just a rather faint change of the surface brightness can be recognized in the left part of the first frame. At 300 fs a distinct bright area indicating a strong increase of the optical reflectivity can be seen which signifies the generation of a supercritical [2] electron-hole plasma. Detailed examination of the reflectivity pattern shows that the high reflectivity area is surrounded by a narrow zone of decreased optical reflectivity at the left edge, signifying areas of subcritical electron-hole density. This effect is due to the spatial distribution of the laser energy across the Gaussian beam profile of the pump pulse, which produces corresponding spatial variation of the electron-hole density: a density maximum in the center of the beam and lower (subcritical) density at the edges.

At a delay time of 1.0 ps a bright spot of uniform optical reflectivity has developed. Quantitative measurements indicate that the optical constants over the bright area correspond to the values of liquid silicon, suggesting that a molten surface is observed. This ultrafast solid-liquid transition is physically distinct from the very much slower normal thermal melting, which is observed when lower laser energy [3] or longer laser pulses are used [4]. Ultrafast solid-liquid transformations which were first observed by Shank et al. [5] in silicon are also seen in gallium arsenide and other semiconductors. Although the physics of these transitions has not yet been fully clarified, there is evidence that the very strong non-equilibrium electronic excitation caused by the intense laser pulse plays a role.

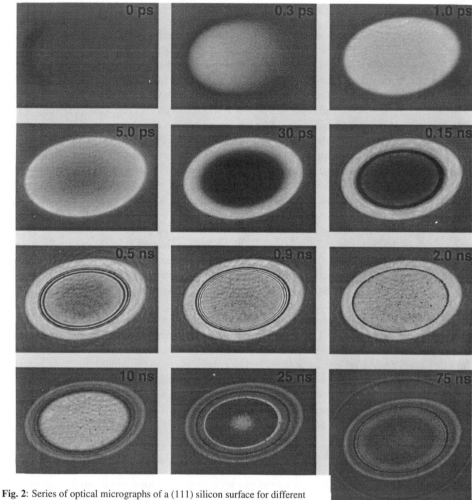

Fig. 2: Series of optical micrographs of a (111) silicon surface for different
delay times between pump and probe. Excitation energy : 0.9 J/cm²;
Frame dimensions: 300×220 μm²;

Let us now shift gears and move in larger time increments. An interesting new phenomenon begins to show up around 5 ps as an apparent decrease of the reflectivity. At 30 ps we observe some dark area which covers a large part of the former high reflectivity region. As more time elapses some black ring develops at the circumference of the dark area, while at the same time the brightness in the inside increases again (150 ps). More detailed measurements showed that a well defined threshold for this phenomenon exists and that the diameter of the dark ring increases with the energy of the laser pulse.

What is observed here is the onset of ablation of surface material, but the physical nature of the ablation process is not yet well understood. Formation of a cloud of droplets has been invoked to explain the dark area [1]. The possibility that this observation can be explained by vaporization of surface material and/or formation and expansion of a plasma has not yet been quantitatively ana-lysed. If ablation is occurring there should be a corresponding change of the surface morphology such as a crater or some depression. As will be shown below, this is not the case in silicon. On the other hand, crater formation correlating with the appearance of the dark cloud has been observed in gallium arsenide.

After 500 ps the dark cloud has almost completely disappeared unveiling again the molten surface. However, the dark ring that marked the circumference of the dark area is left behind, now accompanied by some wavelike structure in its neighborhood. The impression of waves running towards the center of the molten area is even more compelling from the picture recorded at 900 ps. We have considered the possibility of capillary waves using the known values of density, surface tension and viscosity of liquid silicon. Estimates based on the simple dispersion relation of capillary waves lead to oscillation frequencies which seem to be too low to explain our observations. In any case the wavelike structure is observed only during a relatively narrow time interval between 0.5 ns and 2 ns (for 0.9 J/cm^2; the time depends on the laser fluence). As can be seen from the picture taken at 2 ns, only the sharp dark ring is left at this time.

Even more complicated structures appear during resolidification, which starts about 10 ns after excitation for the laser fluence used in this particular measurement. Roughly speaking resolidification proceeds from the periphery of the molten area towards the center. The picture corresponding to 10 ns shows a stage where the outside of the dark ring has solidified, as can be recognized from the low reflectivity, while the inside is still liquid and highly reflecting. Note that at the periphery there is some annulus with a reflectivity slightly higher than the background reflectivity of the unexposed crystalline material. This higher reflectivity is an indication of amorphization.

At 25 ns just a small bright area is left in the center of the laser-excited surface area, indicating that the resolidification is almost complete. The narrow bright ring could be due to delayed solidification or renewed local melting caused by the release of latent heat.

The last picture of this series (75 ns) shows three distinct main features: (i) two amorphous annular areas as discussed above, (ii) the dark ring which will be examined more closely in the following, and (iii) a new feature which can be identified as a shock wave in air propagating away from the laser-excited surface area. It has been verified that this feature disappears when the experiment is repeated under vacuum. In addition, the speed of the shock wave has been measured by comparing the propagation distances from a series of different time frames.

Changes of the surface morphology

The permanent changes of the surface morphology produced by exposure to a single femtosecond laser pulse have been examined by means of optical microscopy and electron microscopy, and also by scanning a mechanical stylus across the laser treated area. Examples of the measured surface profiles for silicon (solid line) and GaAs (dotted line) are shown in Fig. 3, where Δz is the measured change in the direction of the surface normal, and x is the coordinate parallel to the surface. The spatial resolution in the perpendicular direction was 5 nm.

Somewhat surprising, treatment with intense femtosecond laser pulse does not produce craters on the silicon surface. The measured Δz is always positive, indicating some piling up of material over the unperturbed surface level. The peak in the center of the surface profile corresponds to an elevation of $\Delta z = 30$ nm. There are two secondary maxima near $x = \pm 50$ μm which can be readily identified with the dark ring feature discussed above. The relative height of the central peak and the peaks due to the ring vary with laser energy.

Comparison with the dashed curves shows that the situation is quite different for GaAs, where a distinct crater with steep walls is observed. The depth in the center corresponds to 55 nm. The crater is surrounded by a thin wall of about 20 nm in height. This structure appears in the corresponding optical micrographs of GaAs (not shown here) as a sharp dark ring, just like the corresponding feature in silicon.

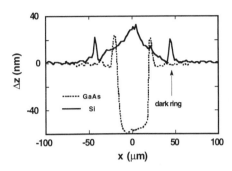

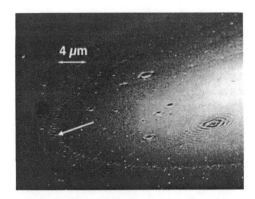

Fig. 3: Examples of surface profiles across the laser-excited spot. The laser fluence was 0.65 J/cm² and 0.6 J/cm² for silicon and GaAs, respectively. The spot diameter for GaAs was about a factor of two smaller.

Fig. 4: Electron micrograph of the surface after exposure to a femtosecond laser pulse. The white arrow marks the feature corresponding to the dark annulus in the optical pictures.

Figure 4 depicts an electron micrograph of a laser-exposed silicon surface. Shown is the section in which the sharp dark ring is seen in the optical pictures. We find that the dark ring consists of blister-like structures having a typical size of a few hundred nm.

Conclusions

Ultrafast time-resolved optical spectroscopy combined with optical microscopy reveals a detailed picture of the physical processes following femtosecond laser excitation of semiconductor surfaces. The first few hundred femtoseconds are dominated by the dynamics of the electron-hole plasma produced by the laser pulse. For laser fluences exceeding about 0.5 J/cm² a surface layer of liquid silicon develops in less than 1 ps. Ablation of surface material is observed on the picosecond time scale. Changes of the surface morphology begin to form in the liquid phase around 1 ns. Resolidification accompanied by the formation of amorphous material at the periphery of the molten area occurs after about 10 ns. Most of the examples discussed here were results for silicon, but with a few specific exceptions the behavior GaAs is very much the same as silicon.

Acknowledgement

The author is indebted to K. Sokolowski-Tinten and J. Bialkowski for their contributions to this work.

References

[1]   M. C. Downer, R. L. Fork, and C. V. Shank, J. Opt. Soc. Am. **B 2**, 595 (1985)

[2]   At 620 nm the critical density in silicon corresponds to a few times 10^{21} cm⁻³

[3]   K. Sokolowski-Tinten, H. Schulz, J. Bialkowski, and D. von der Linde, Appl. Phys. **A 53**, 4564 (1991)

[4]   D. von der Linde, in *Resonances,* ed. by M. D. Levenson, E. Mazur, P. S. Pershan, and Y. R. Shen (World Scientific, Singapore, 1990)

[5]   C. V. Shank, R. Yen, and C. Hirlimann, Phys. Rev. Lett. **50**, 454 (1983)

Materials Science Forum Vols. 173-174 (1995) pp. 7-16

LOW TEMPERATURE DEPOSITION AND CRYSTALLIZATION OF SILICON BY MEANS OF LASER TECHNIQUES

M. Hirose

Department of Electrical Engineering, Hiroshima University,
Higashi-Hiroshima 724, Japan

Keywords: Laser CVD, Polycrystalline Silicon, Layer-by-Layer Growth, Polysilane, Laser Induced Crystallization

Abstract. Saturated adsorption of Si_2H_6 molecules on a cooled substrate held at about -50°C and subsequent irradiation of 193nm laser light in the fluence range 20~50mJ•cm^{-2}/pulse result in atomic layer growth of polycrystalline silicon. It is shown that the photochemical decomposition rate of the first monolayer of Si_2H_6 adsorbed on the growing Si surface is extremely high compared to that of the second or third Si_2H_6 layer. This explains the self-limiting layer-by-layer growth of silicon and the formation of ordered structure at such a low temperature and a low laser fluence. The as-grown film contains about 5% bonded hydrogen and has an average grain size of 260~340Å as revealed by Raman scattering. The other approach to grow polycrystalline silicon is laser induced crystallization of a polysilane film deposited at -110°C from a silane plasma. This film consists of $(SiH_2)_n$ chains terminated with SiH_3. The structure is amorphous and UV light sensitive. 193nm laser irradiation of polysilane at a fluence of 10~20mJ•cm^{-2}/pulse causes desorption of bonded hydrogen in the film, resulting in the formation of a three dimensional network and crystallization. The mechanism is presumably that the photoexcited metastable polysilane with sufficient Si dangling bonds can easily overcome the activation barrier to crystallization. The film is composed of 300Å grains with some amorphous tissue in the matrix. Since the nonirradiated region of polysilane is easily oxidized in air, a fine polysilicon pattern is generated by the laser irradiation onto polysilane through a mask and subsequent air exposure followed by dilute HF etching.

Introduction

In atomic layer epitaxy (ALE) or molecular layer epitaxy, a self-limiting process on the reacting surface controls the growth rate[1,2]. For instance, the strong chemical bonding between group VI (or V) surface atoms and adsorbed group II (or III) organic radicals tends to suppress the multilayer adsorption. Also, Ge ALE has been carried out by using thermal reaction of Ge $(C_2H_5)_2H_2$ on the Ge surface based on a similar mechanism to that of compound semiconductor ALE[3]. A polycrystalline silicon layer has been grown by the excimer laser induced photochemical decomposition of Si_2H_6 at cryogenic temperatures[4]. The silicon atomic layer growth by laser-induced, cryogenic chemical-vapor-deposition (CVD) has also been demonstrated by controlling the adsorption kinetics of Si_2H_6 on a cooled substrate[5]. Highly hydrogenated silicon films such as polysilane $(SiH_2)_n$ groups are found to be sensitive to oxygen exposure or electron beam (EB) irradiation [6]. It has been shown that polysilane is changed to hydrogenated amorphous silicon by EB irradiation as a result of the hydrogen desorption followed by the Si network formation and that a $0.1\mu m$ Si pattern is generated by EB direct writing on the polysilane film [7]. Considering a polysilane bandgap of ~3.1 eV and a Si-H bond energy of 70.4 kcal/mol (3.05 eV), the irradiation of an ArF excimer laser beam with an energy of 6.4 eV could cause the desorption of bonded hydrogen from polysilane and simultaneous three-dimensional silicon network formation and crystallization [8]. This paper reviews some of interesting aspects of the excimer-laser-induced atomic layer growth of polycrystalline Si (poly-Si) at cryogenic temperatures and the formation of poly-Si or microcrystalline Si at room temperature from polysilane grown at cryogenic temperatures.

Atomic Laser Growth of Silicon

Adsorption kinetics of gas molecules on a solid surface depends on the partial pressure of the source gas and the substrate temperature. The Si_2H_6 condensation onto a cooled quartz substrate has been studied by measuring the silicon growth rate as a function of the laser pulse-to-pulse interval by changing the repetition rate of the ArF excimer laser as shown in Fig. 1. It is evident that the Si

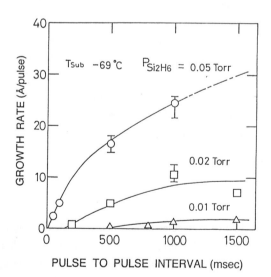

Fig. 1   Silicon growth rate on quartz substrates plotted against pulse-to-pulse interval of laser irradiation at Si_2H_6 partial pressures P_{Si2H6} of 0.01, 0.02 and 0.05 Torr and a substrate temperature Tsub of -69°C. Si_2H_6 (1% in He) gas flow rate was 12 sccm.

deposition occurs after the incubation time τ_i which increases with decreasing the Si_2H_6 partial pressure. No thin-film growth takes place at a laser repetition rate greater than $1/\tau_i$, presumably because two-dimensional silicon nuclei created by the laser-induced dissociation of adsorbed Si_2H_6 on a quartz substrate do not exceed a critical radius when the Si_2H_6 surface coverage is too low. The Gibbs free energy as a function of silicon nucleus radius takes its maximum at the critical radius above which the nucleus is stable and starts to grow. The growth rate per laser pulse saturates with increasing the pulse-to-pulse interval simply because the surface coverage of Si_2H_6 is determined by the Si_2H_6 partial pressure and substrate temperature. Therefore, the monolayer growth can be achieved by controlling the adsorbed layer thickness as shown in Fig. 2, where silicon monolayer thickness of 1.36Å for Si(100) and 1.57Å for Si(111) are indicated. Note that the actual layer thicknesses for (111) are 2.36 and 0.78Å whose average is 1.57Å. The grown film is polycrystalline as confirmed by Raman scattering and X-ray diffraction and hence the atomic layer thickness corresponds to a value between 1.36 and 1.57Å. The temperature dependence of conductivity of an as-grown film with a grain size of ~300Å exhibits an activation energy of 0.55eV which corresponds to one half of the Si bandgap.

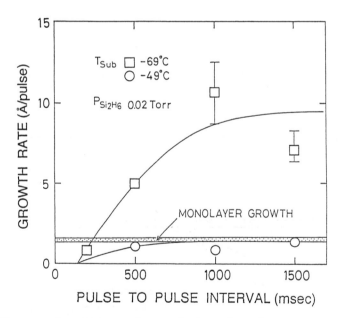

Fig. 2   Silicon growth rate versus pulse-to-pulse interval at substate temperatures of -49 and -69°C and a Si_2H_6 gas flow rate of 12 sccm.

In order to reveal the existence of the self-limiting process in the atomic layer growth, the temperature dependence of the growth rate was measured as shown in Fig. 3. The growth rate per laser pulse shows the plateau region where the atomic layer growth proceeds even at different laser fluences. The growth rate is almost independent of temperature in a plateau region.

The Si_2H_6 coverage is calculated as a function of substrate temperature. A rate equation for the adsorbed Si_2H_6 concentration N under the incident Si_2H_6 flux of G is given by:

$$\frac{dN}{dt} = -\frac{N}{\tau} + cG .$$

(1)

Here, τ is the residence time of Si_2H_6 molecules adsorbed on the surface, being given by:

$$\tau = \tau_0 \exp\left(\frac{E}{kT}\right) ,$$

(2)

where the value of E depends on the Si_2H_6 coverage θ.
In the case of the Si_2H_6 coverage $\theta \leq 1$, θ is described as:

$$\theta = \frac{N}{N_0} = \frac{cG\tau_0}{N_0} \exp\left[\frac{E_a - \theta(E_a - E_b)}{kT}\right] .$$

(3)

Here, E_a is the heat of adsorption of a single Si_2H_6 molecule and E_b is the heat of saturated adsorption.

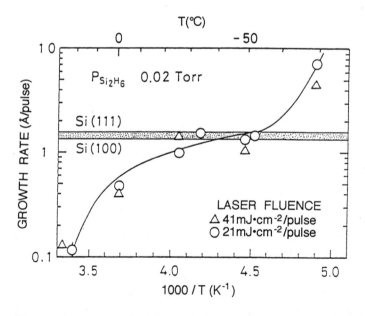

Fig. 3   Silicon growth rate as a function of reciprocal temperature at a repetition rate of 0.66 Hz and a Si_2H_6 gas flow rate of 12 sccm. The solid curve is the calculated growth rate for laser fluence of 21mJ•cm⁻²/pulse.

For Si_2H_6 coverage $\theta > 1$, one obtains:

$$\theta = \frac{cG\tau_0}{N_0} \exp\left[\frac{E_b}{kT}\right] . \tag{4}$$

By adopting $E_a = 0.5eV$ and $E_b = 0.35eV$, the surface coverage is obtained as the solid curve in Fig. 3, where the coverage in the plateau region corresponds to about one monolayer. To explain the measured growth rate, it must be assumed that the first monolayer of Si_2H_6 in contact with the underlying silicon surface is fully decomposed by a single laser pulse. This implies that the photodecomposition rate of the first monolayer Si_2H_6 on silicon surface is significantly enhanced with respect to that estimated from the optical absorption coefficient of an isolated Si_2H_6 molecule at 193 nm which is about 45.5 atom^{-1}•cm^{-1}. The enhancement factor of photodissociation rate for the first monolayer of Si_2H_6 is estimated to be about 40 times larger than isolated Si_2H_6 molecules at a laser fluence of 20 mJ•cm^{-2}/pulse.

Laser Induced Crystallization of Polysilane

Polysilane films were deposited on quartz and c-Si(100) substrates cooled at -110°C by the rf glow discharge decomposition of 3% SiH_4 diluted with H_2 in a capacitively coupled reactor[9]. The total gas pressure and the rf power density were maintained at 0.4 Torr and 0.26 W/cm^2, respectively. A pulsed ArF excimer laser beam (193nm) was used to irradiate the polysilane film with a fluence of 10 or 20 mJ•cm^{-2}/pulse at a repetition rate of 10 Hz in a vacuum camber.

The film deposited at -110°C from the silane plasma is mainly composed of $(SiH_2)_n$ chains terminated with SiH_3 as shown in the infrared absorption spectra (Fig. 4). Obviously, the hydrogen desorption from $(SiH_2)_n$ chains as well as SiH_3 molecular units is promoted by ArF excimer laser irradiation as seen in the figure. This desorption feature is apparently similar to the case of EB irradiation of polysilane [7]. The hydrogen content estimated from the measured integrated absorption intensity of the stretching modes is rapidly decreased down to about 60% of the initial value by the first 200 pulses of the laser irradiation with a fluence per pulse of 20mJ•cm^{-2}, being gradually reduced to the minimum value. Si dangling bonds produced by hydrogen desorption and the optical excitation of the polysilane matrix will change the Si-Si bond configuration and promote three-dimensional Si network formation. The minimum hydrogen content in the laser-irradiated film depends on the film thickness because the penetration depth of 193 nm light in polysilane is limited in the range of tenths of a micron. The significant increase of laser fluence slightly reduces the remaining hydrogen content in the film as indicated in Fig. 5. This implies that the penetration depth of laser beam is smaller than 0.3μm. The hydrogen evolution by laser irradiation was also confirmed by the in situ mass spectrometry. It is interesting to note that when as-deposited polysilane is irradiated by laser pulses at a fluence of 20mJ•cm^{-2}, silicon monomers as well as hydrogen are emitted

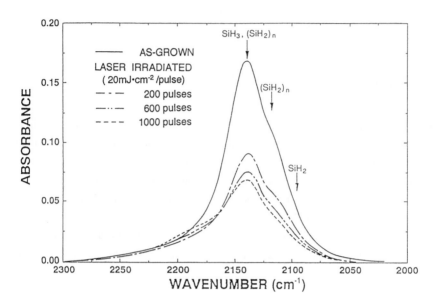

Fig. 4. Infrared absorption spectra due to the stretching modes of SiH_x (x=2, 3) and $(SiH_2)_n$ chains for a 0.3 μm thick polysilane film before and after laser irradiation. The fluence of ArF excimer laser pulse is 20 mJ•cm^{-2}.

from the film surface, while silicon dimer signals can not be detected with a sensitivity of 1% of monomer signal intensities. The mass spectrum of the silicon monomers is very similar to the cracking pattern of SiH_4 except for the significantly more intense signals of Si^+ and SiH^+. This suggests that SiH_3 molecular units which terminates the $(SiH_2)_n$ chains react with neighboring bonded-hydrogen to form SiH_4 molecules. Also, the stronger yields of Si^+ and SiH^+ signals imply the evolution of SiH_2 and SiH_3 molecular units due to the Si-Si bond-breaking in polysilane.

Three-dimensional Si network formation by ArF excimer laser irradiation is directly examined by Raman spectra measured for as-deposited and laser-irradiated films as shown in Fig. 6. In an as-deposited film, the observed peak at ~430cm^{-1} is attributable to the LO phonon mode of the Si skeleton in long-chain polysilane[10]. After 2000 pulses of laser irradiation at a fluence of 20mJ•cm^{-2}/pulse, the sharp, intense peak is clearly observed at ~518cm^{-1} arising from the TO mode of crystallized Si together with a broad yield around at ~480cm^{-1} that is assigned as the TO phonon mode in amorphous Si network. In contrast to this, the Raman spectrum for the EB irradiated film consists of only broad phonon bands at ~480cm^{-1} originating from the amorphous Si matrix. This is attributed to the difference in the energy loss mechanism between electron beam (25 keV, 50 μA/cm^2) and UV photons (193 nm) in polysilane. Namely, UV photons are more effectively absorbed in the surface region of the film than the electron beam. The optical excitation of Si-Si and Si-H bonds in the metastable polysilane promotes the Si dangling bond formation and change in Si atom configuration, leading to

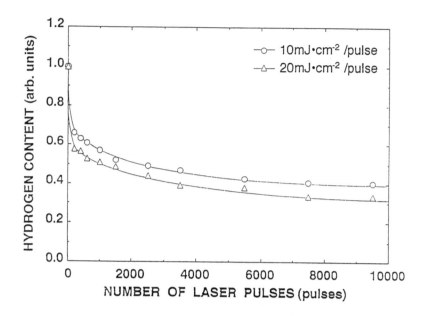

Fig. 5. The integrated absorption intensity of the hydrogen bond-stretching modes plotted as a function of excimer laser irradiation time for a 0.3 μm thick film with a laser fluences of 10 and 20 mJ•cm^{-2}/pulse.

the network crystallization. By taking into account the facts that a TO phonon peak at ~520 cm^{-1} for bulk crystalline Si has a full width at half maximum (FWHM) of 8.3 cm^{-1} and the observed FWHM of polycrystallized Si is ~9 cm^{-1}, the FWHM difference gives an average grain size of ~300 Å. Furthermore, the 2 cm^{-1} red shift of the TO peak for the laser crystallized film suggests that a tensile stress of ~5X10^9 dyne/cm^2 exists in the crystallized region [11].

The chemical stability of the film is dramatically improved by laser irradiation because of the three-dimensional Si network formation promoted by laser-induced hydrogen desorption and crystallization. Consequently, the laser irradiated film immersed in a dilute (0.16%) HF solution exhibits a much slower etch rate than the as-deposited film (Fig. 7). A fine polycrystalline Si pattern can be generated by laser irradiation to polysilane through a Cr mask and subsequent exposure of the film to air for the oxidation of nonirradiated region. The removal of the nonirradiated, oxidized region by a dilute (0.16%) HF gives a poly-Si pattern with a linewidth of 1 μm under the conditions of a mask pattern linewidth of 1 μm and 1000 pulses laser irradiation at a pulse fluence of 10 mJ•cm^{-2} as illustrated in Fig. 8. A little roughness observed at the pattern edges might reflect the spatially inhomogeneous Si network formation in polysilane. This problem can be solved by the improvement of structural homogeneity of an as-deposited polysilane film. Further reduction of the pattern size will be achieved by using 5 to 1 projection optics and a phase shifting mask instead of the contact-type Cr mask.

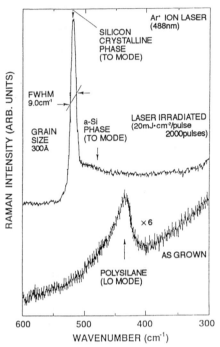

Fig. 6. Raman spectra for as-deposited and laser-irradiated film. The fluence and the number of laser pulses are 20 mJ•cm^{-2}/pulse and 2000 shots, respectively. A 488 nm line from an Ar$^+$ ion laser was used as an excitation source for the Raman scattering.

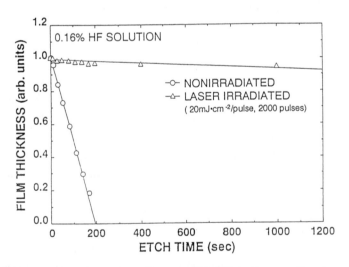

Fig. 7. The film thickness change by a dilute (0.16%) HF etching as a function of etch time for nonirradiated and irradiated films which are exposed to air at room temperature for 1 hr. The fluence and the number of ArF excimer laser pulse are 20 mJ•cm^{-2}/pulse and 2000 pulses, respectively.

0.5μm

Fig. 8   Scanning electron micrograph of a poly-Si pattern obtained by irradiating the film through a Cr mask and subsequent dilute HF etch.

Summary

It is shown that polycrystalline silicon can be grown at temperatures of about -50°C by irradiating ArF excimer laser onto Si_2H_6 adsorbates. Also, it is found that crystallization of metastable polysilane occurs at room temperature by ArF excimer laser irradiation at low fluences. These results indicates very strong interaction of excimer laser light with polymeric species of silicon such as Si_2H_6 and $(SiH_2)_n$.

References

[1]   T. Suntola, Extended Abstracts of the 16th (1984, Int.) Conference on Solid State Devices and Materials, Kobe, 1984 (Business Center for Academic Societies Japan, Tokyo, 1984), p. 647.

[2]   J. Nishizawa, H. Abe and T. Kurabayashi, J. Electrochem. Soc. **132**, 1197 (1985).

[3]   Y. Takahashi, Y. Sese and T. Urisu, Digest of Papers 2nd Micro Process Conference, Kobe 1989 (Business Center for Academic Societies Japan, Tokyo, 1989), p. 138.

[4]   T. Tanaka, K. Deguchi, S. Miyazaki and M. Hirose, Jpn. J. Appl. Phys. **27**, L2149 (1988).

[5]   T. Tanaka, T. Fukuda, Y. Nagasawa, S. Miyazaki and M. Hirose, Appl. Phys. Lett. **56** 1445 (1990)

[6]   S. Miyazaki, H. Shin, K. Okamoto and M. Hirose, Mat. Res. Soc. Symp. Proc. **242**, 681 (1992).

[7]   K. Okamoto, H. Shin, K. Shiba, S. Miyazaki and M. Hirose, Jpn. J. Appl. Phys. **31**, 4441 (1992).

[8]   K. Okamoto, M. Shinohara, T. Yamanishi, S. Miyazaki and M. Hirose, Appl. Surf. Sci. (1994), to be published.

[9]   H. Shin, S. Miyazaki, H. Ichihashi and M. Hirose, Proc. Int. Seminar on Reactive Plasmas (Nagoya, 1991) p. 201.

[10] P. Vora, S. A. Solin and P. John, Phys. Rev. B **29**, 3423 (1984).

[11] Th. Englert, G. Abstreiter and J. Pontcharra, Solid-State Electron. **23**, 31 (1980).

Materials Science Forum Vols. 173-174 (1995) pp. 17-22
© 1995 Trans Tech Publications, Switzerland

FEMTOSECOND-PULSE LASER MICROSTRUCTURING OF SEMICONDUCTING MATERIALS

W. Kautek and J. Krüger

Laboratory for Laser and Chemical Surface Technology,
Federal Institute for Materials Research and Testing, D-12200 Berlin, Germany

Keywords: Femtosecond-Pulse Laser, Subpicosecond-Pulse Laser, Laser-Ablation, Multi-Photon Absorption, Microstructuring, Dry Etching, Semiconductors, Silicon

Abstract. Production of surface structures of < 30 µm diameter with high depth-to-diameter ratio (aspect ratio value, ARV) on semiconductor materials is a delicate task either for mechanical cutting and drilling tools, or even for conventional nanosecond-pulse lasers like e.g. excimer lasers. On silicon wafers, nanosecond-pulse lasers tend to cause *micro-cracks* extending from an annular melting zone, or substantial *disruption*, respectively. The development of intense *ultrashort-pulse laser systems* ($\gg 10^{12}$ W cm^{-2}; < 1 ps) has opened up possibilities for materials processing by *cold ablation*. A *femtosecond-* and a *nanosecond*-dye laser with pulse durations of 300 fs (< 200 µJ) and 7 ns (< 10 mJ), and centre wavelengths at 612 and 600 nm, respectively, both focused to an area of the order of 10^{-5} cm^2, have been applied to (111)-oriented silicon. Because the time is too short (< ps) for significant transport of mass and energy, the beam interaction results in the formation of a thin *plasma layer* of approximately *solid state density* which expands rapidly away from the surface thus avoiding large recoil effects like melt drainage and splashing in contrast to conventional nanosecond-pulse laser technology. Multi-photon absorption contributions affect the penetration depth of radiation in silicon. This nonlinear nature of the mechanism, and phenomena like light channelling and autofocusing lead to an "amplification" of minor random surface inhomogeneities to (chaotic) corrugation features within ablation cavities and grooves.

Introduction

Applications of micromachined components occur in the electronics, aerospace, optics, medical, and photovoltaic industries. Most microsystems are currently made of silicon or similar semiconductors to which adaptations of current microlithography and etching technologies can be applied. These reach aspect ratio values (ARV; ratio between depth to kerf size) typically of only 2:1. However, ARV's of more than 100:1 are required in the future. Nanosecond-pulse lasers therefore play a significant role in ultraprecise micromachining for excavating holes, grooves, or cuts with micrometer and submicrometer tolerances in advanced integrated-circuit (IC) packaging, such as multichip modules (MCM), or in macroscopic devices that require ultratight mechanical specifications [1]. The interaction of strong laser-pulses with silicon targets is also of relevance to the annealing of ion-implanted samples [2], the crystallisation of thin amorphous silicon films [3], the pulsed laser deposition (PLD) of thin films [4], and to the laser-induced damage of Si photosensor arrays [5].

Subbandgap photon energy nanosecond-pulse laser-ablation of semiconductors like GaP [6] led to atom emission indicative of direct photochemical decomposition. A mechanism of vacancy-initiated ablation of semiconductors has been suggested. It consists in the electronic excitation of vacancies, and the gradual bond breaking of weakly bonded atoms around vacancies and vacancy clusters during comparatively long ns laser-pulses. Recent ablative studies demonstrated that thermal processes dominate particularly polymer ablation [7]. Ablation models however, both in the nanosecond- and in the femtosecond-pulse laser regime, are still contradictory, and can only partly explain various experimental results [8-10].

First *in situ* femtosecond-pulse laser second-harmonic generation (SHG) measurements at Si(111) electrodes in contact with aqueous electrolytes [11] directed our attention towards the disruption behaviour of silicon, and the prospects to produce microstructures on this material by subpicosecond-pulse laser treatment [12]. Ablation requires selective ultraprecise removal of thin layers without mechanical or thermal damage of neighbouring substrate regions. Decreasing sizes and film thicknesses let conventional nanosecond-pulse laser technology hit intrinsic limits because heat-affected zones, HAZ $\approx [D\tau]^{-0.5}$ (thermal diffusion coefficient D, laser-pulse duration τ), of the order of 10 μm prevent etching precision [12]. Further, coupling of nanosecond-pulse laser energy into a sample, the ablation process, and the removal of material all take place at the same time resulting in so-called plasma shielding.

It is shown that now commercially available femtosecond-pulse laser technology provides superior ablation quality and precision on silicon, because it can avoid plasma shielding of the laser radiation, and because it minimises both substrate heating and mechanical disruption by strong hot plasma recoil interaction with the substrate. Novel mechanistic aspects are discussed on the basis of a first systematic fs laser-ablation study of Si(111) targets.

Experimental

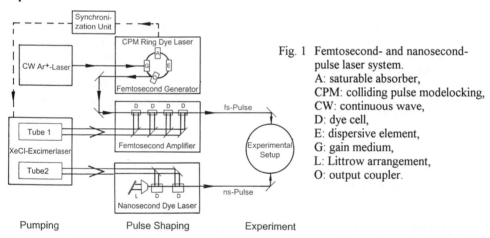

Fig. 1 Femtosecond- and nanosecond-
pulse laser system.
A: saturable absorber,
CPM: colliding pulse modelocking,
CW: continuous wave,
D: dye cell,
E: dispersive element,
G: gain medium,
L: Littrow arrangement,
O: output coupler.

The fs pulses (~100 fs, 612 nm) were generated from a passively modelocked dye laser in the colliding pulse modelocked (CPM) regime (BESTEC, Berlin) pumped by a continuous wave Ar-Ion-Laser (< 2 W, 514.5 nm, model 2025-11, Spectra-Physics; Fig. 1). These low-energy pulses were amplified to pulses with a duration of ~300 fs and a single pulse energy of 200 μJ in four dye cells transversely pumped by a XeCl excimer laser (< 100 mJ, 308 nm, 17 ns, model EMG 150, Lambda Physik, Göttingen). For comparative ns laser-pulse experiments, a commercial excimer laser pumped ns dye laser was used (> 3 mJ, 600 nm, 7 ns, model LPD 3002, Lambda Physik). A 58 mm focal length lens focused the laser beams to spot areas of the order of 10^{-5} cm^{-2}.

N-type Si(111) wafers (phosphorus doped, 10^{18} cm^{-3}; Werk für Fernsehelektronik, Berlin) with 30 mm diameter and 400 μm thickness were used. A computerised x-y-translation stage (LOT, Darmstadt) served to position the wafers perpendicular to the direction of laser incidence. The ablation depth was determined by means of an optical microscope. The precision of this manual focusing method was better than 10 μm. More experimental details are presented elsewhere [12].

Results and Discussion

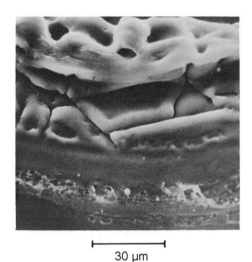

|⊢―――――――⊣|
   30 μm

Fig. 2 Nanosecond (excimer) laser-ablation of Si(111).
$\tau = 17$ ns, 308 nm, 100 pulses, 3.7 Hz, focus $\varnothing \approx 350$ μm, $F = 3.5$ J cm^{-2}.

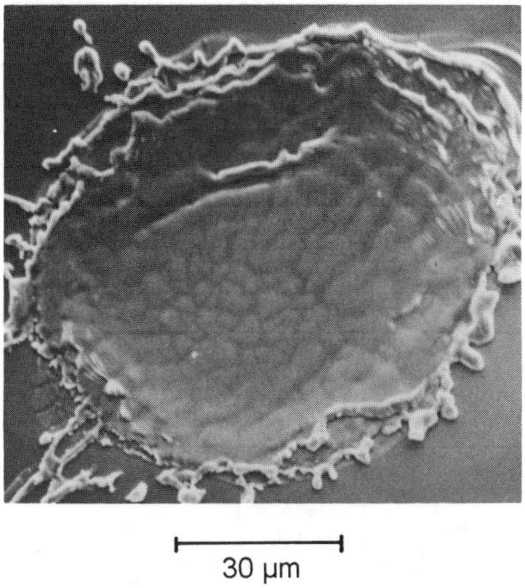

|⊢――――――⊣|
   30 μm

Fig. 3 Nanosecond (dye) laser-ablation of Si(111).
$\tau = 7$ ns, 600 nm, 1 pulse, focus $\varnothing \approx 60$ μm, 30 J cm^{-2}.

Nanosecond-pulses of an excimer laser cause dramatic destruction of the single crystalline Si(111) material within a comparatively wide circumference of the ablation cavity [12] (Fig. 2). Only in a centre region of the laser spot, the ablation fluence threshold (F_{th}) of silicon is surpassed and vaporisation occurs. Outside the inner intense part of the beam profile a halo of concentric waves with a periodicity of the order of 100 μm is generated. This has no connection with optical interference processes, but is due to acoustic shock-waves travelling in a melted surface layer with a thickness corresponding to the HAZ ($\approx [D\tau]^{-0.5} \approx 10$ μm). The penetration depth of ultraviolet laser light in silicon is much smaller ($\alpha_{uv}^{-1} \ll$ HAZ), and therefore does not play any role. Bulk regions of the substrate below the HAZ are also affected by the strong shock-waves (shock-affected zone, SAZ), and exhibit defects. These appear like straight cracks arranged either parallel or tilted by 60° to each other, typical for a (111) oriented surface.

The penetration depth of nanosecond laser-pulses in the visible wavelength range into the indirect semiconductor Si is much greater, $\alpha_{612nm}^{-1} \approx 2.5$ μm, than for the excimer laser radiation, and almost approaches the extension of the HAZ. Thus, both light and heat penetration into the target determine the ablation process (compare ablation of ceramic oxide material by visible ns lasers [13]). Moreover, plasma shielding together with plasma heating by the laser beam probably accelerate the plume expansion to such an extent that considerable drainage of melted material and extensive splashing are caused (Fig. 3 and 4).

The application of visible fs laser-pulses eliminates destruction and splashing (Fig. 5,6). At the end of the 300 fs laser-pulse, the irradiated materials remains still in a condensed phase, and the plasma at the bottom of the crater has not even started to expand. That means that the light-plasma interaction is negligible in contrast to conventional 10-ns-pulses where an interaction depth of the order of 100 μm attenuates the incoming laser beam [14] (Fig. 4).

The average ablation depth per pulse, d, vs. laser fluence, F, with femtosecond laser-pulses is shown in Fig. 7. The ablation depth d varies logarithmically with the laser fluence F (< 2 Jcm^{-2})

according to Beer's law assuming an intensity independent absorption coefficient α:

$$d \approx \alpha^{-1} \ln\left(\frac{F}{F_{th}}\right) \qquad (1)$$

where F_{th} is the ablation threshold fluence. A $F_{th} \leq 0.2$ J cm^{-2} ($I_{th} \leq 0.7 \times 10^{12}$ W cm^{-2}) can be appreciated. This value is less than the threshold at metals (e.g. Au: $F_{th} \approx 0.3 - 0.6$ J cm^{-2}; $I_{th} \approx 1 \times 10^{12}$ W cm^{-2}; [12]). The threshold in transparent media, where visible subpicosecond-laser-pulses induce multi-photon absorption, is even one order of magnitude higher (e.g. human cornea: $F_{th} \approx 1.0$ J cm^{-2}; $I_{th} \approx 3 \times 10^{12}$ W cm^{-2}) [15].

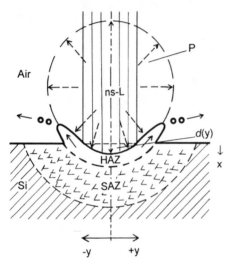 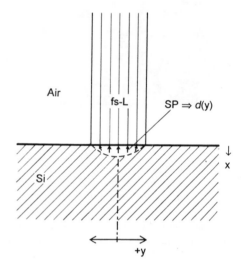

Fig. 4   Schematic model of ns pulse laser-ablation of Si. ns-L: ns laser beam; d(y): ablation depth; HAZ: heat-affected zone; SAZ: shock-affected zone; P: expanding laser-heated plasma.

Fig. 5   Schematic model of fs pulse laser-ablation of Si. fs-L: fs laser beam; SP: solid state plasma (end of pulse) d(y): ablation depth.

Si shows coarse columns of variable size within the spot region (Fig. 6). The pulse duration of subpicoseconds makes the heat-affected zone (HAZ) extremely small (~10 nm) regardless of thermal conductivities. A layer of more than 2.0 µm thickness below the Si surfaces should attain energy densities above the melting threshold. However, no resolidified melt layers of this thickness occur due to plasma recoil drainage out of the cavities (compare ns experiment, Fig. 3). Actually, one observes an optical channelling effect. Surfaces normal to the laser beam direction, i.e. bottoms of the troughs between corrugation features, are preferentially ablated. All other surface orientations exhibit higher reflectivity.

From Fig. 7, an apparent α of the order of 10^5 cm^{-1} is calculated by Eq. 1, which is at variance with the low intensity single-photon value α_1 ($\approx 4 \times 10^3$ cm^{-1} at 612 nm). This contradiction could be resolved by a contribution of two-photon absorption at ultrahigh intensities ($I = F \tau^{-1} > 10^{13}$ W cm^{-2}) expressed by a second order absorption coefficient $\alpha_2 \approx 10^{-9} - 10^{-8}$ cm W^{-1} [10,14], so that

$$\frac{\partial I(x,t)}{\partial x} = -\alpha I(x,t) \qquad \text{and} \qquad \alpha \approx \alpha_1 + \alpha_2 \cdot I. \qquad (2)$$

We suggest that light channelling between the corrugation features (Fig. 6) splits the original beam, and concentrates each partial beam at the bottoms of the cavities ("autofocusing"). Only in these particular locations, I attains high enough values in order to cause non-linear two-photon absorption and a comparatively high absorption coefficient

$$\alpha \approx (\alpha_2 \cdot I) \gg \alpha_1 . \tag{3}$$

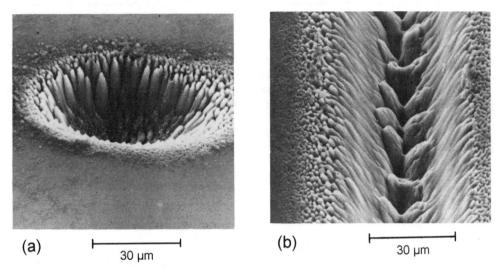

(a) |———————————————|
          30 µm

(b) |———————————————|
          30 µm

Fig. 6 Femtosecond-pulse laser-ablation of Si(111).
   $\tau = 300$ fs, 612 nm, 2.9 Hz, focus $\varnothing \approx 70$ µm. (a) crater, 1000 pulses, $F = 0.8$ J cm^{-2};
   (b) groove, 2 scans, 0.5 µm s^{-1}, ~460 pulses per spot, $F = 1.8$ J cm^{-2}.

This non-linear nature of the mechanism could lead to an "amplification" of minor random surface inhomogeneities to (chaotic) corrugation features observed in the craters and grooves. A detailed study and a discussion of a possible non-linear character of this involved ablation mechanism will be given elsewhere [16].

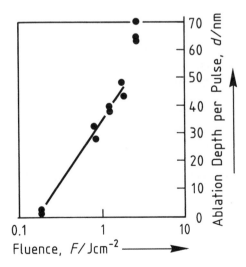

Fig. 7 Ablation depth per pulse vs. laser fluence of the femtosecond-pulse laser-ablation of Si(111) at air. $\tau = 300$ fs, 2.9 Hz, 612 nm. The points represent experimental data averaged over 1000 pulses.

Summary

The interaction of visible *femtosecond laser-pulses* with Si(111) has been studied at peak intensities of more than 10^{13} W cm^{-2}. At such conditions, rapid ionisation of the material occurs and a *thin plasma layer* of approximately *solid state density* is formed during the pulse. This expands rapidly away from the surface *without* any light absorption and *plasma heating*. Therefore, energy transfer to the target material and consecutive *thermal* and *mechanical disruption* are *minimised*.

The *non-linear multi-photon* nature of the visible femtosecond-pulse laser-ablation mechanism, *light channelling*, and *autofocusing* lead to an "amplification" of minor random surface inhomogeneities to (chaotic) *corrugation features* within ablation cavities and grooves.

Femtosecond-pulse lasers *allow precise micromachining of Si wafers* in contrast to conventional nanosecond-pulse (excimer) lasers which cause serious disruption and/or splashing.

Acknowledgements

Technical assistance by S. Pentzien, A. Conradi, and B. Strauß is gratefully acknowledged.

References

[1]   E. Wiener-Avnear, Laser Focus World **29**, 75 (1993).

[2]   H. J. Leamy, G. A. Rozgonyi, T. T. Sheng, and G. K. Celler, Appl. Phys. Lett. **32**, 535 (1978).
      J. M. Liu, R. Yen, H. Kurz, and N. Bloembergen, Appl. Phys. Lett. **39**, 755 (1981).
      D. Haneman and R. J. Nemanich, Solid State Commun. **43**, 203 (1982).
      P. M. Fauchet and A. E. Siegman, Appl. Phys. Lett. **40**, 824 (1982).
      P. S. Peercy, G. J. Calvin, M. O. Thompson, J. W. Mayer, and R. B. Hammon, Physica B **116**, 558 (1983).
      G. Gorodetsky, J. Kanicki, T. Kazyaka, and R. L. Melcher, Appl. Phys. Lett. **46**, 547 (1985).

[3]   J. S. Im and H. J. Kim, Appl. Phys. Lett. **63**, 1969 (1993).

[4]   M. Hanabusa, M. Suzuki, and S. Nishgaki, Appl. Phys. Lett. **38**, 385 (1981).

[5]   O. Muller and R. Joecklé, Mater. Sci. Eng. A **168**, 81 (1993).

[6]   A. Okano, K. Hattori, Y. Nakai, and N. Itoh, Surf. Sci. **258**, L671 (1991).
      K. Hattori, A. Okano, Y. Nakai, and N. Itoh, Phys. Rev. B **45**, 209 (1992).
      A. Okano, A.Y. Matsuura, K. Hattori, N. Itoh, and J. Singh, J. Appl. Phys. **73**, 3158 (1993).

[7]   R. Sauerbrey and G.H. Pettit, Appl. Phys. Lett. **55**, 421 (1989).
      S.V. Babu, G.C. D'Couto, and F.D. Egitto, J. Appl. Phys. **72**, 692 (1992).

[8]   R. Kelly, A. Miotello, B. Braren, A. Gupta, and K. Casey, Nucl. Instr. and Meth. B **65**, 187 (1992).

[9]   G.H. Pettit and R. Sauerbrey, Appl. Phys. A **56**, 51 (1993).

[10] S. Preuss, M. Späth, Y. Zhang, and M. Stuke, Appl. Phys. Lett. **62**, 3049 (1993).

[11] W. Kautek, N. Sorg, and J. Krüger, Electrochim. Acta **39**, (1994), in print.

[12] W. Kautek and J. Krüger, in *Laser Materials Processing: Industrial and Microelectronics Applications*, SPIE Vol. **2207** (1994), in print.

[13] W. Kautek, B. Roas, and L. Schultz, Thin Solid Films **191**, 317 (1990).

[14] S. Küper and M. Stuke, Appl. Phys. B **44**, 199 (1987).

[15] W. Kautek, S. Mitterer, J. Krüger, W. Husinsky, and G. Grabner, Appl. Phys. A **57**, (1994), in print.

[16] J. Krüger and W. Kautek, Appl. Phys. A (1994), in preparation.

Materials Science Forum Vols. 173-174 (1995) pp. 23-28
© 1995 Trans Tech Publications, Switzerland

MODIFICATION OF FOREIGN ATOM CONCENTRATIONS AND PROFILES IN SILICON AND $Si_{1-x-y}Ge_xC_y$ ALLOYS BY LASER CHEMICAL ETCHING

J. Boulmer[1], A. Desmur[1,3], B. Bourguignon[2], J.-B. Ozenne[1], R. Laval[1], A. Aliouchouche[1], D. Débarre[1] and J.-P. Budin[1]

[1] Institut d'Electronique Fondamentale (associé au CNRS), Bât. 220

[2] Laboratoire de Photophysique Moléculaire du CNRS, Bât.213, Université Paris-Sud, F-91405 Orsay, France

[3] Present address: CNET-CNS, Chemin du Vieux Chêne, BP 98, 38243 Meylan, France

Keywords: Etching, Laser Etching, Laser Melting, Laser Desorption, Chemisorption, Si, SiGe, SiGeC, SIMS, EDX

Abstract: Ion implanted silicon (As, Sb, B) samples and $Si_{1-x-y}Ge_xC_y$ films deposited on Si are laser etched in the presence of chlorine in experimental conditions where laser surface melting occurs. We measure the differential desorption and the modifications of concentrations and depth profiles of B, As, Sb and Ge as laser etching proceeds. We find that they depend greatly on the nature of the foreign atom and on the composition of the etched layer. Finally, we observe the formation of shallow doped layers, which are pushed towards the bulk as laser etching proceeds.

Introduction

Laser chemical etching of silicon by chlorine and a pulsed UV laser has been studied extensively /1-8/. Two main classes of experimental situations where laser etches silicon efficiently have been analysed.

The first one involves laser-stimulated reactions of gas phase chlorine (Cl_2 and Cl) at the silicon surface. The ambient chlorine pressure is generally in the range of 10 to 100 mbar, and the laser wavelength lies in the chlorine dissociation spectrum whose maximum is near 300 nm /5,6,8,9/. In this case, the chemisorption and desorption cycles leading to the etching process are activated by the presence of chlorine atoms and by the extra energy delivered to the gas by photodissociation by $\approx$ 4 eV photons and subsequent recombination of chlorine species. Moreover, evidence of an important influence on chemisorption and desorption of the dopant type (n or p) and concentration has been observed by Horiike et al /8/ and was recently confirmed in our laboratory /9/. Finally, in the most favourable experimental situation (highly doped n samples, 308 nm, 100 mbar), etch rates as high as 1 Å/pulse are obtained for laser fluences well under the melting threshold of silicon.

In the second class of experimental conditions, ambient pressure is so low that laser excitation of the chlorine gas cannot play any significant role /1-4/. In this case, etching is due to laser-induced desorption of silicon chlorides, previously created by spontaneous chemisorption, and involves laser-induced surface melting /1,2/. As a consequence, the process requires only very small amounts of chlorine (pressure down to 10^{-7} mbar) and the dependence of the etch rate over laser fluence exhibits a threshold, equal to the melting threshold ($\approx$ 350 mJ/cm^2 at 308 nm). For fluences above 500 mJ/cm^2, we observe that laser etching is self-limited to 0.6 Å/pulse (0.44 monolayer / pulse) /1,2/. As far as surface melting is involved in laser etching, the doping level of silicon has no influence on the etch rate /9/ and we do not observe any measurable dependence of the laser etch rate on the silicon sample composition, even for $Si_{1-x-y}Ge_xC_y$ compounds with x $\approx$ 13% and y=0 or $\approx$ 1%.

Modification of profiles and concentrations induced by laser etching

In the present work, we focus our attention on the physical processes involved in this second class of experimental conditions, where surface melting is required to laser etch the silicon samples. It is well known that laser-induced surface melting (annealing) also allows for the modification of the concentration depth profile of impurities in the melted layer /10-12/. Our purpose is to study the influence of laser-induced desorption and surface melting on samples containing dopants (As, Sb, B) or germanium and/or carbon. We measure the induced modifications of the concentrations and depth profiles of these species during laser etching by chlorine with a XeCl excimer laser.

In the presence of chlorine, laser etching and surface melting simultaneously induce three types of processes which contribute to the modification of the concentration and depth profiles: diffusion in the molten layer, segregation at the liquid/solid moving interface during recrystallisation and laser-induced desorption from the liquid surface, which is responsible for the etching process.

Even at temperatures close to the silicon melting temperature, most impurities have very short diffusion lengths ($<<1$ Å) on the time scale of the laser pulse duration (≈ 10 ns). However, diffusion coefficients are larger by about 7 orders of magnitude in liquid silicon. The diffusion length in 10 ns is about 100 Å in liquid silicon for most dopants, to be compared to melted depths of about 1000 Å in this experiment, allowing for diffusion over the full melted layer in a small number of laser pulses.

Segregation is characterized by a coefficient k_s equal to the ratio n_s/n_l of the concentrations in the solid and liquid phases. During our laser experiments, the interface moves at velocities of about 6 m/s. For most atoms in silicon and at low interface velocities, k_s is < 1, but at such high velocities, k_s becomes close to unity /12/. A large diffusion coefficient and $k_s=1$ leads to a flat distribution of the foreign atoms in silicon after recrystallisation. On the contrary, a segregation coefficient smaller than one leads to an excess of foreign atoms near the surface. Segregation also takes place at solid surfaces /13/ or interfaces /14/ on much longer time scales, resulting in very different concentrations at the surface and in the bulk.

In this work, laser-induced desorption is observed only for laser fluences above the melting threshold. To be desorbed, an atom must be present on the molten surface. The apparent desorption efficacy results from the desorption probability of a surface atom (silicon or foreign atom), the probability for an atom to diffuse from the surface (bulk) to the bulk (surface) and the segregation processes during and after solidification of the melted layer. As a consequence, laser-induced desorption efficacies may vary greatly from one atomic species to another (silicon or foreign atom), so that the laser etching process may induce an increase or a decrease of the foreign atom concentration near the surface.

Experimental procedure

Two types of samples have been studied: Si(100) samples doped by ion implantation (B, As or Sb, implantation dose: 6×10^{15} cm^{-2} at 60 KeV), and 700 Å thick Si$_{1-x-y}$Ge$_x$C$_y$ films, deposited by Chemical Beam Epitaxy (CBE) on Si(100) substrates with $x \approx 0.13$ and $y=0$ or $x \approx 0.13$ and $y \approx 0.01$. An excimer laser (XeCl, 308 nm, pulse duration ≈ 20 ns) is used to anneal or etch the samples. We use laser fluences of 600 mJ/cm^2, well above the surface melting threshold and well below the ablation threshold of 1 J/cm^2.

Etching experiments are conducted in a vacuum apparatus (base pressure $\approx 10^{-7}$ mbar), where the surface is chlorinated by a pulsed molecular beam before each laser pulse /1-3/. Etched depth, obtained after a given number of laser pulses, is measured with a Dektak stylus profilometer. Dopant depth profiles, before and after laser processing, are measured with a SIMS apparatus (Cameca SMI 300) and calibrated with the Dektak profilometer. Germanium concentration in Si$_{1-x}$Ge$_x$C$_y$/Si samples, before and after laser processing, is determined by RBS. Atomic surface densities (at/cm^2) are more conveniently obtained on a SEM coupled with an EDX (Energy Dispersive X-ray analysis) equipment. We were unable thus far to obtain a useful measurement of the carbon concentration and profile.

Experimental results

Modification of surface densities during laser etching. Figure 1 summarizes the measured surface densities of B, As or Sb in Si samples (as obtained from integrated SIMS data), and of Ge in $Si_{0.87}Ge_{0.13}/Si$ or $Si_{0.86}Ge_{0.13}C_{0.01}/Si$ samples (as obtained from EDX data), as a function of the number of laser pulses. The laser fluence is 600 mJ/cm^2. At this fluence, the etch rate is self-limited to 0.6 Å/pulse, as verified for all samples, and the calculated melted depth is 1000 Å. As a consequence, germanium or carbon in SiGe/Si or SiGeC/Si samples is redistributed over a depth larger than the initial one (700 Å) after the first ≈ ten laser pulses. We observe that surface densities decrease very differently from one atom to another as laser etching proceeds. The desorption efficacy increases by a factor of about 60 from boron, which desorbs very inefficiently, to germanium (in the SiGe sample), which desorbs ten times more efficiently than silicon, while the etch rate remains equal to 0.44 monolayer per laser pulse for all samples. To deduce the apparent desorption efficacy from fig. 1, the foreign atom depth concentration profiles have to be measured.

Surface Density (10^{15} Cm^{-2})

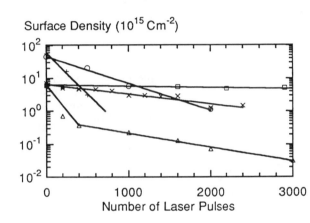

Figure 1: Evolution, as laser etching proceeds, of the measured surface densities of B (squares), As (x) or Sb (triangles) in Si samples, and of Ge in $Si_{0.87}Ge_{0.13}/Si$ (+) or $Si_{0.86}Ge_{0.13}C_{0.01}/Si$ samples (circles)

Modification of dopant concentration profiles during laser etching. Figures 2, 3 and 4 show the evolution of SIMS depth profiles measured before and after etching by an increasing number of laser pulses for silicon samples implanted with boron, antimony and arsenic, respectively. The laser fluence is 600 mJ/cm^2. The depth "zero" on the figures corresponds to the position of the surface before etching. As etching proceeds, the surface is located at increasing values of the depth. Up to now, we do not have enough precise data to conclude about the evolution of germanium profiles. However, our first observations and results published by Kramer et al /16/ indicate that Ge profiles would be almost flat on a depth smaller than its theoretical value (≈ 700 Å instead of 1000 Å).

It readily appears (fig. 2) that boron profiles are almost flat over a depth (125 nm) sligthly larger than the calculated one (100 nm), and unchanged as etch depth increases. After 6000 laser pulses, the sample is etched to a depth of 3600 Å. The etched away layer contained initially about 99% of the implanted atoms; 50% of boron atoms have moved towards the bulk during etching. Moreover, the profiles are very abrupt: the boron concentration decreases by two orders of magnitudes in 100 Å.

The implanted antimony profile (fig. 3) is narrower than that of boron. We have performed etching to depths exceeding the implanted region. As for boron, the antimony atoms are pushed towards the bulk during the etching process, but the loss of dopant is much larger due to the high desorption efficacy of Sb. The profiles are approximately flat,

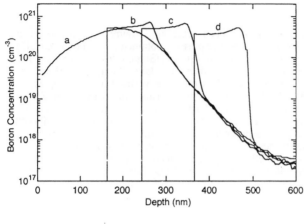

Figure 2: SIMS depth profiles of boron as implanted in silicon (a), and after etching by 2900 (b), 4000 (c), 6300 (d) laser pulses. As laser etching proceeds, the surface is located at increasing values of depth.

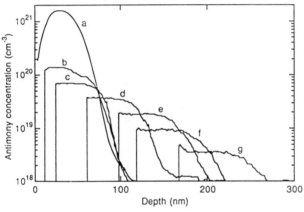

Figure 3: SIMS depth profiles of antimony as implanted in silicon (a), and after etching by 200 (b), 400 (c), 1000 (d), 1600 (e), 2000 (f), 3000 (g) laser pulses.

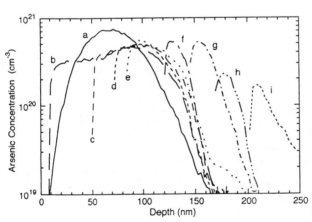

Figure 4: SIMS depth profiles of arsenic as implanted in silicon (a), and after etching by 100 (b), 800 (c), 1200 (d), 1400 (e), 2000 (f), 2400 (g), 2800 (h), 3200 (i) laser pulses.

but less abrupt, probably because of Sb has a lower segregation coefficient than B and As. Finally we observe that the Sb profile width continuously increases during the etching process: its FWHM value varies from 600 Å (200 laser pulses) to 1000 Å (3000 laser pulses), to be compared to the calculated melted depth of 1000 Å.

The implanted arsenic profile (fig. 4) is broader: about 2400 laser pulses are necessary to etch most of the implanted layer. During the first 2400 laser pulses, the dopant profile is approximatly flat, as expected, but its width decreases as etching proceeds. After the etching of the implanted layer, the As profile width stabilizes at ≈ 200 Å (FWHM), which is one-fifth of the calculated value, while the peak concentration of As decreases, due to desorption. The As profile obtained after etching of the implanted layer is very narrow, but not as abrupt as for boron. The peak dopant concentration is as high as 60% of the maximum value of the initial profile.

Desorption efficacy. The apparent desorption efficacy per pulse is a phenomenological coefficient describing the probability for an atom present on the surface to be desorbed, as a result of the competition between laser desorption, diffusion and segregation /15/. This efficacy can be evaluated from the above data, assuming a flat concentration profile over the FWHM width obtained by SIMS (figs 2 to 4), which may vary greatly during the etching process. For Ge, we assumed a constant width of 700 Å. Figure 5 summarizes the apparent desorption efficacies, which are compared to the value for Si obtained from the etching rate (0.6 Å/pulse). It is shown that the apparent desorption efficacies of the foreign atoms vary over a factor of 60 and depend upon the species, the etching step (before or after the implanted layer has been etched), and the composition of the substrate (SiGe or SiGeC). By comparison, the apparent desorption efficacy when the laser anneals the samples (without chlorine) is much smaller: it is not measurable for Si and Ge, and equals 0.04 for As and 1 for Sb, for the first laser pulses.

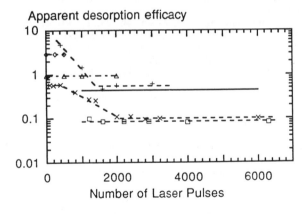

Apparent desorption efficacy

Number of Laser Pulses

Figure 5: Evolution, during the etching process, of the apparent desorption efficacy for Sb (+), As (x) and B (squares) in Si and for Ge in $Si_{0.87}Ge_{0.13}$/Si (diamonds) and Ge in $Si_{0.86}Ge_{0.13}C_{0.01}$/Si (triangles). The horizontal straight line shows the corresponding value for Si.

Conclusion

In the present experimental conditions, laser etching of silicon by chlorine results from surface melting. Desorption and diffusion of atoms are strongly enhanced in the melted layer, and, as a result, a redistribution of dopants is produced together with etching. The magnitude of desorption relative to diffusion and segregation plays a major role in the resulting profiles and concentrations. The physical processes involved in these observations will be discussed in more detail in a forthcomming paper /15/.

The apparent desorption efficacy per laser pulse, in the presence of chlorine, is found to vary strongly -over more than one order of magnitude- as a function of the nature of the foreign atom and also of the composition of the etched layer (case of SiGe and SiGeC). We also observe that the dopant profile width may differ from the theoretical value, depends on the dopant atom, and may vary greatly during the etching process. However, in these experiments, the etching rate remains equal to its self-limited value (0.6 Å/pulse).

After the implanted layer has been laser etched, dopant redistribution, segregation and differential desorption result in the formation of shallow doped layers which are pushed towards the bulk as laser etching proceeds. In these shallow doped layers, peak concentration is equal (B) or almost equal (As) to the peak concentration obtained by implantation.

Acknowledgements

We are indebted to the Centre d'Etudes et d'Education pour les Techniques Avancées de la Microélectronique (CEETAM) for technical assistance and for providing dopant-implanted Si wafers. Its a pleasure to thank CNET-CNS (Meylan, France), with a special attention to J. C. Oberlin, for performing RBS measurements, and to Y. Campidelli and F. Glowacki for their contribution to the realisation of the SiGe and SiGeC samples.

References

/1/ J. Boulmer, B. Bourguignon, J.-P. Budin, D. Débarre and A. Desmur, J. Vac. Sci. Technol. A **9**, 2923 (1991).
/2/ J. Boulmer, J.-P. Budin, B. Bourguignon, D. Débarre and A. Desmur, in: "*Laser ablation of electronic materials*", E. Fogarassy and S. Lazare, editors, Elsevier (1992), p. 239.
/3/ A. Aliouchouche, J. Boulmer, B. Bourguignon, J.-P. Budin, D. Débarre and A. Desmur, Appl. Surf. Sci. **69**, 52 (1993).
/4/ T. S. Baller, J. C. S. Kools and J. Dieleman, Appl. Surf. Sci. **46**, 292 (1990). H. Feil, T. S. Baller and J. Dieleman, Appl. Phys. A **55**, 554 (1992). J. C. S. Kools, T. S. Baller and J. Dieleman, in: "*Photochemical processing of electronic materials*", I. W. Boyd and R. B. Jackman, editors, Academic Press (London, 1992), p.339.
/5/ R. Kullmer and D. Bäuerle, Appl. Phys. A **43**, 227 (1987). R. Kullmer and D. Bäuerle, Appl. Phys. A **47**, 377 (1988).
/6/ W. Sesselmann, E. Hudeczek and F. Bachmann, J. Vac. Sci. Technol. B **7**, 1284 (1989).
/7/ Q.-Z. Qin, Y.-L. Li, P. H. Lu, Z.-J. Zhang, Z.-K. Jin and Q. K. Zheng, J. Vac. Sci. Technol. B **10**, 201 (1992).
/8/ Y. Horiike, N. Hayasaka, M. Sekine, T. Arikado, M. Nakase and H. Okano, Appl. Phys. A, **44**, 313 (1987)
/9/ J. Boulmer, A. Aliouchouche, B. Bourguignon, J.-P. Budin, D. Débarre, A. Desmur and J. Mambou, internal report (1983)
/10/ F. Foulon, in: "*Photochemical processing of electronic materials*", I. W. Boyd and R. B. Jackman, editors, Academic Press (London, 1992), p.257, and references within.
/11/ B. Leroy, Revue Phys. Appl. **21**, 467 (1986).
/12/ R. F. Wood, Phys. Rev. B **25**, 2786 (1982).
/13/ K. Jagannadham and J. Narayan, J. Appl. Phys. **61**, 985 (1987).
/14/ J. M. Moison, F. Houzay, F. Barthe, J. M. Gérard, B. Jusserand, J. Massies and F. S. Turco-Sandroff, J. Cryst. Growth **111**, 141 (1991).
/15/ A. Desmur, B. Bourguignon, J. Boulmer, J.-B Ozenne, J.-P. Budin, D. Débarre and A. Aliouchouche, J. Appl. Phys., submitted.
/16/ K.-J. Kramer, S. Talwar, E. Ishida, K.-H. Weiner and T.-W. Sigmon, Appl. Surf. Sci. **69**, 121 (1993)

Materials Science Forum Vols. 173-174 (1995) pp. 29-34
© 1995 Trans Tech Publications, Switzerland

PROPERTIES OF RECRYSTALLIZED AMORPHOUS SILICON PREPARED BY XeCl EXCIMER LASER IRRADIATION

I. Ulrych[1], K.M.A. El-Kader[1], V. Cháb[1], J. Kocka[1], P. Prikryl[2], V. Vydra[3] and R. Cerný[3]

[1] Institute of Physics, AVCR, Cukrovarnická 10, 162 00 Prague 6

[2] Institute of Mathematics, AVCR, Zitná 25, 115 67 Prague 1

[3] Department of Physics, Faculty of Civil Engineering, Czech Technical University, Thákurova 7, 166 29 Prague 6, Czech Republic

Keywords: a-Si:H, Recrystallization, Mathematical Model, Excimer Laser, TRR, SEM

Abstract. A mathematical model of a XeCl-excimer-laser induced nonequilibrium melting and solidification is applied to a-Si:H including recrystallization. Experimental time-resolved reflectivity curves are fitted using some material quantities of pc-Si (latent heat, thermal conductivity) as free parameters which correspond to relaxation and changing grain size processes during a surface irradiation. The melting time is well reproduced for the first three shots, proving irradiation results in a fine-grain crystallization process ($< 0.1\,\mu m$). This is confirmed by SEM observation. The irradiated layers possess a luminescence spectrum identical to that of porous Si. The intensity of the luminescence increases with energy density of laser pulses.

Introduction

Technological interest has initiated remarkable research work for obtaining high quality polycrystalline Si (pc-Si) layers deposited onto inexpensive glass substrate. Many theoretical and experimental studies have appeared dealing with the crystallization process of a-Si induced by pulse laser irradiation (e.g., [1–4]). Pulsed laser annealing of thin a-Si films results in rapid heating and subsequent cooling. Depending on pulse duration and its energy density, either a-Si or pc-Si are formed [5]. It has been reported recently that various phases of thin Si films processed by an excimer laser irradiation possess a luminescence, the spectrum of which is identical to that of a luminescence spectrum observed on porous Si prepared by an etching in HF based solution (see, e.g., [6]). In this paper, both experimental and theoretical analyses are performed to study the kinetic and thermodynamic processes in a pulsed-laser-irradiated a-Si:H thin film.

Mathematical model

Our model of phase change processes in a-Si induced by pulsed laser irradiation is based on the experimental work from [1–4]. Accounting for the recent experimental findings [4], we propose the following mechanism of phase change processes in the initial heating phase. The solid material is heated without a phase transition until the surface reaches the temperature T_{ac}, which is high enough to initiate the solid-state explosive crystallization (XC) of amorphous silicon. Absorption of the laser energy together with the release of latent heat of a-Si $\to$ pc-Si transition leads then to the melting of the pc-Si surface. With a further temperature increase, evaporation from the surface also may become important [7].

From the mathematical point of view, the described process can be generally classified as a four-phase problem with three moving boundaries. It can be simplified as in the case of a conversion of the classical two-phase Stefan problem into a one-phase problem. Three phases, l-Si, pc-Si and a-Si are treated explicitly, and the fourth phase, Si vapor, is expressed by means of boundary conditions for l-Si. Supposing one-dimensional heat conduction to be the dominant mode of energy transfer (see [8–9], for a detailed justification of this treatment), we obtain a heat conduction problem with a volume source term.

As in our previous work [7] we consider all phase changes to be nonequilibrium, which follows from the rapidity of the process. Therefore on every phase interface we have one heat balance condition and one kinetic condition. The interface response functions in kinetic conditions are not known either from experimental studies or from molecular dynamics simulations (except for the condition at the liquid/vapor interface). Therefore we use the most simple way in this case and choose the velocities of the interfaces to be linear functions of their temperatures. The latent heat of melting of pc-Si is also not exactly known from the experimental studies. We assume here that $L_{m,pc} = L_{m,a} + L_{ac}$, where $L_{m,a}$ is the latent heat of melting of amorphous silicon, and L_{ac} the latent heat of the a-Si → pc-Si transition.

Later in the laser pulse the position of the l-Si/pc-Si interface reaches its maximum value due to the decrease of the absorbed laser energy. The melting process is then stopped and a crystallization from the melt begins. Since the substrate is pc-Si previously recrystallized from a-Si, we can also assume that the product of the crystallization from the melt will be pc-Si. However, the grain size of the newly formed pc-Si may differ from that of pc-Si substrate, since the process is running at a different temperature and velocity of the interface than the previous a-Si → pc-Si transition. Therefore the values of thermal conductivity K_{pc} and other thermophysical parameters may be different from those used before. Also, the latent heat of crystallization, L_{pc}, released during this process, and the equilibrium temperature of the l-Si → pc-Si transition, T_{pc}, do not generally need to be equal to $L_{m,pc}$ and $T_{m,pc}$, resp.

After some time, the l-Si/pc-Si interface reaches the surface and the crystallization is completed. In a general case the following structure of the sample is finaly formed. Just below the surface there are two pc-Si layers with possibly different grain sizes, then comes the remaining a-Si and the substrate (usually quartz).

This layer arrangement constitutes the initial state for the second laser shot. Modeling the phase transitions during the second shot in the same place on the sample, we assume that there is no further crystallization in the pc-Si layers. The remaining a-Si layer, however, can recrystallize, providing the pc-Si/a-Si interface reaches the temperature T_{ac}. Thus, after the surface of the sample reaches the equilibrium melting temperature $T_{m2,pc}$ (which we consider to be equal to the equilibrium crystallization temperature of pc-Si from the melt, T_{pc}, used in the previous laser shot), the surface begins to melt and an analogous system of model equations as in the solidification phase of the first pulse is solved. A similar situation occurs in the subsequent crystallization from the melt and in eventual further laser pulses.

Experimental

The a-Si:H samples were prepared by a radio frequency glow discharge deposition on the 7059 Corning glass with the following parameters: deposition temperature ∼ 250°C, working pressure ∼ 20 Pa, SiH_4 flow rate ∼ 30 sccm, deposition rate ∼ 1μm/hour.

The samples were irradiated under vacuum and oxygen atmospheres (10^{-6} mbar) by XeCl excimer laser pulses (LPX LAMBDA 250iC, 308 nm, 28ns FWHM, Gaussian shape). The homogeneous part of the beam was transmitted through a rectangular diaphragm giving the energy density uniform to 10%. The energy density on the sample surface was controlled by focusing the beam with a fused silica lens. The experimental setup is described in detail elsewhere [10].

The morphology of the samples was studied by an electron microscope (JEOL SUPER-PROBE 733). Figs. 1, 2 show the surface images of the thin (230 nm) and thick (3μm) a-Si:H layers after the irradiation. The images of thick samples exhibit a finer structure in contrast to thin ones, where a characteristic morphology is already developed after the first laser pulse. The morphology originates from surface tension of the molten layer. There are no composition changes or crystallites bigger than 100 nm in both types of the samples. The crystalline character of the irradiated layer was deduced from the electrical conductivity measurements which showed $\sim$ 3-6 order increase after the irradiation.

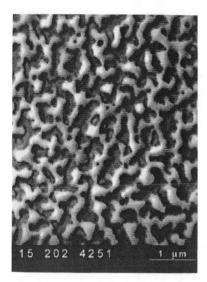

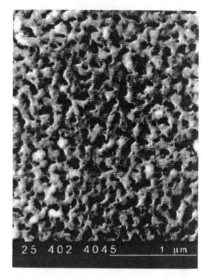

Fig. 1 SEM image of a typical sample with a 230nm thick a-Si layer after three XeCl laser shots.

Fig. 2 SEM image of a typical sample with a 3μm thick a-Si layer after three XeCl laser shots.

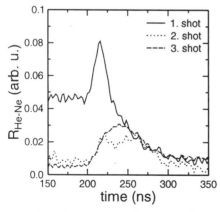

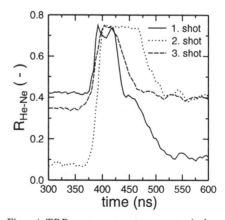

Fig. 3 TRR measurements on a typical sample with a 230nm thick a-Si layer for the first three consecutive shots of XeCl laser, $E = 0.25\,Jcm^{-2}$.

Fig. 4 TRR measurements on a typical sample with a 3μm thick a-Si layer for the first three consecutive shots of XeCl laser, $E = 0.44\,Jcm^{-2}$.

For the thin samples (see Fig. 3), TRR spectra exhibited one sharp maximum followed by a rapid decrease during and after the first pulse above $\sim 0.2\,J/cm^2$ which reflects the melting of the layer followed by a recrystallization. After the first shot, the a-Si/pc-Si recrystallization process was apparently finished, but the changes of grain size were expected during the subsequent shots. Similar processes were also identified in thick samples (see Fig. 4). The main difference is that TRR spectra possess a three-peak structure which is connected with repeated cycles of melting and recrystallization from the liquid phase. The oscillatory behavior was observed during the first pulse only. As in the previous case, the changes in the melt duration between the second and third shots are connected with changes in the grain structure of the polycrystalline matrix. Note that the changes were related to the energy density in a pulse. All the effects described above were observed for energy densities higher than the melting threshold of $E \sim 0.12 - 0.15\,J/cm^2$. Below this value of E, a crystallization from the solid state was identified.

For the thick samples, the intensity and the wavelength of photoluminescence were studied depending on the energy density of a laser pulse. For all energy densities, the wavelength spectrum was identical, but the intensity of the photoluminescence increased with energy density (see Fig. 5) up to $\sim 0.7\,J/cm^2$. The photoluminescence was time stable.

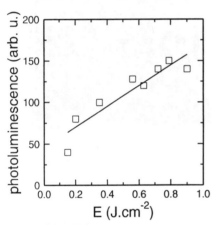

Fig. 5 Photoluminescence as a function of the energy density of XeCl laser for a typical sample with a $3\mu m$ thick a-Si layer.

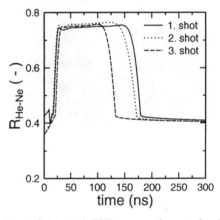

Fig. 6 Calculated TRR curves of a sample with a $3\mu m$ thick a-Si layer for the first three consecutive shots of XeCl laser, $E = 0.44\,Jcm^{-2}$.

Numerical simulation

In the numerical calculations, we used our mathematical model to simulate the experimental situations described in the previous section. Since, to our knowledge, no thermophysical or optical data for pc-Si have so far been published, we had to use estimated values in the model calculations. Some material quantities, K, L_m, L_{ac}, T_{ac}, were supposed to be free parameters in our calculations.

First, we simulated the phase change processes in $3\,\mu m$ a-Si layers on the quartz substrate for three consecutive shots in the same spot on the sample surface. The main criterion in our fitting procedure was to achieve an agreement with the experimental data in the melt duration within the margin of $\pm 5\%$ for all three shots. The results of our computational experiments can be summarized as follows:

First shot: $T_{ac} = 1200\,K$, $L_{a,pc} = 100\,J/g$, $L_{m,pc} = 1420\,J/g$, $T_{m,pc} = 1550\,K$, $L_{pc} = 1400\,J/g$, $K_{a \to pc} = 0.05\,W/cmK$, $K_{l \to pc} = 0.02\,W/cmK$
Second shot: $L_{m,pc} = 1400\,J/g$, $T_{m,pc} = 1450\,K$, $L_{pc} = 1500\,J/g$, $K_{a \to pc} = 0.05\,W/cmK$, $K_{l \to pc} = 0.02\,W/cmK$
Third shot: $L_{m,pc} = 1500\,J/g$, $T_{m,pc} = 1520\,K$, $L_{pc} = 1600\,J/g$, $K_{a \to pc} = 0.05\,W/cmK$, $K_{l \to pc} = 0.04\,W/cmK$.

The values of thermal conductivity K of the pc-Si layer crystallized from the liquid were found to be very low in our numerical experiments. This leads us to conclude that no large-grained pc-Si layer could appear closest to the surface, which is in good agreement with our SEM results.

Numerical simulations of TRR measurements (see Fig. 6) show that in the first pulse, the shapes of TRR curves do not agree very well with our experimental results. The reason apparently lies in the fact that the model does not assume either the existence of the buried molten layer during the solidification phase or the heterogeneous structure of the resulting pc-Si layer nearest the surface.

For the second and third laser shots, the agreement between the measured and calculated TRR curves is much better. This can be regarded as evidence that the resulting structure of pc-Si becomes more homogeneous and regular.

Fig. 7 shows the development of Si phases during the first three laser shots. The upper curves are the positions of the pc-Si/a-Si interface $Z_2(t)$, the lower curves the positions of the pc-Si/l-Si interface $Z_1(t)$. It can be seen from Fig. 7 that for the energy density E=const. for the three pulses, the direct a-Si $\to$ pc-Si transition can run during the first pulse only. In the subsequent shots, the temperature of the pc-Si/a-Si interface is too low to initiate the recrystallization.

The values of free parameters resulting from our fitting procedure were used in the numerical simulations of our second set of experiments, for the 230 nm layers of a-Si:H. Fig. 8 shows that the numerical TRR curve can approximate (except the oscillations in the initial heating phase) the experimental data for the first shot, including the shape of the curve. However, this agreement is much worse in the subsequent shots. This may be attributed to the heterogenity of the resulting structures which is much more important here than after irradiation of the 3 μm a-Si samples.

Comparing our experimental and computational results with other authors, we can see that our mathematical model can predict quite well the experimental results by Im et al [4]. This might be caused by better time resolution of the TRR measurements in [4] than in our experiments.

Conclusions

Experimental and computational analyses of phase change processes in a-Si:H irradiated by a XeCl excimer laser revealed a strong dependence of the resulting structure on the a-Si:H layer thickness, the type of substrate and number and energy density of laser pulses. A good agreement between experimental and simulated melt durations showed a capability of our model to deal with multiple phase transitions occurring during the irradiation. Photoluminescence measurements gave evidence that laser annealing is a promising tool for the preparation of new forms of light-emitting silicon.

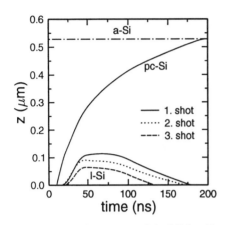
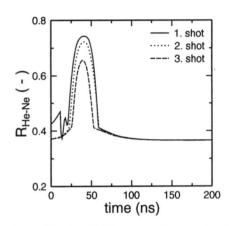

Fig. 7 Calculated positions of the l-Si/pc-Si and pc-Si/a-Si interfaces vs. time in a sample with a $3\mu m$ thick a-Si layer for the first three consecutive shots of XeCl laser, $E = 0.44\,Jcm^{-2}$.

Fig. 8 Calculated TRR curves of a sample with a 230nm thick a-Si layer for the first three consecutive shots of XeCl laser, $E = 0.25\,Jcm^{-2}$.

Acknowledgement

This paper is based upon work supported by the Grant Agency of the Czech Republic, under grant # 202/93/2383, and by the Grant Agency of the Czech Academy of Sciences, under grant # 110433.

References

[1] M. O. Thompson, G. J. Galvin, J.W. Mayer, P. S. Peercy, J. M. Poate, D. C. Jacobson, A. G. Cullis, N. G. Chew, Phys. Rev. Lett. **52**, 2360 (1984).

[2] W. Sinke and F. W. Saris, Phys. Rev. Lett. **53**, 2121 (1984).

[3] D. H. Lowndes, S. J. Pennycook, G. E. Jellison, Jr., S. P. Withrow, D. N. Mashburn, J. Mater. Res. **2**, 648 (1987).

[4] J. S. Im, H. J. Kim, M. O. Thompson, Appl. Phys. Lett. **63**, 1969 (1993).

[5] T. Sameshima and S. Usui, J. Appl. Phys. **70**, 1281 (1991).

[6] C.H. Lin, S.C. Lee and Y.F. Chen, Appl. Phys. Lett. **63**, 902 (1993).

[7] R. Černý and P. Přikryl, Computational Solution of a Moving Boundary Problem with Two Phase Interfaces, to appear in ZAMM **74**, No. 6 (1994)

[8] R. Černý, R. Šášik, I. Lukeš, V. Cháb, Phys. Rev. B **44**, 4097 (1991).

[9] R. Černý, I. Lukeš, V. Cháb, R. Šášik, Thermochimica Acta **218**, 173 (1993).

[10] Z. Dohnálek, V. Cháb, I. Lukeš, R. Šášik, Proceedings of "Laser Advanced Materials Processing 92", Vol. 1, p. 275, Nagaoka, Japan 1992.

Materials Science Forum Vols. 173-174 (1995) pp. 35-40
© 1995 Trans Tech Publications, Switzerland

KINETICS OF LASER-INDUCED SYNTHESIS AND CRYSTALLIZATION OF THIN SEMICONDUCTORS FILMS

K. Kolev and L.D. Laude

Department of Materials and Processes, University of Mons - Hainaut,
7000 Mons, Belgium

Keywords: Laser-Synthesis, Thin Films, Chalcogenides, Kinetics

Abstract

Thin film stacks of Sb and Se layers are successively deposited in 2:3 atomic proportion onto a glass substrate. During irradiation of these multilayered films by a CW Ar^+ laser beam, the kinetics of the resulting Sb_2Se_3 formation is traced via the time dependence of the optical reflectivity of the irradiated zone, which is simultaneously measured at 633 nm using a low power He-Ne laser. XPS and electron diffraction characterizations of the products of irradiation are performed at various stages of laser processing which helps delineating the three essential phases of the compound formation : atomic interdiffusion, solid state synthesis and crystallization. The specific advantages of laser induced growth of semiconducting thin films are discussed.

Introduction

There has been considerable interest in the development of amorphous and polycrystalline thin films of chalcogenide semiconductors for electro-optical and photovoltaic applications, waveguiding materials, photoacoustic devices and image recording media. This interest is further strengthened by the fact that the materials are physically stable and relatively cheap. Among the methods for producing chalcogenide semiconductor films, laser-induced synthesis (L.I.S.) is a promising new one. In a previous paper [1], the present authors have studied the possibility for LIS of Sb_2Se_3, one of the least investigated of the V-VI compound semiconductors.
In this paper, materials characterization by electron diffraction and X-ray photoelectron spectroscopy is added to a study of the kinetics of CW-laser induced growth of Sb_2Se_3 thin films to underline the specific advantages of the synthesis process.

Experimental

Thin layers of (polycrystalline) antimony and (amorphous) selenium are successively deposited onto a glass substrate by electron gun evaporation at a vacuum better than 10^{-4} Pa. The substrates are ultrasonically and glow-discharge cleaned microscope glass plates. The layer thicknesses are controlled during evaporation by a vibrating quartz monitor in order to maintain (within ± 1%) the overall atomic proportion at 2:3 between the two elements. To perform the synthesis, the multilayer sandwich film samples (Fig. 1) are irradiated in air with CW Ar⁺ laser, operating at λ = 514 nm. During irradiation, synthesis develops. Since the resulting product is eventually semiconducting, in contrast with the initial film which is metallic, large variations of the film optical properties may be anticipated. Upon compound synthesis such variations are here traced with time by recording the optical reflectivity of the film at a given wavelength. For this purpose, low power He-Ne laser beam is pointed at the center of (and well within) the Ar⁺ laser irradiated spot, under a few degrees incidence. The back-reflected He-Ne beam intensity is measured as a function of time with a photodiode. This time-resolved reflectivity measurement (λ = 633 nm) is used to analyse the kinetics of the laser-induced compound formation at fixed laser power values. The same irradiated sandwich films are independently characterized by electron diffraction and by X-ray photoelectron spectroscopy (XPS) for which case films obtained after restricted irradiation exposures have been analyzed to corroborate the time-resolved reflectivity measurements.

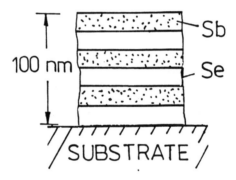

Fig. 1 : Initial multilayer sandwich film

Results and Discussion

The time dependence of the 633 nm reflectivity R(t) is shown in Fig. 2 for a sample having an initial multilayer structure. After the beginning of Ar⁺ laser CW-irradiation, R(t) is seen to decrease abruptly within some 4 s. from the start. After this initial decrease, R(t) is observed to increase rapidly and to reach a maximum at R ≈ 35% after some 10 seconds. Upon further maintaining the Ar⁺ laser irradiation, another reflectivity decrease is observed. This coincides with the development of a circular transparent zone at the center of the irradiated spot on the film surface, i.e. the area which is actually probed by the He-Ne beam.

The evolution of the phases which are successively present at the surface of the irradiated sandwich films is more closely traced with XPS. For this purpose, irradiation is interrupted after various periods of time and the resulting samples may then be XPS-analyzed. This analysis may be restricted to either the Sb levels or Se levels. Since the Se 3d levels have proved to be practically unsensitive to the formation of covalent bonds between Se and Sb atoms [2], XPS analysis is here focused on the Sb 3d levels. These spectra are restricted to the energy range 522 to 543 eV, i.e. in the immediate vicinity of the Sb 3d levels, localized at 528.1 eV (3d 5/2) and 537.6 eV (3d 3/2) binding energies for metallic antimony. Fig. 3 shows XPS spectra of a multilayer film sample before irradiation (Fig. 3a) and after irradiation for 4, 10 or 20 s. (Fig. 3b, c, d, respectively). After the first seconds of irradiation, the observed decrease of R (Fig. 2) may

be associated with progressive Sb exhaustion in massive metal state. In effect, the XPS lines
(Fig. 3b)

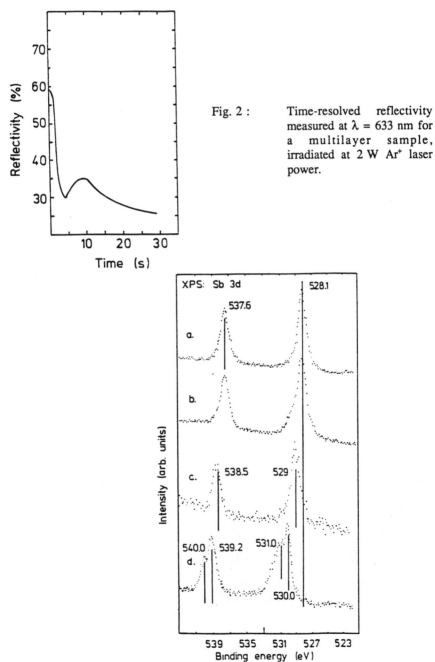

Fig. 2 :   Time-resolved   reflectivity
measured at $\lambda = 633$ nm for
a   multilayer   sample,
irradiated at 2 W Ar⁺ laser
power.

Fig. 3 :   Sb 3d XPS-lines from the surface of a Sb/Se multilayer film before
irradiation (a) and after 4 s. (b), 10 s. (c) or 20 s. (d) irradiation at
2 W Ar⁺ laser power.

which correspond to levels 3d 3/2 and 3d 5/2 are still centered at the same energies as those detected for metallic Sb (Fig. 3a) with only a marginal broadening to higher energy of the 3d 5/2 peak. This is indicative of an alloying process rather than a compound formation at, or very near, the surface of the sample. In contrast, after 10 seconds (Fig. 3c), the 3d lines are both chemically shifted to higher energies by $\Delta E_0 = 1.0$ eV. An electron diffraction analysis of the resulting film has been carried out. The obtained pattern (Fig. 4) shows the usual rings which are characteristic of a polycrystalline material. Identification of the rings is shown in Table I which clearly evidences for the formation of the crystalline (ortorhombic) Sb_2Se_3 compound alone. Some texturing is responsible for differences in intensity at d = 3.75, 2.86 and 1.98 Å, respectively. The XPS chemical shift ΔE_0 may thus correspond to the formation of Sb-Se bonds in the immediate vicinity (0 to 15 Å) of the free surface of the film, i.e. within the depth of XPS analysis. After 20 s. irradiation (Fig. 3d), the XPS 3d levels now appear at even higher energies. Clearly the XPS peaks are split. After appropriate deconvolution, they may be delineated at 530 eV and 531 eV for Sb 3d 5/2 and 539.2 and 540.4 eV for Sb 3d 3/2. The pair of peaks at 530-539.2 eV is shifted by $\Delta E_1 = 1.7 \pm 0.2$ eV and the second pair at 531-540.4 eV by $\Delta E_2 = 3.0 \pm 0.2$ eV. These two different chemical shifts ΔE_1 and ΔE_2 are sufficiently well resolved to be associated with two distinct Sb configurations. Note that they are also different from ΔE_0. On the other hand, the time occurence of these two simultaneous shifts corresponds to the second decrease of R(t) in Fig. 2. This means that the Sb_2Se_3 compound, which was evidenced after 10 s. irradiation (Fig. 4) has now disappeared from the film surface and from within (at least) 15 Å from the surface. The irradiation being performed in air, a possible oxidation of Sb_2Se_3 could be responsible for the Sb_2Se_3 disappearance. This seems to be confirmed by the literature [2] in which the 3d 5/2 binding energy is quoted at 529.6 eV in Sb_2O_3 and 530.6 eV in Sb_2O_5. These values compare reasonably well with XPS peak positions in Fig. 3d, within experimental uncertainties.

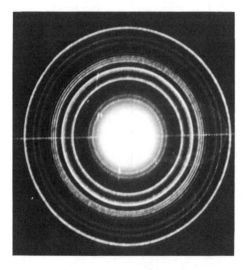

Fig. 4 :　Electron diffraction pattern from a multilayer Sb/Se sample after 10 s. irradiation at 2 W Ar$^+$ laser power.

Table I

Indexed electron diffraction pattern of the observed laser-synthsized film (A) and expected orthorhombic Sb_2Se_3 (B)[1] compound.

A		h	k	l	B	
I/I_0	d (Å)				d (Å)	I/I_0
10	8.26	1	1	0	8.28	8
10	5.88	0	2	0	5.89	25
10	5.22	1	2	0	5.25	55
1	4.13	2	2	0	4.14	10
70	3.75	1	0	1	3.76	12
50	3.68	3	1	0	3.682	16
10	3.29	0	2	1	3.297	12
80	3.15	2	1	1	3.162	75
50	2.86	2	2	1	2.868	100
50	2.77	3	0	1	2.776	60
50	2.70	3	1	1	2.703	20
40	2.60	4	2	0	2.608	20
30	2.51	3	2	1	2.513	30
20	2.30	4	1	1	2.303	12
40	2.18	2	5	0	2.184	35
30	2.16	5	2	0	2.164	20
5	2.06	4	4	0	2.070	10
40	2.01	5	0	1	2.010	30
100	1.98	0	0	2	1.989	25

[1]JCPDS International Center for Diffraction Data USA, 1980.

It is now interesting to reconsider comparatively the R(t) behaviour given in Fig. 2 and XPS-spectra, shown in Fig. 3. Obviously, five collaborating processes are developing in this multilayer sandwich film upon CW Ar^+ laser irradiation : a) atomic interdiffusion, b) compound synthesis, c) crystallization, d) eventual compound reduction and e) antimony oxydation. Though the order of these processes follows logically from a) to e), it may not be clear whether some time or/and space overlap between them may take place effectively within the volume of the film, at one time or another. Would this overlap be the case, and given the respective (and dissimilar) kinetics of these processes, difficulties would follow in a proper modelling of R(t). An attempt to solve the problems of this R(t) modelling will be proposed in another article.

Conclusion

It has been shown in this work that good quality Sb_2Se_3 films can be produced by laser irradiating sandwich films composed of a superposition of elemental layers in stoichiometric atom proportions. A distinct advantage is obtained in constructing the film from a multiplication of single elemental layers the thickness of which does not exceed 150 Å. In that case, the three essential phases of the compound formation (atomic interdiffusion, solid state synthesis and crystallization) are forced to develop at once within the whole volume of the film.

References

[1]   K. Kolev and L.D. Laude, Appl. Surf. Sci. **54**, 358-361 (1992)

[2]   Perkin-Elmer : *Handbook of X-ray Photoelectron Spectroscopy*, (Physical Electronics Publications, USA 1982), p. 92 and 120.

Materials Science Forum Vols. 173-174 (1995) pp. 41-52
© 1995 Trans Tech Publications, Switzerland

SURFACE PATTERNING AND THIN-FILM FORMATION BY PULSED-LASER ABLATION

D. Bäuerle, E. Arenholz, J. Heitz, S. Proyer, E. Stangl and B. Luk'yanchuk

Angewandte Physik, Johannes-Kepler-Universität Linz, A-4040 Linz, Austria

Keywords: Pulsed-Laser Ablation, Surface Patterning, Thin Film Formation

Abstract. In this paper we present recent results on laser-induced surface modifications, surface patterning by ablation, and thin-film formation by pulsed-laser deposition. Different types of structure formation and the modelling of UV-laser ablation are also discussed.

Definition of terms

Material removal under the action of laser light is denoted as laser ablation. If the process takes place under quasi-equilibrium conditions it is also denoted as laser-induced (thermal) vaporization and very similar to conventional evaporation techniques. New phenomena are observed with short high-intensity laser pulses. In this case material removal takes place under *non*-equilibrium conditions and is often denoted as pulsed-laser ablation. This regime is considered throughout the present paper.

Pulsed-laser ablation can tentatively be classified into thermal, photophysical, and photochemical ablation. The microscopic mechanisms involved in the ablation process depend on the laser parameters (photon energy, fluence, and dwell time of the laser light), the type of excitation, the dissipation of the energy within the solid, etc.

Thermal ablation is based on laser-induced heating, melting and (thermal) vaporization. Here, the dissipation of the excitation energy, which is characterized by the thermal relaxation time τ_T , is so fast that the detailed excitation mechanism becomes irrelevant.

In *photophysical* ablation, non-thermal excitations (electron-hole pairs, electronically excited species, etc.) directly influence the ablation rate.

Photochemical (photolytic) laser ablation is based on direct non-thermal bond breaking.

Irrespective of the fundamental mechanisms involved, laser ablation has been demonstrated to be a powerful tool in surface micro-patterning of hard, brittle, and heat-sensitive materials and in the fabrication of thin films with complex stoichiometry. The latter technique is denoted as *pulsed-laser* deposition (PLD).

Surface Modifications, Structure Formation

Materials that are irradiated with short high-intensity laser pulses show a number of common features. Congruent surface ablation is observed only if the laser fluence, ϕ , exceeds a certain threshold fluence, ϕ_{th} . Correspondingly, experimental observations on different materials can be classified into regimes $\phi < \phi_{th}$, $\phi \approx \phi_{th}$, and $\phi > \phi_{th}$. Clearly, laser ablation can be performed in

vacuum.   Nevertheless,   reactive   and   even   non-reactive   atmospheres   may
considerably affect the rate and chemistry of the ablation process.
     Interactions below Threshold. With laser fluences $\phi < \phi_{th}$, changes in
surface morphology and microstructure, the generation of defects, surface
roughening, and the depletion of one or several components of the material are
frequently observed [1]. With certain systems, a tail in the ablation rate
appears [2]. Sub-threshold and near-threshold irradiation can also result in the
formation of different types of coherent and non-coherent structures.
     The most well-known coherent structures are the so-called *ripples* which have
been investigated in detail for semiconductors and metals for laser wavelengths
between 1 and 10 μm [3]. Recently, ripples were also observed with polymers [4].
Figure 1 shows new results for polyethylene terephthalate (PET, DuPont Mylar, 50
μm thick) obtained with 308 nm excimer-laser radiation ($\phi \approx 26$ mJ/cm^2, N = 3000,
5 pps). Here, the periodicity of ripples, Λ, is plotted versus the angle of
incidence for p- and s-polarization. For p-polarization the experimental data
can be fitted by $\Lambda = \lambda/ [n^2 - \sin^2\theta]^{1/2}$ and for s-polarization by $\Lambda = \lambda/ [n -
|\sin \theta|]$ where θ is the angle of incidence. The fit parameter n can be
interpreted as refractive index of the material in which the scattered light
propagates. For 308 nm and 248 nm laser radiation the fit yields n = 1.17 ± 0.01
and n $\approx$ 1.22, respectively. Uniform sample heating during irradiation diminishes
the minimum number of pulses necessary for first observation of ripples.
AFM-analysis gave direct evidence that ripple formation is dominated by material
flow. The tops of ripples are elevated over the non-irradiated surface while the
valleys are below this surface. Thus, at least with PET-samples and with the
wavelengths investigated, neither evaporation nor ablation is responsible for
these structures. Typical ripple amplitudes are 100 - 300 nm.
     The Threshold Fluence ϕ_{th} . Significant ablation is observed only above a
certain threshold fluence, ϕ_{th} . With most inorganic insulators, ϕ_{th} is around
0.5 to 2 J/cm^2. With organic materials it is, typically, between 0.01 and 1
J/cm^2. Clearly, these numbers give only the order of magnitude for standard
experiments using UV-laser pulses with nanosecond duration.
     For fluences $\phi \approx \phi_{th}$ many of the surface modifications already mentioned for
subthreshold fluences become more pronounced. Additionally, non-stoichiometric
ablation is observed [5]. Recently, it has been found that some types of surface
roughening of polymers are related to (non-coherent) dendritic surface
structures that develop after single-pulse excimer-laser irradiation [6]. An
example of such a dendrite is shown in Fig. 2 [7].
     The number of dendrites, the length of arms, and the number of bifurcations
all increase with time. Figure 3 shows the lateral growth rate of dendrites
versus time. The dendrites seem to be polymeric crystallites that grow within
the amorphous surface layer (ℓ_α is some 0.1 μm thick) formed during irradiation
of the (semi-crystalline) polymer foil. Compared to bulk crystallization in PET
[8] the growth velocities of dendrites (at room temperature) are very high. The
laser-induced modification of the polymer surface which gives rise to dendrite
formation and to possible changes in chemical composition is responsible for the
significant increase in adhesion force observed for metal films. Figure 4 shows
recent results achieved with CoNi films on PET [9].
     Another phenomenon observed in this fluence regime is the increase in
electrical conductivity of PI surfaces. Depending on the ambient atmosphere,
this conductivity change may exceed 15 orders of magnitude [10].
     Congruent Ablation, $\phi > \phi_{th}$. High power short laser pulses cause short
interaction cycles resulting in congruent ablation of small material volumes. In
this regime, the thickness of the layer ablated per pulse, Δh, should be of the
order of the thermal penetration depth $\ell_T \approx [D_T \tau_\ell]^{1/2}$ or the optical penetration

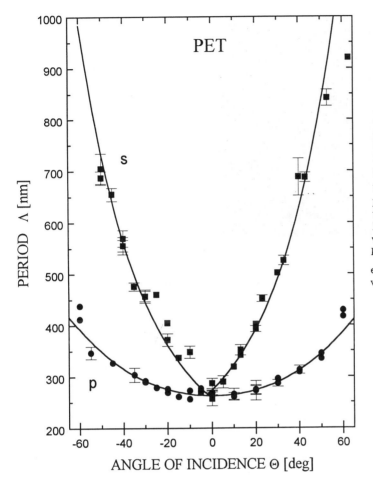

Fig. 1: Period, Λ, of ripples observed on PET versus angle of XeCl-laser-beam incidence, θ ($\phi < \phi_{th}$). Different branches refer to p- and s-polarization of the laser light. The grating vector of ripples, k_r, is $\perp$ to the electric field vector, E.

depth, $\ell_\alpha = \alpha^{-1}$, depending on which is larger, i.e.

$$\Delta h \approx \max(\ell_T , \ell_\alpha) .$$ (1)

With this condition, the dissipation of the excitation energy beyond the volume ablated during the pulse and the segregation of the single material compounds is suppressed. It is evident that (1) is a crude estimation for ideal conditions only. With many materials, this condition can be reasonably well fulfilled with UV-laser light and nanosecond pulses. This can be made plausible from a simple estimation of the characteristic times involved in the process. Let us assume purely thermal ablation and one-dimensional heat flow. For surface absorption, the time to reach the effective vaporization temperature, T_v^{eff}, on the target

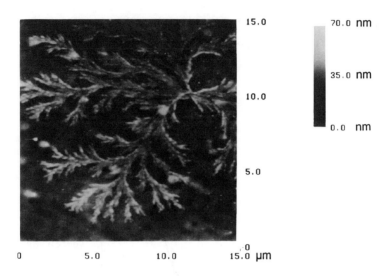

Fig. 2: AFM-picture of a surface dendrite on PET. Irradiation was performed in vacuum with a single KrF-laser pulse (ϕ = 41 mJ/cm^2).

surface can be estimated from

$$t_v \approx \frac{\pi}{4D} \left\{ \frac{\kappa\,\theta_v^{eff} \cdot \tau_\ell}{\phi\,[1 - R]} \right\}^2 \qquad (2)$$

The effective temperature rise $\theta_v^{eff} \equiv \Delta T^{eff} = T_v - T(\infty) + \Delta H_v/c_p$ takes into account, in a crude way, the enthalpy of vaporization. With laser parameters typically employed in pulsed-laser ablation, $\phi \approx 5J/cm^2$ and $\tau_\ell \approx 20$ ns we find for a metal target such as Cu and a non-metallic ceramic target such as a high temperature superconductor (HTS) values of t_v of about 10^{-8} and 10^{-11} s, respectively. Clearly, this estimation ignores laser-plasma interactions. At least the latter value seems to be small enough to avoid material segregation. This is supported by experimental findings.

With visible- and infrared-laser radiation, this condition is often more difficult to fulfill.

Ablation Rates. The ablation rate is often defined as the thickness of the layer ablated per laser pulse, W [µm/pulse]. It should be noted, however, that other definitions are frequently used as well, for example the mass loss per pulse, etc. In most experiments, W is an *average* rate which is obtained by dividing the total ablated thickness by the (large) number of laser pulses employed. W depends on the quantum energy and fluence of photons, the width of the laser focus, the heat or optical penetration depth, internal stresses, etc. The heat penetration depth depends on the thermal diffusivity of the material and the laser-pulse length and pulse shape. The optical penetration depth depends on the absorption coefficient and thereby on the laser wavelength. With the fabrication of deep holes, W becomes dependent also on the number of laser

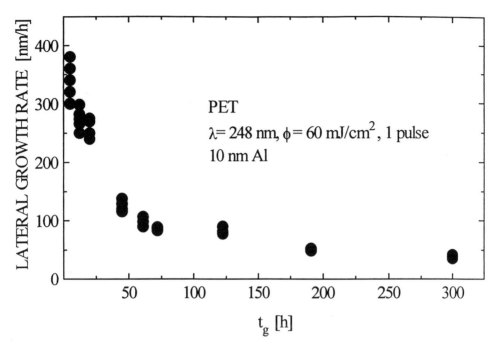

Fig. 3: Lateral growth rate of dendrites at T ≈ 300 K versus time, t_g (single laser pulse, ϕ = 60 mJ/cm^2, 10 nm Al coating).

pulses.

Surface Patterning

Laser-induced surface patterning is a single-step process which can be based on structural or chemical transformations of surfaces, on material removal by pulsed-laser ablation, etc. The resolution achieved in these processes is in the sub-micrometer range. This technique has been described extensively within the literature [1].

Surface patterning by pulsed-laser ablation can be performed by direct focusing of the laser light onto the substrate surface, by direct masking, or by laser-light projection. This has been demonstrated in particular for inorganic insulators, high-temperature superconductors (HTS), organic polymers, and biological tissues. The ablation rates are, typically, between some 0.01 μm/pulse and several μm/pulse. The corresponding laser fluences are between 0.1 J/cm^2 and several J/cm^2. Reasonable suppression of material damage can be achieved as long as (1) is fulfilled. With (thermally) well conducting heat sensitive materials ns pulses may be too long for high resolution micro-patterning. An example are thin films of Y-Ba-Cu-O. Here, almost damage free patterns with lateral widths as small as 4 μm can be fabricated only with femtosecond laser pulses [11,12]. Figure 5 shows the critical current density j_c(77K) versus the width of Y-Ba-Cu-O lines on a (100) MgO substrate fabricated

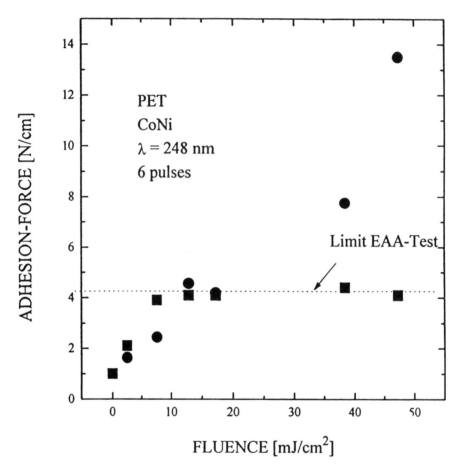

Fig. 4: Enhancement in adhesion force versus laser fluence. Laser irradiation was performed prior to metal coating. The data were derived from two different peel tests. ■ Conventional inverse EAA peel test. ● Improved inverse EAA peel test (after [9]).

by different patterning techniques [12]. Dry (!) patterning with fs pulses yields almost the same results as wet-chemical etching. Similar results can be achieved with ns pulses if the film is protected by a thin layer of photoresist.

With many materials and with certain experimental conditions, ablation results in smooth surfaces while in other cases, rough surfaces or periodic structures are obtained. Experimentally, some of these surface structures have been shown to be related to interference phenomena, to thermocapillary waves, to internal stresses, etc. In the latter case, structure formation can be avoided by adequate pretreatment of the sample. In many other cases, however, the experimental results cannot be explained adequately along these lines [6, 13].

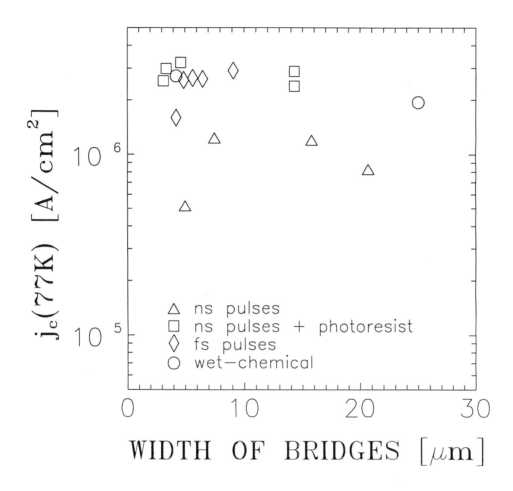

Fig. 5: Critical current density (at 77 K) versus the widths of Y-Ba-Cu-O lines (film thickness ≈ 2500 Å) on (100) MgO. Symbols refer to different techniques: wet-chemical etching o, ns-KrF-laser-light projection with □, and without △ a photoresist layer, and fs-KrF-laser-light projection ◊ (after Ref.[12]).

Pulsed-Laser Deposition (PLD)

Thin-film formation by *pulsed-laser* deposition (PLD) is achieved by condensing the material ablated from a solid or liquid target on a substrate [14]. The techniqe is fast, cheap, and versatile. It permits to fabricate from a *single* target thin stoichiometric films of multicomponent materials with areas up to several square centimeters.

    Among the materials studied in most detail are compound semiconductors, insulators, and high-temperature superconductors (HTS). Due to the strong forward direction of the plasma plume, films produced by PLD have a characteristic thickness profile [15].

Pulsed-laser deposition permits to synthesize components that can be prepared either not at all or not in single-phase by solid-state reactions or by standard evaporation techniques. This unique possibility is related to the lower substrate temperatures that can be employed in PLD, to the type and energy of species involved in this process, and to the short turn around times. Let us consider three examples:

Reactive deposition of $YBa_2Cu_3O_7$ in N_2O atmosphere permits to lower the substrate temperature from, typically, $730°C$ to about $600°C$ [16].

Pulsed-laser deposition permits to extend the solid-solution range for substitutions in compounds. For $YBa_{2-x}Sr_xCu_3O_7$ standard ceramic techniques permit to stabilize the (pure) 123-phase only for Sr contents x < 1.2 [17,18]. Pulsed-laser deposition permits to extend this range up to x = 1.8 [19].

Single-phase $LuBa_2Cu_3O_7$ [20] and $TmBaSrCu_3O_7$ [21] films were successfully prepared in oxygen atmosphere by PLD from stoichiometric ceramic targets. With $LuBa_2Cu_3O_7$ the substrates employed were (100)MgO and (100)$SrTiO_3$ at a temperature $T_s \approx 690°C$. With (100)$SrTiO_3$ substrates, the zero resistance temperature for c-axis oriented films was 90 K. The transition width was well below 1K. The critical current density of such films at zero magnetic field, $j_c(83K)$, exceeded 10^6 A/cm^2. On (100)MgO substrates the transition temperature

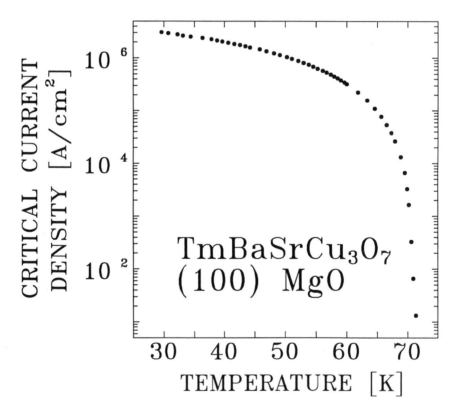

Fig. 6: Temperature dependence of the critical current density of single-phase $TmBaSrCu_3O_7$ films fabricated by PLD.

was $T_{co} \approx 85$ K.

Single phase $TmBaSrCu_3O_7$ films on (100) MgO substrates were fabricated by employing deposition parameters similar to those used with $YBa_2Cu_3O_7$ and $TmBa_2Cu_3O_7$ ($T_s \approx 750°C$; $p(O_2) \approx 1$ mbar). The maximum values achieved were $T_{co} = 71$ K and $j_c(57$ K$) \approx 4 \cdot 10^5$ A/cm^2 (see Fig. 6). For comparison, with Tm-123 films on (100) MgO substrates we find $T_{co} \approx 88$ K and $j_c(77$ K$) \approx 2.5 \cdot 10^6$ A/cm^2 [22]. It should be emphasized that with standard techniques it has *not* been possible to synthesize *single*-phase $LuBa_2Cu_3O_7$ and $TmBaSrCu_3O_7$.

A material where standard techniques did *not* permit to synthesize the 123 phase at all is $LuBaSrCu_3O_7$. By means of PLD, however, pure 123 phase $LuBaSrCu_3O_7$ films have been grown in N_2O atmospere at $T_s \approx 600°C$. The zero resistance temperature achieved was $T_{co} \approx 54$ K and the critical current density $j_c(45$ K$) \approx 10^5$ A/cm^2 [20].

Theoretical Description of Pulsed-Laser Ablation

Pulsed-laser ablation by infrared laser radiation is considered to be based on purely thermal mechanisms. The situation is different for UV-laser ablation. Here, the relative importance of thermal and non-thermal mechanisms is still under discussion, in particular with organic polymers where the UV-photon energy is comparable with the bond breaking energy.

Among the arguments frequently used in favor of a mainly photochemical process are: (i) The observation that UV laser-induced ablation of heat-sensitive materials such as organic polymers, can be performed almost without any damages of the remaining material, and in particular without indications for melting. (ii) The non-equilibrium between the translational, vibrational, and rotational temperatures of ablated products. (iii) The large number of species with high translational energies. (iv) The differences in the composition of species obtained with UV- and IR-laser ablation. (v) The very high temperatures that would be necessary to explain the experimentally observed ablation rates on the basis of a purely thermal model. On the other hand, for a simple photochemical process, one would expect the ablated thickness to be dependent only on the total laser fluence (dose), and not on the laser-beam intensity and the dwell time separately, as found experimentally. The most serious argument against a purely photochemical process, however, is the Arrhenius-type behavior of the ablation rate observed for certain materials near the ablation threshold [2]. Last but not least it should be emphasized that the dominating ablation mechanism will depend on the particular material under investigation and the laser parameters employed.

The difficulty in the interpretation of the experimental results is closely related to the complexity of the optical excitation and energy-dissipation mechanisms involved in the ablation process.

It is evident that the role of thermal and non-thermal mechanisms in pulsed-laser ablation is closely related to the different relaxation channels and the corresponding times involved. The situation is even more complicated. The laser-light intensities employed in pulsed-laser ablation yield high densities of excited species, induce multiphoton (non-linear) excitations, etc. This will not only result in changes of the characteristic relaxation times derived from experiments using UV-lamp irradiation but, additionally, will open up new relaxation channels. Clearly, these changes in excitation and energy-dissipation processes depend not only on the laser-beam intensity but on the duration of the laser pulse, τ_ℓ, as well. In the total energy balance, we have to take into account excited species that leave the material surface before they transfer their excitation energy, or part of it, to the bulk. Non-radiative relaxation processes will result in a laser-induced temperature rise, the generation of internal stresses, defects etc. It is evident that all of these processes influence each other and depend on both the laser parameters and the physical properties of the particular material. Because of the complexity of the problem and because of the lack of reliable data on relaxation times, local temperatures, etc., theoretical models can only try to describe common features observed in laser-induced material ablation.

Recent careful experimental investigations on the ablation of polyimide (PI) can neither be described on the basis of a purely photochemical process nor on the basis of a purely photothermal process [2]. They can, however, be described on the basis of photophysical ablation [13,23,24]. An important feature of this mechanism is the lower activation energy for the desorption of electronically excited species with respect to electronic ground-state species. The details of this model have been described in [23,24]. It can explain the following

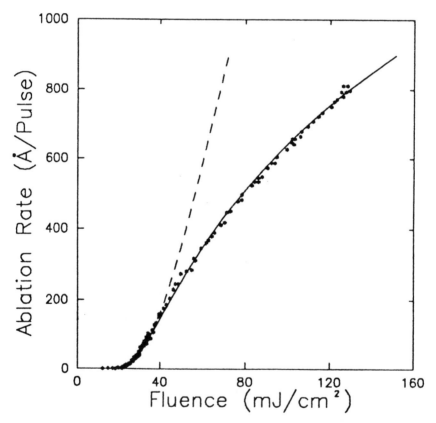

Fig. 7: Ablated layer thickness per pulse for polyimide (PI) versus laser fluence (248 nm KrF, $\tau_\ell \approx 15$ ns). Experimental data have been taken from [2]. Full curve is calculated from interpolation formula (3) with A = 116618 [Å/pulse], B = 220 [mJ/cm^2], α_o = 0.447 α. Dashed curve represents the interpolation given in [2].

experimental results in UV-laser ablation of polymers:
- The Arrhenius-type behavior of the ablation rate for small fluences (below the ablation threshold).
- The anomalously small activation energies, which are significantly smaller than bond-breaking energies.
- The relatively small surface temperatures together with high ablation rates.
- The absence of thermal instabilities in UV-laser ablation.

Figure 7 shows experimental data taken from [2]. These data can be described by the interpolation formula given in [25]

$$\phi = B \exp [\alpha_o \, \Delta h] \, \ln^{-1} \left[\frac{A}{\Delta h} \right] , \tag{3}$$

where A, B, α_o are constants. This equation is valid for nanosecond laser ablation within a wide range of fluences $0 < \phi \le 4 \, \phi_{th}$. An important prediction of the model is the low surface temperature for ablation. For example, with polyimide, KrF-laser radiation, and $\phi \approx \phi_{th}$ this temperature is below 2000 K [23]. Recently, direct measurements of the surface temperature did yield $T_s \approx$ 1660 K [25].

Summary

Laser-induced surface modifications and micro-patterning by ablation is increasingly applied in different fields of modern technology, in particular in microelectronics and micromechanics. Pulsed-laser deposition permits to fabricate thin films of compounds that cannot be synthesized at all or not in single phase by standard techniques. For an understanding of the fundamental mechanisms involved in pulsed-laser ablation, the analysis of experimental results on the basis of model calculations is of great importance. Here, photophysical processes related to the desorption of both electronically excited and ground state species seem to be important.

Acknowledgements: We wish to thank the "Fonds zur Förderung der wissenschaftlichen Forschung in Österreich" for financial support. One of us (B.L.) would also like to thank the "Russian Fund of Fundamental Research".

References

[1] D. Bäuerle: *Chemical Processing with Lasers*, Springer Series in Materials Science **1** (Springer-Verlag 1986); and Appl. Phys. B **46**, 261 (1988).
D. Bäuerle, J. Heitz, W. Ludorf, P. Schwab, X.Z. Wang: In the Proceedings of "*In-situ Patterning: Selective Area Deposition and Etching*", ed. by R. Rosenberg, A.F. Bernhardt and J.G. Black, Mat.Res.Soc.Symp.Proc. Vol. **158** (1990), p. 451.
D.C. Paine and J.C. Bravman (Eds.): *Laser Ablation for Materials Synthesis* Mat.Res.Soc.Symp.Proc. Vol. **191** (Pittsburgh, Pennsylvania 1990);
J.C. Miller and R.F. Haglund,Jr. (Eds.): *Laser Ablation - Mechanisms and Applications*, Lecture Notes in Physics, Vol **389** (Springer, Berlin 1991).
E. Fogarassy and S. Lazare (Eds.): *Laser Ablation of Electronic Materials - Basic Mechanisms and Applications,* Proc. E-MRS, Vol **4** (North-Holland Elsevier 1992);
E. Arenholz, V. Svorcik, T. Kefer, J. Heitz and D. Bäuerle, Appl. Phys. A **53**, 330 (1991).
[2] S. Küper, J. Brannon and K. Brannon, Appl. Phys. A **56**, 43 (1993).

[3] For an overview see: S.A. Akhamanov, V.I. Emel'yanov, N.I. Koroteev and
     V.N. Seminogov, Sov. Phys. Usp. **28**, 1084 (1985) and references therein.
[4] P.E. Dyer and R.J. Farley, Appl. Phys. Lett. **57**, 765 (1990);
     M. Bolle, S. Lazare, M. Le Blanc and A. Wilmes, Appl. Phys. Lett. **60**, 674
     (1992).
[5] J. Heitz, X.Z. Wang, P. Schwab, D. Bäuerle and L. Schultz, J. Appl. Phys.
     68, 2512 (1990).
[6] J. Heitz, E. Arenholz, D. Bäuerle, H. Hibst, A. Hagemeyer and G. Cox, Appl.
     Phys. A **56**, 329 (1993).
[7] J. Heitz, E. Arenholz, D. Bäuerle and K. Schilcher, to be published.
[8] F. van Antwerpen and D.W. van Krevelen, J. Polym. Sci. Polym. Phys. Ed. **10**,
     2423 (1972).
[9] A. Hagemeyer, H. Hibst, J. Heitz and D. Bäuerle, J. Adhesion Sci. Technol.
     8, 29 (1994);
     J. Heitz, E. Arenholz, D. Bäuerle, H. Hibst and A. Hagemeyer, Appl. Phys. A
     55, 391 (1992).
[10] E. Arenholz, J. Heitz, M. Wagner, D. Bäuerle, H. Hibst and A. Hagemeyer,
     Appl. Surface Science **69**, 16 (1993);
     M. Schuhmann, A. Sauerbrey and M.C. Smayling, Appl. Phys. Lett. **58**, 428
     (1991).
[11] P. Schwab, J. Heitz, S. Proyer and D. Bäuerle, Appl. Phys. A **53**, 282
     (1991).
[12] S. Proyer, E. Stangl, P. Schwab and D. Bäuerle, to be published in Appl.
     Phys. A (1994);
     R.M. Schalk, G.S. Hosseinali, H.W. Weber, S. Proyer, P. Schwab,
     D. Bäuerle and S. Gründorfer, Phys. Rev. B **49**, 3511 (1994).
[13] B. Luk'yanchuk, N. Bityurin, S. Anisimov and D. Bäuerle, Appl. Phys. A **57**,
     449 (1993).
[14] For an overview see, e.g., D. Bäuerle, Appl. Phys. A **48**, 527 (1989).
[15] S.I. Anisimov, D. Bäuerle and B.S. Luk'yanchuk, Phys. Rev. B **48**, 12076
     (1993).
[16] P. Schwab and D. Bäuerle, Physica C **182**, 103 (1991);
     P. Schwab, A. Kochemasov, R. Kullmer and D. Bäuerle, Appl. Phys. A **54**, 166
     (1992).
[17] A. Ono, T. Tanaka, H. Nozaki and Y. Ishizawa, Jpn. J. Appl. Phys. **26**,
     1687 (1987).
[18] B.W. Veal, W.K. Kwok, A. Umezawa, G.W. Crabtree, J.D. Jorgensen, J.W.
     Downey, L.J. Nowicki, A.W. Mitchell, A.P. Paulikas and C.H. Sowers, Appl.
     Phys. Lett. **51**, 279 (1987).
[19] P. Schwab, X.Z. Wang, S. Proyer, A. Kochemasov and D. Bäuerle,
     Physica C **214**, 257 (1993).
[20] P. Schwab, X.Z. Wang and D. Bäuerle, Appl. Phys. Lett. **60**, 2023 (1992).
[21] E. Stangl, S. Proyer, B. Hellebrand and D. Bäuerle, to be published.
[22] E. Stangl, S. Proyer and B. Hellebrand, to be published in
     Physica C (1994).
[23] B. Luk'yanchuk, N. Bityurin, S. Anisimov and D. Bäuerle, Appl. Phys. A
     57, 367 (1993).
[24] B. Luk'yanchuk, N. Bityurin, S. Anisimov and D. Bäuerle: In *Excimer Lasers*,
     ed. by L.D. Laude, NATO ASI Series (Kluwer Academic Publishers, 1994).
[25] D.P. Brunco, M.O. Thompson, G.E. Otis and R.M. Goodwin: In *Laser Ablation
     in Material Processing: Fundamentals and Applications,* ed. by B. Braren,
     J.J. Dubowski and D.P. Norton, MRS Symp. Proc., **285** (Pittsburgh,
     Pennsylvania 1993) p. 151.

Materials Science Forum Vols. 173-174 (1995) pp. 53-58
© 1995 Trans Tech Publications, Switzerland

LASER PATTERNING OF CuInSe₂ /Mo/SLS STRUCTURES FOR THE FABRICATION OF CuInSe₂ SUB MODULES

L. Quercia[1], S. Avagliano[1], A. Parretta[1], E. Salza[2] and P. Menna[1]

[1] Ente per le Nuove Tecnologie, l'Energia e l'Ambiente (ENEA), Centro Ricerche Fotovoltaiche,
I-80055 Portici (Napoli), Italy

[2] ENEA, Centro Ricerche Energia, I-00060 Casaccia (Roma), Italy

Keywords: Laser Back Scribing, CuInSe₂ (CIS) Integrated Sub Module Fabrication, Molybdenum Patterning, CuInSe₂ Patterning

Abstract. In this work we show the viability of laser scribing as a tool for selective patterning of CuInSe₂(CIS)/Molybdenum (Mo)/Soda Lime Silica (SLS) thin film structures. The dependence of scribing quality on the basic process parameters is investigated by optical microscopy, talystep, SEM and microprobe analysis.

Two different configurations are used for Mo-patterning: the frontal scribing whereby the laser beam interacts first with the air/Mo interface and the back scribing [1] with the laser beam passing through the glass slide and interacting with SLS/Mo interface. Back scribing can result in a high quality patterning with no droplets or swelling, but only the Mo layer is removed and the SLS is left untouched. The Mo removal is due to a laser induced stress detachment. Using the frontal scribing, it is also possible to obtain good quality scribing where evaporation is the main mechanism for Mo removal. To obtain good quality scribing, it is necessary to gain accurate control of the focus position and of laser beam characteristics. When a laser with a poor gaussian profile was used, high swelling and large droplets were always present.

Optimum scribing parameters for both CIS/SLS and Mo proved to be roughly equivalent. Consequently, selective scribing of an individual layer was not possible using Q-switched operation. Using cw laser power output the relatively low melting and evaporation temperatures, low thermal diffusivity, optical reflection and high absorption coefficient of the CIS layer allow for local heating and consequent evaporation of the CIS whilst leaving the Mo layer unaffected. Scribes are performed on CIS/Mo/SLS layers and the resulting selective removal of the CIS layer is shown.

Introduction

The fabrication process for integrated solar CIS modules requires some steps to electrically connect the thin film layers. For serial connected modules, having substrate/Mo/CIS/CdS/ZnO structure, three interconnection steps are required: Mo, CIS/CdS and ZnO contact scribing. The scribes in the Mo and ZnO layers electrically isolate the top and bottom segment contacts from the adjacent segment. The scribes through the adsorber layer, with subsequent ZnO deposition, form the interconnect between the top contact of one segment and the bottom contact of the adjacent segment. Light is transmitted through the ZnO top transparent contact into the CIS layer where it is absorbed and converted into electrical energy. The photo generated current flow across the CIS/ZnO junction, passing through the ZnO, and across the ZnO/Mo contact interconnects the next stripe cells.

To interconnect the cells different techniques have been employed [2,3], but till now the more promising is the laser scribing techniques. This has been already applied on the first step of the CIS module's fabrication, but on this subject no information has been published.

Recently, it has been reported [4] that the Mo/SLS directly scribing yielded poor results due to the extensive cracking, flaking and burning along the scribe edges. To overcome these problems, a specular CTO layer, between Mo and SLS layers, has been interposed . Anyway, no laser processing data has been reported.

The selective removal, via laser, of each layer is quite difficult with the results that shorting, material mixing and other problems are commonly met.

The principle issues to be faced are:

1) The CIS module is a multi layer structure which consists of different materials that need to be

patterned separately and in the correct order without damaging the other layers;
2) The layers are thin, (1-2 μm) so that the cutting quality must be very high to avoid swelling phenomena and particulate residues which adversely affect the module characteristics.

Laser Apparatus

The scribing apparata we used consist of two different AO Q-switched, cw pumped, Nd-YAG lasers with a wavelength of 1064 nm. The first one is placed in Casaccia (Rome) and is mainly devoted to crystalline silicon processing. The second one which is placed in Portici (Naples), is mainly devoted to the scribing of a-Si based solar cells.

The first laser has a pulse width equal to 100 ns FWHM, with a beam profile that approach the pure fundamental gaussian profile (TEM_{00}). The second one had a pulse width of 200 ns FWHM and a poor beam quality. The spatial distribution of intensity had a camel hump like profile.

Both the lasers have a closed-circuit TV viewing system for the precise visual alignment, a focalisation system with X-Y-Z micro positioning of the condenser lens and a completely automated X-Y table for the movement of samples with position accuracy of 6 μm.

Outside the cavity, and after the beam-expander, a spatial filter removes the ghosts due to the scattering from dust particles. To reduce the laser beam intensity without loss of stability, induced by output power decreasing, neutral density filters have been used.

The 90° beam deflection is obtained by using a dichroic mirror mounted on an adjustable assembly to direct the beam onto the axis of the objective's lens for optimal focusing apparatus performance. The focusing apparatus is composed of a multielement lens system; this solution reduces longitudinal and transverse spherical aberrations. The focal lengths are 46.4 mm and 27.2 mm, the relative F-numbers of the objective's lens are 1.36 and 1.81, respectively. Three micrometers control both the laser focus positions (Z axis) and the laser beam centring in the focusing lens (X-Y axis).

Scribing Conditions

The laser scribing operating conditions for the groove's formation has been searched with the aim that the scribed layer be removed completely and the underlying one to be left unchanged. The parameters optimised to reproduce the different scribing steps are:
i) the distance of the focusing plane from the working one (defocus distance, Δf));
ii) average laser power (P_a);
iii) the specimen speed (V_y);
iv) the pulse repetition rate (RR).

In a four-factor, two-levels factorial experiment with interactions, the ANOVA analytical techniques (1% of accuracy) has shown that the defocus position does affect the scribing quality. The defocus provides added control of the amount of heat generated in the working area and simply through his variation we obtain different groove's width.

The specimen's speed and pulses frequencies are correlated: knowing the single pulse effective spot size, the laser repetition rate and specimen speed must be adjusted to achieve a continuous scribed line. Practically, the following relationship: Effective Spot Size > V_y/RR must be satisfied to overlap discrete pulses.

The single pulse effective spot size depends on the material thermal properties, as well as the optical spot size and spatial intensity distribution across the beam. Moreover, the edge definition of the grooves depends primarily on target thermal properties and pulses duration.

Experimental Work and Discussion

In consequence of the lack of information we have made much experimental work for determining the best set of the four parameters to obtain good quality scribing.

So far the ranges of parameters that we explored are:

P_a = 10 mW - 5 Watt   RR = 1 - 10 kHz V_y = 20 - 200 mm/s   Δf = 0 - 5 mm

We scribed about 16 Mo/SLS 7x3.5 cm^2, 2 Mo/SLS 7x7 cm^2 and 6 CIS/Mo/SLS 7x1.5 cm^2 samples with more than 600 scribing to study the scribe quality dependence by the above parameters. The groove quality is measured by cutting selectivity, lack of debris, absence of swelling phenomena and low extension of heat-affected zone (HAZ). The groove's morphology and the removal mechanism are discussed by optical microscopy, talystep, SEM and microprobe analysis. Here we will show only 10 scribings, everyone is that with the best set of the four parameters.

Molybdenum patterning. For Mo, two different configurations are investigated: the frontal scribing whereby the laser beam interacts first with the air/Mo interface and the back scribing with the laser beam passing through the glass slide and interacting with SLS/Mo interface. Fig. 1 shows a schematic picture of the back-scribing process.

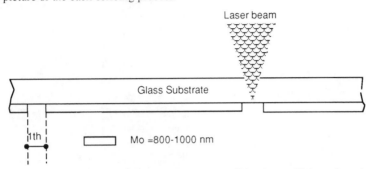

Figure 1: Back-scribing approach applied to the Mo laser scribing layer. Enlarged sectional view.

The scribe quality obtainable using the back scribing is very good, without droplets or swelling. The talystep analysis confirms that all Mo is removed and SLS is unaffected. Fig. 2 shows the SEM cross section of a typical back scribing, suggesting that the mechanism involved in Mo removal is a laser induced stress detaching. The partially detached scribing edges, composed of brittle residues, can be easily removed after cleaning.

Using the other configuration, Mo evaporation seems to be the main mechanism for Mo removal. The droplets showed in Fig. 3 can be removed after cleaning. This happens only if optimised laser parameters are used (see Table 1 sample F4.2). Otherwise, the produced droplets have a strong adhesion to SLS and are not removable.

As just shown, it is possible to obtain good quality scribing but it is necessary to have an accurate control of focus position (tolerance 50 μm with 27 mm focal objective lens), particularly in the frontal scribing configuration.

It is important to observe that the laser beam characteristics (width, spatial distribution of intensity, propagation) are fundamental for Mo scribing quality. In fact, using a laser with a poor Gaussian profile, we did not obtain the same results. The best one is that denoted by F6.1 (see Table 1) where high swelling ($\approx$4-5 μm) and large droplets are always present. It has been impossible to obtain a laser induced detaching even with the back scribing configuration.

After these results, a laser upgrade has been realised in Portici to obtain the proper beam profile.

At this aim, we have forced the laser to operate in low-order mode using a pinhole inside the cavity (ϕ=1.4 mm). In such a way, the beam profile is approximately gaussian. As a consequence, recently, it has been possible to perform very high quality back-scribing patterning on a 7x7 cm^2 Mo/SLS structure (see Table 1 FM02 sample). Two adjacent Mo stripes are electrically isolated.

Attempts to determine specific region in the parameter's space for "evaporation" and "laser detaching" scribing has been unsuccessful. Looking at Table.1 it is difficult to discern different region associated to different removal mechanism. The power density, pulse duration and laser wavelength involved in this study are related to evaporation connected with plasma effects. In this regime, the vapour tends to become supersaturated and it evolves from the surface. Condensing droplets of sub micrometer size then lead to absorption and scattering. It is possible during the frontal scribing to observe the vapour plume and sometimes droplets with a sub micrometer size are present also inside the groove. Rough estimates of the thickness that is possible to evaporate with the power density used give values of 0.4 μm, about one half of the Mo thickness. Also considering that metal reflectivity (50% for the cold Mo film) decreases with temperature, ignoring the thermal losses, it is likeable to suppose that much of the Mo removed leaves the groove as melt rather as vapour. It's well established [5] that the fraction of beam fluxes absorbed in metals increase above a certain threshold power density; such an effect can be produced by formation of an LSCW (Laser Supported Combustion Wave) in the evolving vapour. Associated vapour pressure is of the order of tenth of atmospheres (overwhelming the radiation pressure), they seem responsible for the melt ejection in the frontal scribing and the detachment in the back scribing. In conclusion, frontal scribing and back scribing physical removal mechanisms are not easily differentiable in spite of the clear different typology.

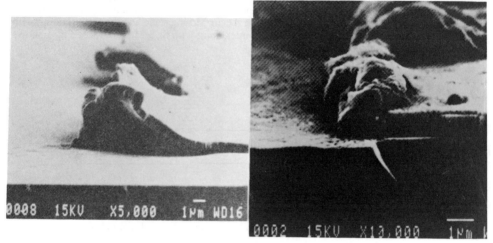

Figure 2: Mo back scribing on sample F1.2   Figure 3: Mo front scribing on sample F4.2
        reported in Table 1   reported in Table 1

Table. 1: Molybdenum/glass laser scribing data

	F1.2 back	F4.2 front	F6.1 S3(1) front	FM02 back
Lab.	Casaccia	Casaccia	Portici	Portici
Structure	Mo/SLS	Mo/SLS	Mo/SLS	Mo/SLS
Thickness (μm)	1.1	1.1	1.0	1.0
Vy (mm/s)	30	30	30	50
I (A)	23.9 a	23.9 a	11	14.5
Rep. Rate (kHz)	2.6	2.6	2	5
Pa (W)	0.48	0.48	1.1	0.62
Defocus (μm)	-200	+400	-500	+900
HAZ width (μm)	50	60	80	60
Groove Depth (μm)	1.1	1.1	1.0	1.0
Overlap (%)	77	81	81	83
Pp (W)	1800	1800	2750	620
Power Density (W/cm^2)	9×10^7	6.4×10^7	5.5×10^7	2.2×10^7
Pulse Energy (mJ/pulse)	0.18	0.18	0.55	0.13
Energy Density (J/cm^2)	9	6.4	10.9	4.6
Substrate Damage	no	no	light	no

Patterning of CuInSe$_2$ on Mo/SLS. With the aim to know the best parameters for CuInSe$_2$ scribing, we performed scribing on a CIS/SLS structure (see Table 2 samples M222 S8(2) and S4(4)). Optimum scribing parameters for both CIS/SLS and Mo/SLS proved to be roughly equivalent. Consequently, selective scribing of an individual layer proved impossible using Q-switched mode. All attempts to obtain selective scribing of CuInSe$_2$ with Q-switched beam failed. It was only possible either to remove both Mo and CIS layers together, or to have a partial evaporation of CIS leaving a solidified mash containing Mo, Cu, In and Se. A microprobe analysis showed that the solidified mash has a stoichiometry poor in In and Se respect to the original polycrystalline film. High peak power used in Q-switched mode (in the range of kWatt) does not discriminate between the two materials.

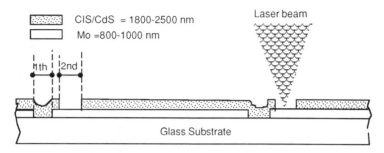

Figure 4: Front-scribing of CIS/Mo/SLS structure. Enlarged sectional view with groove in CIS layer for successive ZnO/Mo electrical coupling.

The solution to this problem is to work in cw power output mode.

CIS optical and thermal properties are compatible with a low power intensity evaporation. The high absorption coefficient ($>10^4$ cm^{-1}) and low reflectivity (9% for the films scribed) at 1.16 eV photon energy, allows a good laser-beam/material coupling. The low melting and evaporating temperatures ($\approx$900-1000°C) and the low thermal diffusivity ($\approx$0.05 cm^2/s) don't require high energy and limit his dispersion. In this way, it is possible to find the parameters for which CIS evaporates and Mo-film remains unaffected (Fig. 5). Droplets of molten, resolidified and brittle residues of CIS (Fig. 6) can be easily detached by cleaning after scribing. After cleaning, the scribing is very good and all the semiconductor film is removed. Microprobe analysis showed that only Mo is present in the groove and the original stoichiometry of CuInSe$_2$ along the scribing edge is preserved. It is important to observe that all the HAZ is removed with the droplets, leaving an edge made of CIS heat unaffected. Eventual Se loss would increase the material conductivity leaving to have an increase in a short circuit current that adversely affect the module performances.

For cw-CIS-scribing the beam quality it is not so critical as for Mo scribing, henceforth, the patterning with poor beam profile is acceptable.

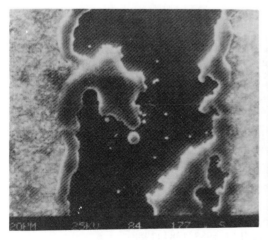

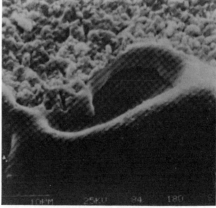

Figure 5: CIS/Mo/SLS selective groove on sample M217A S11(4) reported in Table 2

Figure 6: Fig 5 detail showing the scribing edge with brittle residues easily detachable

Table 2. CuInSe$_2$/Molybdenum/glass laser scribing data

	M217A S11(4)	M217A S5(4)	M222B S8(2)	M222B S4(4)	M217B S4(4)	M221B S2(4)
Lab	Casaccia	Casaccia	Casaccia	Portici	Portici	Portici
Structure	CIS/Mo/SLS	CIS/Mo/SLS	CIS/SLS	CIS/SLS	CIS/Mo/SLS	CIS/Mo/SLS
Thickness (μm)	1.8/0.5	1.8/0.5	2.4	2.4	1.6/0.58	2.3/0.95
Vy(mm/s)	50	50	30	50	150	45
I (A)	14 a	17.8 a	22 a	10.6	10.6	9.9
Rep. Rate (kHz)	cw	8	2.6	10	5	cw
Pa (W)	0.95	0.70	0.12	1.6	1.45	0.9
Defocus(μm)	-	+300	-100	0	+1500	+1500
HAZ width (μm)	100	40	40	100	210	120
Depth (μm)	1.8	1.5	2.5	2.5	1.5	2.3
Overlap (%)	-	84	71	95	86	-
Pp (W)	-	875	230	800	1450	-
Power Density (W/cm^2)	1.2x10^4	7x10^7	3.7x10^7	1.0x10^7	4.2x10^6	0.8x10^4
Pulse Energy (mJ/pulse)	1.9	0.087	0.046	0.16	0.29	2.4
Energy Density (J/cm2)	24.2	7	3.7	2	0.8	21.8
Substrate Damage	no	no	light	light	no	no

Conclusions

The possibility to obtain optimum selective laser scribing of Mo/SLS and CIS/Mo/SLS structures is demonstrated. The process is scalable without problem on a 7x7 cm^2 sub module.

The scribe quality is critically related to the control of either the focus position and the beam profile.

Back scribing approach is an attractive technique for Mo scribing due to laser induced stress detachment removal mechanism. The final groove quality is high because the SLS is unaffected, the HAZ is totally removed, the swelling and the debris are absent.

The cw power laser output operation mode is used to differentiate the two film layers in the CIS/Mo/SLS structure for a selective scribing. It is possible to obtain good quality scribing and to leave the remaining CIS film chemically unaffected.

Acknowledgements

All the samples used in this work have been made by the Institute fur Physikalische Elektronik of Stuttgart University The work has been partially supported by the CEC under a Joule contract No.JOU2-CT92-0141 (EUROCIS II).

References

[1]   S. Avagliano, M.L. Addonizio et al., Proc. 12th E.C.PVSEC 11-15 April 1994, Amsterdam
[2]   K. Mitchell, C. Eberspacher, et al., Proc. IEEE 1988, 1384 (1988).
[3]   J. Ermer, C. Frederic, et al., Proc. IEEE 1990, 595 (1990).
[4]   R. Arya, J. Fogleboch, et al., NREL/TP-413-3759 (1993).
[5]   M. von Allmen: *Laser-Beam Interactions with Materials,* Springer Ser. Mat. Sc.Vol.2 (Springer-Verlag, Berlin, Heidelberg 1987).

Materials Science Forum Vols. 173-174 (1995) pp. 59-66
© *1995 Trans Tech Publications, Switzerland*

UV-EXCIMER LASER ABLATION PATTERNING OF II-VI COMPOUND SEMICONDUCTORS

P.H. Key, D. Sands and F.X. Wagner

Department of Applied Physics, University of Hull, Hull, HU6 7RX, UK

Keywords: Excimer Lasers, Ablation, Patterning, ZnS, CdTe, II-VI Compounds

ABSTRACT. Pulsed excimer laser ablation characteristics of ZnS, CdTe, and ZnSe crystals have been studied using 248 nm and 308 nm radiation in vacuum and in argon at pressures up to 2×10^3 mbar. The depth of material removed per pulse is shown in most cases to hold a Beer's Law relationship to laser fluence (energy/unit area). The threshold fluence for these materials is typically in the range 120-150 mJ cm^{-2} in vacuum, and is found to be increased by raising the ambient pressure. We have exploited the low threshold fluence in vacuum to pattern epitaxial thin films of CdTe using a conformal mask of conventional photo-resist which has been exposed and developed in the normal way. Blanket exposure causes both the exposed CdTe and the photo-resist to be ablated but the relative ablation rates of the two materials allows shallow features to be etched into the CdTe. The maximum depth we have achieved is in excess of 600 nm.

INTRODUCTION

The properties of pulsed excimer lasers make them ideal tools for industrial processing of a wide range of materials, including semiconductors [1]. At the most common wavelengths (193 nm to 308 nm), most semiconductors have very high absorption coefficients so that the radiation is typically absorbed within tens of nanometres from the surface. The high output powers, large beam areas, high beam uniformity, and short pulse lengths all mean that very high surface temperatures can be reached over large enough areas for a process to be considered commercially viable. Pulsed excimer laser recrystallisation of amorphous silicon on glass is being considered by many laboratories around the world as one method of achieving fully integrated active matrix addressed liquid crystal displays [2].

Pulsed laser processing is best suited to applications where, for one reason or another, the temperature of the substrate or wafer should be kept as low as possible. II-VI compounds are one such case. The high diffusivity of many impurities, particularly group III dopants such as In, leads to the requirement that the bulk temperature should be as low as possible and the processing time short. In this paper we define the conditions for excimer laser processing of the II-VI compounds ZnS, CdTe, and ZnSe at

248 and 308 nm. We describe first the ablation characteristics of these materials and then demonstrate a method of patterning them based on conventional photo-lithography. To illustrate the technique, we show some simple patterns etched into a thin film of epitaxial CdTe on InSb.

EXPERIMENTAL

The lasers used in these experiments were a Questek 2200 series operating at 248 nm (pulse length 25 nm) and an in-house built device operating at 308 nm (pulse length 10 nS), each with their own optical imaging system. The imaging optics consisted of an aperture to select an area of uniform energy distribution from the output beam and a single uv-quartz lens to form an image of the aperture at the surface of the target material, which was located in an evacuable cell. The laser output was fixed to give a maximum fluence at the target of approximately 1.2 J cm-2 and lower fluences were obtained by introducing neutral density filters into the beam path. The cell was pumped to a base pressure of ≤ 10-5 mbar and back filled to the required pressure with argon, the pressure being monitored with a suitable gauge for the range (Penning, Pirani, capacitance, Bourdon). Argon was used as the filling gas to avoid, as far as possible, any target-ambient gas reactions during irradiation.

For studies of the ablation characteristics, deep ablation craters were formed using a range of laser fluences and a range of argon pressures by irradiating the target with many hundreds of pulses. The depth of the craters was measured using an optical microscope with a calibrated fine focus (depth resolution of 1 μm), giving an average depth measurement error of ≈ 3% over the range of depths measured. From these measurements the average etch-depth per laser pulse was calculated for each fluence/pressure combination. The fluence was calculated from the image dimensions at the target and the laser energy measured with the target removed.

For ablation patterning, 308 nm radiation was used. The same optical system and vacuum cell were employed, but the target was a thin epitaxial layer of CdTe on InSb which had been coated with photo-resist. The photo-resist (BPRS150 from OCG Chemicals Ltd) was patterned and developed in the normal way using a chrome-on-glass mask and a standard mask aligner. These patterns were etched into the CdTe by ablation by exposing the sample in vacuum to a relatively large area beam (circular cross-section of approximate area 1 mm^2) so that both the CdTe and the photo-resist were irradiated.

RESULTS

Figure 1 shows a semi-log plot of the etch-depth per laser pulse at 248 nm as a function of fluence for ZnS in argon over the pressure range 0.01 mbar to 2000 mbar. Each line can be fitted to the expression;

$$d_e = \frac{1}{\alpha} \cdot \ln\left(\frac{F}{F_T}\right) \qquad (1)$$

which is derived from Beer's law of absorption, where d_e is the etch depth. Here α is the effective absorption coefficient and F_T is the threshold fluence for significant ablation. This relationship between etch-depth per laser pulse and laser fluence has been shown to hold for a large number of materials in the regime of true ablation [3], that is to say for fluences in excess of F_T. F_T is defined by extrapolating the curve back to zero ablation rate and is seen to depend markedly on the ambient pressure. The data have been fitted to a common gradient, from which α is obtained. For CdTe, shown in figure 2, ablation is also described by equation 1, but the threshold fluence is seen to be largely independent of pressure below 0.5 bar.

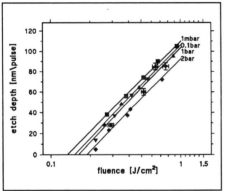

Figure 1 The etch-depth per pulse for ZnS at 248 nm. Selected data are shown with their error bars

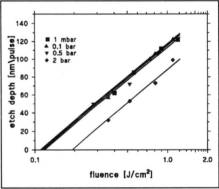

Figure 2 The etch-depth per pulse at 248 nm for CdTe

Figures 3 and 4 show the ablation characteristics of ZnS and CdTe at 308 nm. For ZnS the data is very similar to that shown in figure 1, but with a smaller pressure dependence. For CdTe, there is an essential difference between 308 nm and 248 nm; a change in the slope of the curves at 400 mJ cm-2 which divides the characteristics neatly into a high fluence and low fluence regime. In the low fluence regime, equation 1 holds and a pressure dependence in the threshold fluence is observed. In the high fluence regime, equation 1 also holds but the effective absorption coefficient is different and exhibits a slight variation with the external pressure.

The threshold fluence, F_T, as a function of ambient gas pressure for ablation of ZnS and CdTe are shown in figures 5&6 respectively. It can be seen that at low pressures ($\leq$ 1 mbar) the threshold fluence is constant for each material, but at higher pressures the threshold for ablation is increased significantly. It follows, therefore, that the onset of ablation may be controlled to a certain extent by varying the ambient gas pressure. The curves delineate the two possible processing

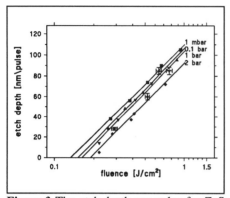

Figure 3 The etch-depth per pulse for ZnS at 308 nm

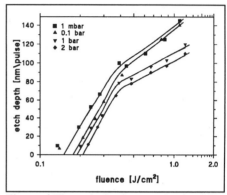

Figure 4 The etch-depth per pulse for CdTe at 308 nm. The change in α at 400 mJ cm^{-2} is clear.

regimes of ablation (above the curve) and surface treatment such as annealing (below the curve). For ZnSe, the behaviour lies between that of ZnS and CdTe. The

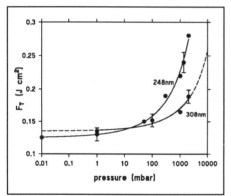

Figure 5 The threshold fluence as a function of pressure for ZnS at 248 and 308 nm.

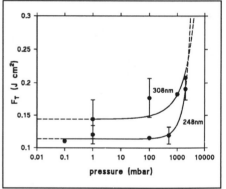

Figure 6 The threshold fluence for CdTe as a function of pressure at 248 and 308 nm

threshold fluence for ablation is very similar to both the above materials, but the pressure dependence of the threshold fluence is not as marked as in the case of ZnS.

The patterns were formed by blanket exposure of a relatively large area of an epitaxial layer of CdTe coated with photo-resist with a predefined pattern. In order to improve the adhesion of the photo-resist, some CdTe layers were baked at 363 K for twenty minutes prior to deposition of the photo-resist while others had no treatment. The photo-resist layer was baked at 363 K for 15 minutes prior to exposure through the mask and baked following development of the pattern to harden the photo-resist. This last step is considered essential since this use of the photo-resist layer is not standard, and the photo-resist needs to withstand as much as possible the incident laser radiation. The maximum post-development baking temperature was 393 K for 30 minutes.

For all samples prepared without pre-deposition baking, poor adhesion of the photo-resist layer was observed. The irradiated photo-resist was removed with the first laser pulse, leaving the underlying CdTe exposed. For photo-resist deposited onto pre-baked CdTe, there is a much greater resistance to ablation. We have etched features into CdTe at 120 mJ cm^{-2} using up to 64 pulses and still retained a photo-resist covering. Figure 7 shows an optical micrograph of the patterned area. The dark circles and small squares are the patterns in the photo-resist (the smallest circles are 140 µm in diameter) which are still visible after 64 shots. Figure 8 shows the bottom left-hand corner of figure 7 (140 µm diameter circle plus 14 µm sized square features) after the residual photo-resist has been removed by dissolution in acetone. The patterns in the CdTe are clearly discernible, the ablated material being the lighter. The estimated depth is in excess of 600 nm based on an average ablation rate of 10 nm per pulse at this fluence.

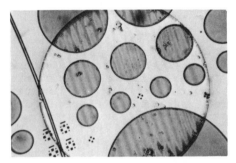

Figure 7 64 pulses at 120 mJ cm^{-2}. The ablated region is the lighter circular region.

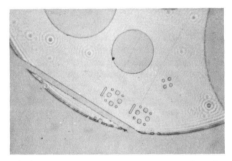

Figure 8 A close up of figure 7 after the photo-resist has been removed.

Figure 9 shows the effect of 50 pulses at 150 mJ cm^{-2} on another part of the sample. The letters UMB are clearly identifiable. The photo-resist is nearly ablated through to the CdTe and at this stage has been damaged to the extent that this residual photo-resist cannot be removed. However, patterns can be etched into the CdTe successfully at this fluence provided fewer pulses are used. Figure 10 shows the effect on the photo-resist of a single pulse at 200 mJ cm^{-2}. The high fluence has caused the photo-

Figure 9 50 pulses at 150 mJ cm^{-2}. Near total ablation has damaged the photo-resist.

resist to melt and splash onto the surrounding CdTe. However this has not affected the pattern produced (figure 11) which has an estimated depth of 30 nm.

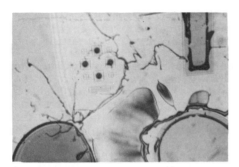
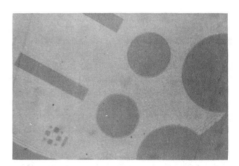

Figure 10 1 pulse at 200 mJ cm^{-2} has melted the photo-resist.

Figure 11 The damage to the photo-resist has not affected the pattern, which is well defined though shallow.

DISCUSSION

The detailed mechanism of ablation of these materials will not be discussed here. Arguments have been presented elsewhere for the mechanisms involved, which we believe to be thermal in nature [5]. In the case of CdTe it is almost certain that melting occurs during ablation, but there is the possibility that at fluences close to threshold at low pressure, sublimation also plays a part. For ZnS, with a much higher melting point of $\approx$2100 K, sublimation at $\approx$1300 K is probably much more significant. We believe that this is responsible for the greater dependence of the threshold fluence on pressure for this material. We have reached no firm conclusions regarding ZnSe yet.

The dependence of the threshold fluence on pressure makes possible a variety of laser-based processes for II-VI compounds. We have already investigated annealing of implanted epitaxial CdTe films and shown that fluences as high as 180 mJ cm^{-2} are needed to restore the excitonic features [6]. This is above the threshold fluence in vacuum and external pressures are required to prevent ablation. We are also investigating the effect of annealing in reactive ambients on the surface composition. In this paper we have exploited the low threshold fluences to create patterns by ablation using conventional photo-resist as a conformal mask. It is essential that the photo-resist adheres well to the substrate and pre-deposition baking is found to be crucial in this respect. With good adhesion, the photo-resist is ablated but shallow features can be etched as long as the photo-resist is not damaged so much that the residue cannot be removed, as is the case if the resist is nearly ablated through to the underlying II-VI material.

The patterns created are well-defined with, as far as we are able to tell from Scanning Electron Microscopy (SEM), straight walls. Features etched with several pulses, as in figure 7, exhibit some rounding of the corners on square features. The reason for this lies in the post-deposition baking of the photo-resist, which has the effect of converting the small squares into pyramidal structures as the resist shrinks slightly. The coverage at the corners is inferior to the bulk, and some ablation occurs.

Clearly, the greater the number of shots the greater the loss of resolution on sharp features. This limitation aside it is possible to etch features into CdTe with depths as little as 10 nm or as high as 600 nm. Similar results can be expected for ZnS because of the similarity in ablation thresholds and rates.

CONCLUSION

The present work has demonstrated quite clearly that a method of patterning II-VI materials using pulsed excimer lasers is feasible. However, the process has not yet been optimised. Refinements in the post-development baking temperature or a different, more durable and perhaps thicker resist might enable much better definition on deeper structures. For very shallow structures, the technique has the advantages that very precise depths can be achieved. It is also possible to envisage an integrated process involving both annealing and etching in the same environment.

ACKNOWLEDGEMENTS

The authors would like to thank P Monk and B L Tait for their assistance with the lasers and D E Ashenford and D Wright for their help in supplying epitaxial CdTe and the photolithography.

REFERENCES

[1] I W Boyd, *Laser processing of thin films and microstructures*, (Springer-Verlag, Berlin 1987).

[2] S D Brotherton, Microelectronic Engineering **15**, 333-340 (1991).

[3] J E Andrew, P E Dyer, D Forster & P H Key, Appl. Phys. Lett. **43**, 717 (1983).

[4] P D Brewer, J J Zinck & G L Olson, Mat. Res. Soc. Symp. Proc. **191**, 67 (1990).

[5] P H Key, D Sands, F X Wagner, Unpublished.

[6] P H Key, D Sands, F X Wagner, Unpublished.

Materials Science Forum Vols. 173-174 (1995) pp. 67-72
© 1995 Trans Tech Publications, Switzerland

FABRICATION AND CHARACTERISATION OF N$^+$ DOPED LASER CRYSTALLISED POLYSILICON

E. Al-Nuaimy, J.M. Marshall, T.E. Dyer, A.R. Hepburn and J.F. Davies

Electronic Materials Centre, Department of Materials Engineering,
University College of Swansea, Singleton Park, Swansea SA2 8PP, UK

Keywords: Spin-on-Dopant, Laser Doping, Laser Crystallisation, Structural Properties, Electronic Properties

Abstract. Laser crystallised doped polysilicon has been prepared from hydrogenated/de-hydrogenated amorphous silicon by excimer laser crystallisation. In one case the precursor has been doped during deposition, while in the other the layer is doped using a spin-on-dopant during the crystallisation process. Structural and electronic properties of these materials were investigated using electron microscopy, optical absorption and DC electrical conductivity. This study reveals that the grain size is dependent on hydrogen content and the means of dopant incorporation. The electrical conductivity increases by orders of magnitude upon crystallisation and is found to be further enhanced by doping. Highly doped layers are found to have a temperature dependence characteristic of impurity band conduction. The absolute conductivity magnitudes are dependent upon the means of dopant incorporation and the hydrogenated state of the layer, with PECVD incorporated dopants producing the highest conductivities.

Introduction

International research interest in polysilicon TFTs continues to grow. These devices are key elements for active matrix liquid crystal displays, linear image sensors, and printing heads. Polysilicon TFTs require shallow and high concentration doping for source and drain formation. However, with conventional ion implantation followed by heat treatment in a furnace for times of the order of minutes to hours, it is difficult to achieve a high dose with low-energy doping. In addition, long-term thermal processing is incompatible with cheap glass substrates and detrimental to the final quality of the deposited gate insulator (self-aligned technology). To overcome these problems, novel techniques such as laser doping from the gas [1] or solid [2] phase and laser annealing of low-energy implanted ions [3] have recently been investigated. In all these experiments pulsed laser irradiation is used to melt locally the amorphous or polycrystalline Si layer and produce liquid-phase incorporation of the dopant and its electrical activation [4-7].

Another process known as "Excimer-Laser Doping using Spin-On Dopant" [8-10] has also received much attention because it is a simple process precluding the expense of gas phase doping or ion-implantation equipment. In this study, the spin-on-doping process has been carried out and the results compared with materials in which the dopant is incorporated during the gas phase. While the conductivities achieved are somewhat below those accessible utilising gas phase doping, the utility of the process is demonstrated and the values obtained are found to be satisfactory for device applications.

Experimental

Sample preparation. All precursor specimens were prepared using a conventional capacitively coupled PECVD reactor operating at 13.56 MHz. Samples, 150 nm (undoped) and 100 nm (doped), were deposited on Corning 7059 glass for structural and electronic characterisation, and on polished crystalline silicon wafers for infrared measurements. PECVD deposition conditions were; flow rates (SiH_4 - 25 sccm, PH_3 - 32 sccm), pressure (0.5 torr undoped, 0.6 torr doped), power density 80 $mWcm^{-2}$ and substrate temperature (T_d) 250 °C. The various de-hydrogenation and doping procedures carried out are detailed in Table 1.

Sample	Dehydro-genated	Spin-on-Doped	Gas Phase Doped (3000vppm PH_3)
Type 1	No	No	No
Type 2	820K	No	No
Type 3	No	Yes	No
Type 4	820K	Yes	No
Type 5	No	No	Yes
Type 6	Yes	No	Yes

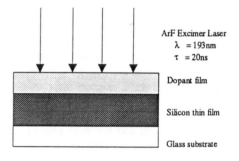

ArF Excimer Laser
λ = 193nm
τ = 20ns

Dopant film

Silicon thin film

Glass substrate

Table 1. De-hydrogenation and doping processes carried out on a-Si:H precursor thin films.

Fig. 1. Schematic representation of the laser doping process.

Dehydrogenation is conducted in an N_2 flow furnace at reduced pressure. During crystallisation the film is melted in ambient atmosphere by a high power ArF laser (193nm radiation (τ=20ns), Lambda-Physik LPX-150E). Fig 1 shows a schematic representation of the laser doping process. For laser spin-on-doping (type 3,4), the emulsion (P509, Filmtronics Inc.) containing phosphorus dopants (2×10^{21} atom/cc) is introduced on to the surface of the silicon films by spin coating. The emulsion is then dried to become a hard film after baking at 100°C for 30 minutes. The laser doping is conducted simultaneously with the crystallisation process and the crystallisation time thought to be sufficient to allow the dopant to diffuse through the whole crystallised layer [9]. A similar procedure is carried out for laser crystallisation of gas phase doped material (type 5,6).

Characterisation. Grain size and microstructure are observed by Transmission Electron Microscope (TEM), and samples prepared using a lift-off technique. Vibrational modes are investigated by Fourier Transform Infra-Red (FTIR) spectroscopy in the range 4000-400 cm^{-1} using a Perkin-Elmer 1720X FTIR spectrometer. The dark DC conductivity is measured in the range 150-450K using standard techniques.

Results and Discussion

Structural Properties. Figs 2 (a,b) illustrates typical planar bright-field Transmission Electron Microscopy (TEM) and Transmission Electron Diffraction (TED) micrographs for type 1 and type 2 polysilicon. Fig 2(a) implies a fine microstructure with a maximum grain size of ~ 0.1 μm (the maximum resolvable limit of the instrument). In addition, the associated TED pattern reveals an almost continuous (111) ring, indicating that the grains are randomly oriented The TEM and TED micrographs of fig 2(b) demonstrate the larger grain size of the type 2 material. It can be argued that this is due to a reduction in structural relaxation in the dehydrogenated material which results in enhanced crystal growth rates and reduced nucleation rates [11].

The crystallisation threshold for type 2 polysilicon is ~135mJcm^{-2}, and for type 1 polysilicon ~125mJcm^{-2}. This difference in the crystallisation threshold is due to the reduced hydrogen content in the a-Si, which reduces the rate of crystallisation [12].

Fig 2(a): TEM/TED micrographs for ArF excimer laser crystallised type 1 polysilicon.

Fig 2(b): TEM/TED micrographs for ArF excimer laser crystallised type 2 polysilicon.

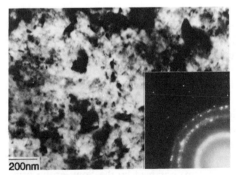

Fig 3: TEM/TED micrographs for n$^+$-doped (type 3) polysilicon.

Fig 3 illustrates typical plan-view bright-field TEM/TED micrographs for type 3 polysilicon. A larger grain size is observed, as compared to both type 1 and type 2 films. This is tentatively attributed to the effect of the spin-on-dopant as a thermal insulator during the crystallisation/dopant activation stage. Computer simulation studies are currently underway to investigate this possibility.

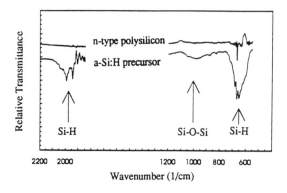

Fig 4: FTIR transmittance spectra for n-type (type 3) polysilicon and a-Si:H precursor (T_d=250°C).

Atomic bonding configurations were investigated by Fourier tranform infra-red (FTIR) spectroscopy. Fig 4 shows the infra-red transmittance for as-deposited a-Si:H at a substrate temperature of 250°C, and a type 3 polysilicon film, as fabricated using the spin-on-dopant technique. For a-Si:H the IR spectrum [13] exhibits broad features around 2000 and 630 cm^{-1} associated with the stretching and rocking vibration modes between Si-H species respectively. After the laser spin-on-doping process, it is evident that all resolvable modes attributable to Si-H interactions disappear from the spectrum.

Electronic Properties. Fig 5 illustrates the temperature dependence of the DC conductivity (σ_{dc}) for the a-Si:H precursor, and type 1 and type 2 excimer (ArF) laser crystallised polysilicon. The conductivity is calculated from the ohmic regime of the I-V characteristic. The transition to the polycrystalline phase is reflected in a 5-6 orders of magnitude increase in room temperature conductivity, together with an associated reduction in the high temperature activation energy. In spite of the different grain sizes (Fig.2(a,b)), the temperature dependencies for both type 1 and type 2 polysilicon are remarkably similar. However, type 2 polysilicon exhibits a higher conductivity than type 1 material, with a corresponding reduction in the conductivity activation energy from 0.39→0.11 eV over a temperature range of ~200 K. A significant deviation from Arrhenius behaviour at lower temperatures is also observed. This phenomenon is often attributed to hopping conduction in amorphous semiconductors.

Fig 6 shows that laser doped (type 3/4) polysilicon films have significantly higher conductivites than their intrinsic counterparts, ~ 3×10^{-2} (Ω.cm)$^{-1}$ for type 3 samples and 9×10^{-3} (Ω.cm)$^{-1}$ for type 4 samples. σ_{dc} for both types of film is temperature independent, which suggests a change to impurity band conduction in heavily doped films. The type 4 laser doped polysilicon shows higher values for conductivity than type 3 material. This may be due to hydrogen passivation of donor atoms [14], or to the larger grain size in dehydrogenated polysilicon.

The magnitude of the conductivity is significantly lower than that obtained when the dopant is incorporated directly into the film from the gaseous phase. At present it is unclear why the conductivities of the type 5/6 material should be significantly higher than those of the type 3/4 material. It may be that the dopant activation in the gas phase doped material is significantly more efficient. In Fig 6, the extremely high conductivity (1300 Ω/cm) obtained by the Sony group using very heavily doped (20000 vppm PH$_3$) precursors [15] is also shown.

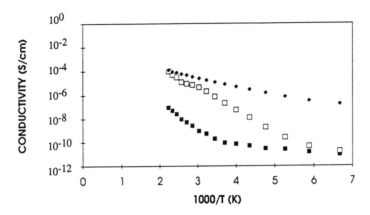

Fig 5 - D.C conductivity as function of inverse temperature for ■ - a-Si:H precursor, □ - type 1 polysilicon, ◆ - type 2 polysilicon.

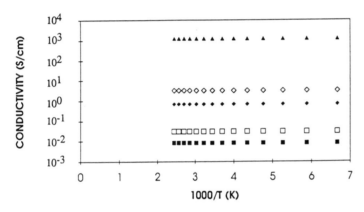

Fig 6 - D.C conductivity as function of inverse temperature for ▲ - 20000 vppm P-doped 20nm thick crystallised film [18], ◊ - type 6 material, ◆ - type 5 material, □ - type 4 material, ■ - type 3 material.

Conclusions

We have successfully fabricated n+ polysilicon layers using both the spin-on-dopant technique and gas phase doping. Crystallisation of the a-Si:H precursor has been carried out (simultaneously with laser doping) and spatially selectively using an ArF excimer laser. TEM/TED studies have confirmed the presence of microcrystallites in the doped samples. A larger grain size is observed in the doped samples as compared to undoped samples. This is tentatively attributed to the effect of the spin-on-dopant as a thermal insulator during the crystallisation/dopant activation stage. The doped films exhibit a significant conductivity enhancement as well as temperature invariance, which may be assigned to impurity band conduction. The magnitude of the conductivity is significantly lower in laser spin-on-doped material than that obtained when the dopant is incorporated directly into the film from the gaseous phase.

Acknowledgements

The authors would like to thank M. Williams for assistance with the TEM characterisation, and W.I. Jones (Department of Chemistry) for access to the FTIR spectrometer. We also wish to acknowledge the Iraqi government for financial assistance to one of us (E. Al-Nuaimy). This work was carried out with partial support from SERC contracts GR/G49999 and GR/H71819.

References

[1] P. Coxon, M. Lloyd and P. Migliorato, Appl. Phys. Lett. **48**, 1785 (1986).

[2] T. Sameshima, M. Hara and S. Usui, Jpn. J. Appl. Phys. **28**, 1789 (1989).

[3] G. Kawachi, T. Aoyama, T. Suzuki, A. Mimura, Y. Ohno, N. Konishi and Y. Mochizuki, Ext. Abstr. 22nd Int. Conf. on *Solid State Devices and Materials*, Sendai, 1990, p. 971.

[4] T.F. Deutsch, in: Laser Diagnostics and Photochemical Processing for Semiconductor Devices, Eds. R.M. Osgood. S.R.J. Brueck and H.R. Schlossberg, *Mater. Res. Soc. Symp. Proc.*, Vol. **17** (North-Holland, New York, 1983), p. 225.

[5] E. Fogarassy, D.H. Lowndes, J. Narayan and C.W. White, J. Appl. Phys. **58**, 2167 (1985)

[6] T.W. Sigmon, in: Photon, Beam and Plasma Stimulated Chemical Processes at Surfaces, Eds V.M. Donnelly, I. P. Herman and M. Hirose, *Mater. Res. Soc. Symp. Proc.* , Vol. **75** (Materials Reasearch Society, Pittsburgh, PA, 1987), p. 619

[7] T. Sameshima and S. Usui, J. Appl. Phys. **62**, 711 (1987).

[8] E. Fogarassy, R. Stuck, J.C. Muller, M. Hodeau, A. Wattiaux, M. Toulemonde and P. Siffert, in: Proc. 3rd EC *Photovoltaic Energy Conf.*, Ed. W. Palz (Reidel, Dordreclt, 1980) p. 639.

[9] K. Sera, F. Okumura, S. Kaneko, S. Itoh, K. Hotta and H. Hoshino, J. Appl. Phys. **67**, 2359 (1990).

[10] A. Slaoui, M. Elliq, H. Pattyn, E. Fogarassy, S. de Unamuno and R. Stuck, in: *Photons and Low Energy Particles in Surface Processing*, Eds. C. Ashby, J.H. Brannon and S. Pany. *Mater. Res. Soc. Symp. Proc.*, Vol. **236** (Materials Research Society, Pittsburgh, PA, 1992).

[11] K. Nakazawa and K. Tanaka, J. Appl. Phys. **68**, 1029, (1990).

[12] S.D. Brotherton, D.J. McCulloch, J.B. Clegg, and J.P. Gowers, IEEE.Trans.Elect.Dev, **40**, 407, (1993).

[13] P.J. Zanucchi, in: *Semiconductors and Semimetals (Part B)*, Ed. J.L. Pankove, (Academic Press Inc, London, 1984).

[14] M. Stutzmann, in: *New Physical Problems in Electronic Materials*, Eds. M. Borissov, N. Kirov, J.M. Marshall and A. Varek, (World Scientific, 1991).

[15] T. Sameshima, M. Hara and S. Usui, in: *Polycrystalline Semiconductors II*, Eds. J.H. Werner and H.P. Strunk, (Springer-Verlag, Heidelberg, 1991).

Materials Science Forum Vols. 173-174 (1995) pp. 73-80
© 1995 Trans Tech Publications, Switzerland

PULSED LASER ABLATION: A METHOD FOR DEPOSITION AND PROCESSING OF SEMICONDUCTORS AT AN ATOMIC LEVEL

J.J. Dubowski

Institute for Microstructural Sciences, National Research Council of Canada,
Ottawa, Ontario, K1A OR6, Canada

Keywords: Pulsed Laser Ablation, Epitaxial Growth, Semiconductor Quantum Wells and Superlattices, Laser-Assisted Dry Etching, Atomic Level Processing

Abstract. The results of pulsed laser ablation are reviewed with a focus on the feasibility of this approach for atomic level processing. We have developed a pulsed laser evaporation and epitaxy (PLEE) method, with two lasers used for simultaneous or sequential ablation of different targets, for the deposition of quantum wells and superlattices of some II-VI semiconductor materials. The unique potential of PLEE exists in bandgap engineering of ternary structures with an arbitrarily changing chemical composition. Another new and rapidly developing application of pulsed laser ablation is low-fluence laser etching. We have been exploring this approach with a laser-assisted dry etching ablation (LADEA) system for the microfabrication of various structures in materials viable for advanced microelectronics and optoelectronics. Using InP it has been demonstrated that the process of material removal can be carried out at rates well below one atomic layer per UV excimer laser pulse. An investigation of the surface morphology of LADEA-processed material indicates the feasibility of the method in achieving an atomic level control.

Introduction

The application of lasers for the growth and in-situ processing of thin films of different materials has attracted steadily growing attention during recent years. The advances made in laser technology, especially excimer laser technology, have significantly contributed to this interest. The results of pulsed laser ablation for the deposition of high-T_c superconductors, semiconductors, metals, ferroelectrics and diamond-like films have been reported in materials of numerous symposia [1-4] and reviewed by several researchers, e.g., [5-8]. Pulsed laser ablation deposition has become increasingly attractive in the deposition of quantum wells and superlattices of various materials [5,7, 9-11]. The ever growing requirements of modern microelectronics and optoelectronics and the demand of an increased level of integration with a simultaneous reduction of the costs of fabrication have stimulated a research of novel technologies where lasers, x-rays and/or particle-beams are used to induce, enhance or stimulate the process of thin film fabrication. One of the advantages in the application of the laser for thin film processing is that it is a non-invasive method. In contrast with the beams of electrons, ions or x-rays, the laser beam is delivered to the processing chamber through a designated window and without interfering with other internal fabrication hardware. This may prove to be especially critical in systems operating in chemically reactive ambients. Laser processing of semiconductors has been demonstrated, for instance, in preparation of atomically clean silicon surfaces [12-15], etching and selective deposition of GaAs [16-18], patterned desorption of GaAs in a MOCVD reactor [19] and laser-assisted chemical beam epitaxy for selective growth of GaAs and InP [20].

We have undertaken an effort to explore the potential of laser-assisted epitaxial growth and laser in-situ processing with the aim to fabricate optoelectronic devices which are viable for the information technology industry. The investigations concerned some II-VI semiconductors and III-V materials such as InP, GaAs and CBE-grown InGaAsP quaternaries. The intention of this paper is to review some results of laser ablation carried out in our laboratory and illustrate the feasibility of pulsed laser ablation in the <u>deposition</u> and <u>removal</u> of semiconducting structures with a near atomic layer resolution. In the first part of the paper, a pulsed laser evaporation and epitaxy (PLEE) method, which has recently been complemented by an in-situ fast-nulling ellipsometer, has been discussed for the deposition of quantum wells and superlattices consisting of CdTe and $Cd_{1-x}Mn_xTe$ layers. In the second part, we discuss excimer laser-assisted dry etching ablation (LADEA) of III-V semiconductors in an Cl_2/He environment.

Experimental Results and Discussion

Pulsed laser evaporation and epitaxy. A sequential of simultaneous ablation of different targets in the pulsed laser evaporation and epitaxy (PLEE) system has been carried out with Nd:YAG and excimer XeCl lasers. Details of this system have been published previously [21] and here only a brief outline is given. A schematic concept of PLEE in depicted in Fig. 1. Two laser beams are delivered to different targets independently and their output characteristics (switching, laser fluence, pulse frequency, number of pulses delivered per step) are computer controlled. An investigation

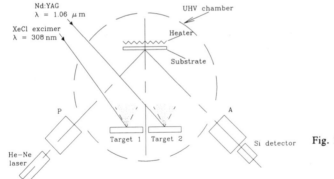

Fig. 1. A schematic diagram of the PLEE system with a fast-nulling ellipsometer implemented for *in-situ* monitoring of the deposition process.

of the structural and optical properties of PLEE-grown CdTe and $Cd_{1-x}Mn_xTe$ thin films, multiple quantum wells and superlattices [7] have indicated the feasibility of this method in achieving an excellent quality material with properties comparable to those of materials obtained with other advanced methods of epitaxial growth. The flexibility of laser ablation in the growth of thin films has resulted in the successful fabrication of superlattices, such as $CdTe/Cd_{1-x}Hg_xTe$ [5], $Ca_{1-x}Sr_xCuO_2$ [9] and ZnS/ZnSe [10,11]. These results have been obtained with single-laser systems. We believe that the application of a two-laser or multi-laser system has the advantage of being more tunable to the requirements of ablation of different materials. Such an approach enables bandgap engineering with an arbitrarily variable composition of ternary compounds [21]. Also, the time required for switching between ablation of different targets can be made practically negligible in a multi-laser system such as PLEE. This is especially important for the fabrication of high quality heterojuctions in which interfaces are known to act as impurity gettering centres.

Inherent with laser ablation is the ability to control deposition at rates of an atomic monolayer, or less, per laser pulse. This offers precision in the film thickness control even with lasers running at high repetition rates and over a long period of time. It is worth mentioning however that a stable ablation rate may not be achieved instantaneously and the process of target erosion has to be monitored in detail [6,22]. By ablation of CdTe and $Cd_{1-x}Mn_xTe$ targets with a Nd:YAG and an XeCl lasers and maintaining the growth temperature at 260 °C, we have been able to maintain

Table 1
Structural parameters of (001)CdTe-Cd$_{1-x}$Mn$_x$Te MQW and SLs illustrating a PLEE-induced atomic level growth.

Sample	N	Well thickness [nm]		Barrier thickness [nm]
	pairs	SIMS	Opt.	SIMS
CCM-109	22	8.8	11.9	22.5
CCM-102	49	5.4[a]	6.3	2.7
CCM-207	41	2.0[b]	2.6	4.5

[a]   From TEM results
[b]   Estimated value

deposition rates at levels such that 1000 or more laser pulses were needed to complete one monolayer of material. Structural parameters of some of the PLEE-grown CdTe-Cd$_{1-x}$Mn$_x$Te multilayer samples have been listed in Table 1. The properties of these samples, which have been established with SIMS depth profiling and x-ray diffraction, and verified by optical measurements [23,24], indicate that an atomic level growth has been achieved. In the case of sample CCM-207, a 2.6 nm thickness of the (001)CdTe well implies that it constitutes 8 atomic layers. In this particular case it took about 6000 pulses of an Nd:YAG laser to complete one well structure. It is interesting to note that a small interface mixing ($\sim$ 0.5 nm) has been concluded from a magnetoreflectivity investigation [25] of PLEE-grown CdTe/Cd$_{1-x}$Mn$_x$Te (x $\approx$ 0.10) heterojunctions.

In addition to in-situ RHEED measurements, which proved to be very useful in the optimization of the conditions leading to the epitaxial growth [21], we have recently implemented in our PLEE system a fast-nulling ellipsometer (Waterloo Digital Electronics, EXACTA 2000). In Fig. 2 is shown an example of the ellipsometric plot (dependence of the polarizer angle on the analyzer angle) obtained during the growth of a CdTe buffer layer and a series of Cd$_{1-x}$Mn$_x$Te-CdTe multiple quantum well structures. The flat portion, A-B, of the plot is believed to be related to the formation of a $\sim$ 1 nm thick transient structure at the (001)CdTe/(001)GaAs interface [7]. After completing the growth of $\sim$ 0.5 μm thick buffer layer, point C on the plot, the structure consisting of 50 pairs of Cd$_{1-x}$Mn$_x$Te(x $\approx$ 0.20)-CdTe has been grown in this case. Preliminary results indicate the usefulness of this instrument in monitoring an atomic level growth and its feasibility for an in-situ investigation of the dynamics of the growth process. A detailed analysis of the ellipsometric data obtained during the PLEE growth of MQWs and SLs will be published in future [26].

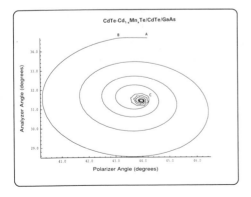

Fig. 2. An example of the ellipsometric plot obtained during the growth of a CdTe buffer layer and a series of Cd$_{1-x}$Mn$_x$Te(x $\approx$ 0.20)-CdTe multiple quantum well structures.

Laser-Assisted Dry Etching Ablation. A schematic diagram of the laser-assisted dry etching ablation (LADEA) system is shown in Fig. 3. An XeCl excimer laser (Lumonics HX-460) with a custom made holographic diffuser (HD) has been employed for these experiments. The application

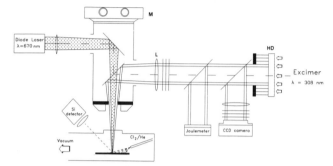

Fig. 3. A schematic diagram of the laser-assisted dry etching ablation system. The diode laser is used for diffused laser light scattering experiments.

of the diffuser enabled production of a more uniform beam with a nearly 'top-hat' beam profile. The beam profile and laser fluence were monitored respectively with a CCD camera and a Joulemeter (Molectron-2000). The laser beam was focused on samples with a 52 cm focal length lens (L). An elliptical laser spot on the surface of etched samples had dimensions ~ 500 μm x 800 μm. A low-resolution inspection of the processed surfaces has been carried out with a microscope (M) equipped with a rocking objective. We have applied an in-situ characterization of the etched surfaces by measuring diffused laser light scattered (LLS) from the surface. This was carried out with a diode laser operating at 670 nm which was delivered to an excimer laser etched spot with the use of a mirror equipped with an x-y micro-manipulator. The spot size of the probing beam focused on the sample was approximately 100 μm in diameter. A standard lock-in detection technique has been applied for the measurements of the intensity of LLS signal.

Etching was carried out in a 5 % mixture of Cl_2 buffered with He. The gas pressure, p, in the chamber was regulated with a leak valve in the range of 10 - 200 mtorr. A gas injector placed in the vicinity of the etched wafer has been applied to provide efficient chlorination of the sample surface. The samples were so called 'epi-ready', S-doped (001) InP wafers from Crismatec Inc. No additional surface cleaning was applied before mounting a sample on a stainless steel holder. We also investigated etching of a multilayer structure of InP/$Ga_{1-x}In_xAs$ (x=0.47) grown by chemical beam epitaxy. The samples were held at room temperature during LADEA experiments.

It has been found [27] that the etch rate of InP in the LADEA process is independent of Cl_2/He pressure for p > 15 mtorr. Also, it has been established that the etching rate for laser fluences (F) between 30 and 60 mJ/cm^2 was almost independent on the laser fluence and, for laser fluences F > 60 mJ/cm^2, it increased rapidly to more than 0.4 nm per pulse at 80 mJ/cm^2. In this paper we discuss the results obtained at a pressure of 20 mtorr and F $\approx$ 50 mJ/cm^2 which results in etching rates of about 0.03-0.08 nm/pulse. Figure 4 shows an example of a Dektak profile of a crater in InP

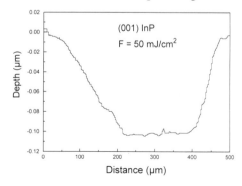

Fig. 4. A cross section profile of a LADEA fabricated crater in (001) InP. No mask has been applied in this process.

etched with about 1500 pulses delivered at 2 Hz. It can be seen that LADEA enables uniform removal of material with the surface of an etched crater smooth to better than ~ 10 nm. It should be pointed out that flat bottom craters have not always been obtained with the current experimental setup, and results were strongly dependent on the stability of the laser and profile of the laser beam. A scanning electron microscopy (SEM) investigation which has been complemented by an atomic force microscopy (AFM) revealed that the samples with craters shallower than 100 nm had featureless surface morphology on a near atomic scale. An example of an AFM image observed for a sample with a 120 nm thick layer removed in the LADEA process is shown in Fig. 5. A slightly corrugated surface can be seen in this case.

The corrugated surfaces are due to an inhomogeneous deposition of laser energy on the sample surface. It has been indicated in literature [28] that the nature of this inhomogeneous deposition is due to the interference between the laser light scattered on the surface and the primary laser beam. If a linearly polarized laser beam irradiates the sample at an angle Θ with respect to the normal, a grating of period $\Lambda = \lambda/[n(1 \pm \sin\Theta)]$ can be formed (n is the index of refraction of an ambient surrounding the sample). The scattering of the excimer laser light is strongly influenced by surface defects such as micro dust particles. This leads to a more non-uniform development of laser induced periodic surface structures (LIPS). A cross sectional AFM scan which is shown in Fig. 5 illustrates a case of a non-uniform structure consisting of weakly developed corrugations 'imprinted' on the sample surface. The spacing of corrugations measured between points a-b and e-f (half period) is approximately 300 nm, and 280 nm in the vicinity of the scattering center (c-d). Thus, it appears that the periodicity of corrugations decreases as the development of LIPS takes place. It has been reported [29,30] that the amplitude of wet-etched LIPS in GaAs grows exponentially until a maximum depth of 50-60 nm is reached. Consequently, a dominant orientation of LIPS may be lost as the laser light is scattered on large size surface irregularities and a disorganized surface structure is formed. It is reasonable to believe that LADEA fabrication of surface structures in InP follows the same pattern. We have demonstrated that a 0.25-μm-period grating can be developed on the surface of a LADEA processed wafer if a sufficiently thick layer of material (about 200 nm) is etched away [31]. It is worth mentioning that the sample in Fig. 5 has been etched with a linearly polarized laser beam having the electric field vector approximately parallel to the direction of the scan (a-b-c-d-e-f). One can expect that rotation of the direction of linear polarization or scrambling of the polarization should lead to smooth surfaces even after the removal of a substantially thick layer of material.

The results of LLS experiments have been found to be consistent with SEM and AFM investigations. The intensity of the LLS signal remained constant or slightly decreased as the material was etched up to a depth of about 150-200 nm. The decrease in the intensity of the LLS

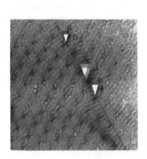
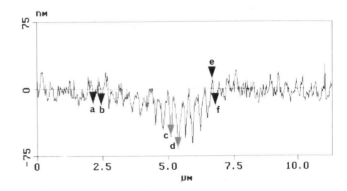

Fig. 5. An atomic force microscopy image and a cross scan of a 120 nm deep crater fabricated with LADEA in (001) InP.

signal appears to be due to an initial smoothing of the surface as a result of etching and/or anisotropic scattering of diffused light as a result of the formation of ordered surface corrugations. As the etching continued, LIPS became more well defined and deeper until their amplitude reached the maximum value. The resulting surface scattered the probing laser beam more efficiently until the saturation of the LLS signal was obtained [27]. Measurements of the LLS signal during LADEA of multilayer structures indicate the feasibility of this method to monitor the ablation of individual layers at rates approaching an atomic resolution. This is illustrated in Fig. 6 where the LLS signal obtained during LADEA of a CBE-grown InP/GaInAs structure is plotted as a function of etched depth. The structure consists of a 15 nm thick cap layer of InP and 5 layers of $Ga_{1-x}In_xAs$ (x=0.47) of thicknesses systematically increasing from about 2 nm near the surface to 15 nm. The $Ga_{1-x}In_xAs$ layers are separated by 15 nm thick layers of InP. Different etch rates can be expected for InP and $Ga_{1-x}In_xAs$ however we have not addressed this problem in this paper. An average etch rate of 0.03 nm/pulse has been observed during the etching of this structure with the laser fluence, repetition rate, and Cl_2/He pressure, 45 mJ/cm^2, 2 Hz and 20 mtorr, respectively. Indicated with arrows in Fig. 6 are five maxima. They can be distinguished from the monotonically increasing LLS signal

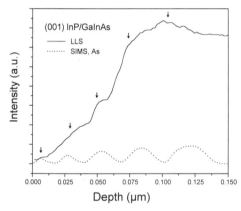

Fig. 6. A comparison between the intensity of a diffused laser light scattering signal (solid line), observed during LADEA of a InP/Ga$_{0.53}$In$_{0.47}$As structure, and that of a SIMS As signal (dotted line) obtained in an independent experiment. The maxima of the LLS signal coincide with those of the SIMS As profile.

which ultimately saturates after the ablation of a 125 nm thick layer. A secondary ion mass spectroscopy (SIMS) investigation which was carried out for the same sample corroborated the LLS results. A dotted line in Fig. 6 corresponds to a SIMS As signal measured during an O_2^+ ion beam etching of the same structure in a Cameca IMS 4f system. It can be seen that five maxima, corresponding to etching of $Ga_{1-x}In_xAs$ in the SIMS experiment, have their positions coinciding with those in the LLS signal. Therefore, it is reasonable to believe that the LLS maxima correspond to etching/ablation of $Ga_{1-x}In_xAs$ layers. A slight discrepancy in the SIMS and LLS measured spacings may be due to an error in the Dektak measurements of the depths of the etched craters. There is still much room for improvements in both LADEA and LLS, but we consider these results encouraging for the field of research on laser ablation for characterization and processing with atomic level resolution.

Summary

Pulsed laser ablation has been reviewed with a focus on the application of this method for the deposition of thin films (quantum wells and superlattices) and processing of semiconductor structures with atomic level resolution. Although the method of pulsed laser ablation has been in use for almost three decades [32] its potential for in-situ processing and fabrication of various structures has been appreciated only recently. This is a result of the development in understanding of the process of laser-surface interaction and of the continuous improvements in the characteristics of lasers suitable for microprocessing. A plethora of relevant problems remain to be solved, however some features of pulsed laser ablation, such as an ability to grow multilayer epitaxial structures with components of several atomic layers thickness, or to remove material by laser-assisted etching with similar resolution, have been demonstrated. The development of various methods for in-situ characterization of material deposited and/or processed with the pulsed laser ablation method also plays an increasingly important role. The laser here, again, proved to be an invaluable tool in both fast-nulling ellipsometry and laser light scattering experiments which have been applied in this work.

Acknowledgements

It is with a great pleasure that I would like to acknowledge the help provided by Jim Thompson during various stages of the development of the pulsed laser evaporation and epitaxy system. I thank Jim Leslie and Steve Buchanan from Waterloo Digital Electronics for furnishing PLEE with the ellipsometer, Lisa Fukunaga from Digital Instruments for AFM measurements, Alain Roth for providing a CBE-grown structure and Stephen Rolfe for SIMS measurements.

References

[1] D.C. Paine and J.C. Bravman (eds.), *Laser Ablation for Materials Synthesis*, Mat. Res. Soc. Symp. Proc., vol. 191 (1990).
[2] E. Fogarassy and S. Lazare (eds.), *Laser ablation of electronic materials*, Eur. Mat. Res. Soc. Mon., vol. 4, North-Holland, Amsterdam, 1992.
[3] B. Braren, J.J. Dubowski and D. Norton (eds.), *Laser Ablation in Materials Processing: Fundamentals and Applications*, Mat. Res. Soc. Symp. Proc., vol. 285 (1993).
[4] I.W. Boyd, *Laser-Assisted Fabrication of Thin Films*, SPIE Proc., vol. 2045 (in print).
[5] J.T. Cheung and H. Sankur, Critical Rev. Solid St. Mater. Sci. **15**, 63 (1988).
[6] J.J.Dubowski, Chemtronics **3**, pp. 66, 1988.
[7] J.M. Wrobel and J.J. Dubowski, in: *Semiconductor Interfaces and Microstructures*, ed. by Zhe Chuan Feng, World Scientific Publishing Co. Pte. Ltd., Singapore, 1992, pp. 216-237.
[8] J. Dieleman, E. van de Riet and J.C.S. Kools, Jpn. J. Appl. Phys. **31**, 1964 (1992).
[9] M. Kanai, T. Kawai and S. Kawai, Mat. Res. Soc. Symp. Proc., vol. 285, 257 (1993).
[10] J.W. McCamy, D.H. Lowndes, J.D. Budai, G.E. Jellison, Jr., I.P. Herman and S. Kim, Mat. Res. Soc. Symp., Vol. 285, 471 (1993).
[11] Y. Rajakarunanayake, Y. Luo, B.T. Adkins and A. Compaan, Mat. Res. Soc. Symp. Proc., vol. 285, 477 (1993).
[12] D.M. Zehner, C.W. White and G.w. Ownby, Appl. Phys. Lett. **36**, 56 (1980).
[13] R.Tsu, D. Lubben, T.R. Bramblett and J.E. Greene, J. Vac. Sci. Technol. **A9**, 233 (1991).
[14] D.M. Zehner, C.W. White, B.R. Appleton and G.W. Ownby, Mater. Res. Soc. Proc. **Vol.4**, p. 683 (1982).
[15] H. Liu, J.C. Roberts, J. Ramdani, S.M. Bedair, J. Farari, J.P. Vilcot and D. Decoster, J. Cryst. Growth **107**, 878 (1991).
[16] P.A. Maki and D.J. Ehrlich, Appl. Phys. Lett **55**, 91 (1989).
[17] S.M. Bedair, J.K. Whisnant, N.H. Karam. M.A. Tischler and T. Katsuyama, Appl. Phys. Lett. **48**, 174 (1986).
[18] Q. Chen, J.S. Osinski and P.D. Dapkus, Appl. Phys. Lett. **57**, 1437 (1990).
[19] J.E. Epler, D.W. Treat, H.F. Chung and T.L. Paoli, SPIE vol. 1043, 36 (1989).
[20] R. Iga, H. Sugiura and T. Yamada, Semicond. Sci. Technol. **8**, 1101 (1993).
[21] J.J. Dubowski, Acta Physica Polonica **A80**, 221-244 (1991).

[22] S.R. Foltyn, R.E. Muenchausen, R.C. Estler, E.Peterson, W.B. Hutchinson, K.C. Ott, N.S. Nogar, K.M. Hubbard, R.C. Dye and X.D. Wu, Mat. Res. Soc. Symp. Proc., vol 191, 205 (1991).

[23] J.J. Dubowski, A.P. Roth, E. Deleporte, G. Peter, Z.C. Feng and S. Perkowitz, J. Cryst. Growth **117,** 862-866 (1992).

[24] A.P. Roth, R. Benzaquen, P. Finnie, P.D. Berger and J.J. Dubowski, SPIE Proc., vol. 2045, (in print).

[25] A. Lewalski, The Khoi Nguyen, P. Kossacki, J.A. Gaj and J.J. Dubowski, Acta Phys. Polon. **A84,** 571-574 (1993).

[26] H. Tran et al., in preparation.

[27] J.J. Dubowski, A. Compaan and M. Prasad, Symposium on Photon-Assisted Processing of Surfaces and Thin Films, E-MRS Spring Meeting, Strasbourg, France, May 1994.

[28] A.E. Siegman, IEEE J. Quantum Electronics, vol. QE-22, 1384 (1986).

[29] A.I. Khudobenko, V.Ya. Panchenko, V.K. Popov and V.N. Seminogov, SPIE Proc., vol. 1723, 7 (1991).

[30] H. Kumagai, K. Toyoda, H. Machida and S. Tanaka, Appl. Phys. Lett.**59,** 2974 (1991).

[31] J.J. Dubowski, Proc. of the 3rd Int. Conf. on Advanced Materials, Ikebukuro, Japan, September 1993 (in print).

[32] J.F. Ready, *Effects of High-Power Laser Radiation,* Academic, New York, 1971.

Materials Science Forum Vols. 173-174 (1995) pp. 81-92
© *1995 Trans Tech Publications, Switzerland*

VACUUM ULTRAVIOLET DEPOSITION OF SILICON DIELECTRICS

I.W. Boyd

Electronic and Electrical Engineering, University College London,
London WC1E 7JE, UK

Keywords: Ultraviolet, Vacuum Ultraviolet, Photochemistry, Photodeposition, Photo-CVD, Dielectrics, Silicon Dioxide, Silicon Nitride, Silicon Oxynitride, Excimer Lamp, Thin Films, Multilayers

Abstract. The use of vacuum ultraviolet (VUV) light generated from a new type of excimer lamp to initiate the deposition of dielectric thin films in a photo-chemical vapour deposition process is reviewed. Compared with other UV lamps, these pseudo-continuous light sources can provide high photon fluxes (more than a few watts) over large areas. The growth rates and physical properties of the films deposited, which include silicon dioxide, silicon nitride, and silicon oxynitride, when SiH_4, N_2O and NH_3 are used, are reported. A layered combination of silicon oxide and silicon nitride, produced in the same reactor at temperatures below 300°C, is also described. The technique is found to produce good quality films at reasonable growth rates, and also offers very good control of the stoichiometry in the case of SiO_xN_y film deposition.

Introduction

In recent years, a general requirement for low temperature processing has evolved, especially given the emergence of a growing number of temperature-sensitive film technologies, including solar cell structures, advanced silicon and compound semiconductor microelectronics, and optoelectronic devices and displays. Of the film growth processes under investigation, photo-induced processing has received considerable attention [1,2,3] since the processed surface is not subjected to damaging ionic bombardment as can be the case in plasma assisted systems [4,5].

Photochemical vapour deposition (photo-CVD) techniques have used various light sources from powerful lasers [6,7] to low wavelength lamp devices [8,9], the former generally offering very high fluences and the latter high photon energies. However, lasers are generally very expensive and costly to operate, and traditional vacuum ultraviolet (VUV) lamps (λ<200 nm) generally offer very limited fluences and efficiencies. Recent advances have brought about the development of new high power sources (5W output can be easily obtained) requiring low operating and maintenance costs, which can provide pseudo-continuous fluxes of VUV radiation over large areas and at high efficiency (typically around 10%) [10,11].

In this paper, after briefly describing the underlying principles involved in excimer lamp sources, their use in depositing thin dielectric films will be described. In particular, the growth of silicon oxide, silicon nitride, silicon oxynitride and ONO multilayers will be discussed.

Excimer Lamp Source and Deposition Apparatus

The radiation sources used in this study, i.e. dielectric barrier discharge (DBD) or excimer (or silent) discharge lamps, have been specially designed and constructed in-house, as they are not currently commercially available. Their operating principle relies on the radiative decomposition of excimer states created by a silent discharge in a high pressure (few hundred mbar) gas column [12]. In a dielectric barrier discharge lamp, one or both electrodes must be electrically insulated, e.g. covered by a dielectric. A high voltage (7 to 10 kV) and high frequency (100 to 500kHz) supply is then applied to the device causing an arc discharge to occur. The charge build-up on the dielectric surface immediately decreases the field in the discharge gap and extinguishes the arc. The total duration of a typical arc is of the order of a few nanoseconds and several arcs can be formed quasi-simultaneously yet randomly across the entire surface of the dielectric at a frequency equal to two times the driving frequency. Each individual current filament is described as a microdischarge because of the short time duration and low electrical energy it involves. The self-extinguishing feature of this type of discharge enables the use of high (≈500 mbar) gas pressure without causing any sputtering of the electrodes and therefore erosion and contamination problems common with traditional arcs are eliminated.

In the case of rare gas halide mixtures circulating between the electrodes, and in particular for the mixtures of ArF, KrF and XeCl, the well-known excimer laser frequencies of 192, 248 or 308 nm are obtained. For excited molecular complexes in pure rare gases, lower wavelength continua are generated at 126, 146 and 172 nm for argon, krypton and xenon respectively. The bandwidth associated with these emissions is typically around 10-20 nm.

The pumping mechanism involved in the formation of excimers in the case of xenon, the gas used in our current studies, is illustrated by Fig. 1[13], while Fig. 2 shows a simplified potential energy diagram for this system. Note the absence of any bound ground level for the Xe_2 dimer, which indicates splitting when emission occurs. Therefore self-absorption of the radiation by the gas phase is completely absent in this system [14].

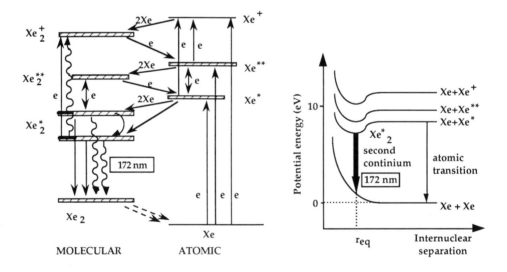

Figure 1: *Simplified pumping scheme for Xe*   Figure 2: *Simplified band-diagram of Xe*

The schematic of Fig. 1 summarises the following operations (1) - (6). Energetic electrons present in the microdischarges excite and ionise the xenon atoms:

$$e^- + Xe \rightarrow Xe^* + 2\ e^- \tag{1}$$

$$e^- + Xe \rightarrow Xe^+ + e^- \tag{2}$$

At pressures above 50 mbar (relevant to our case) the formation of molecular ions is rapid and subsequently leads to the formation of excited neutrals:

$$Xe^+ + 2\,Xe \rightarrow Xe_2^+ + Xe \tag{3}$$

$$Xe_2^+ + e^- \rightarrow Xe^{**} + Xe \tag{4}$$

The creation of the rare gas dimer, Xe_2^*, then occurs through the three body reaction of an electronically excited rare gas atom Xe^* with two others in the ground state:

$$Xe^* + Xe + Xe \rightarrow Xe_2^* + Xe \tag{5}$$

The Xe_2^* excimer consequently dissociates into two Xe atoms, thereby radiating a 172 nm photon

$$Xe_2^* \rightarrow 2\,Xe + h\nu \ (172\ nm) \tag{6}$$

The lamp system used in this study contained the Xe within a cylindrical dielectric cell covered with a metallic mesh acting as the ground electrode. An inner dielectric-covered electrode was supplied by an external discharge voltage [15]. The microdischarges induced upon breakdown occurred in the so-called discharge gap separating the two electrodes and the Xe, and the photons were emitted through the outer dielectric, chosen to be Suprasil® quartz which is known to be transparent to the radiation generated. The lamp output power, measured to be around 3 W (corresponding to 20 mW/cm2), was determined using actinometric techniques. The generic experimental set-up used is sketched in Fig. 3.

The radiation generated is passed through a MgF_2 window into the reaction chamber where the host substrate, which could be heated to a temperature of 400°C, was placed. P-type silicon (100) samples up to one inch in diameter, cleaned in a propanol solution in ultrasonic bath, were used in these present studies. The cell was evacuated to 10^{-6} mbar. and filled with mixtures of SiH_4, NH_3, O_2, and N_2O. Ar was used as the purge gas. Thermocouple control ensured that, although essentially negligible, the low intensity of the lamp did not increase the surface temperature of the samples during film growth. A full description of this reactor is described elsewhere [16].

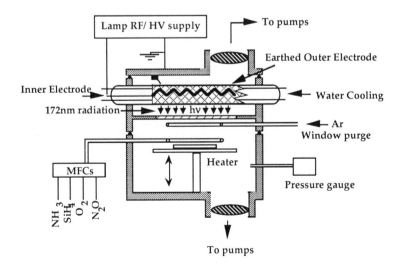

Figure 3: General experimental arrangement used for photodeposition

Dielectric Thin Film Deposition

Deposition of silicon dioxide from silane and nitrous oxide mixtures [17], and of silicon nitride from silane and ammonia [18] have been successfully deposited using this apparatus. Silane is transparent to the 172 nm radiation. Therefore, since only the oxidising and reducing agents exhibit a significant absorption cross section, σ, at this wavelength they are photochemically dissociated with this lamp [19]. The following primary reactions are initiated:

$$NH_2O + hv \rightarrow N_2 + O \qquad\qquad (\sigma = 2\ atm^{-1}cm^{-1}) \quad (7)$$
$$NH_3 + hv \rightarrow NH_2 + H \qquad\qquad (\sigma = 50\ atm^{-1}cm^{-1}) \quad (8)$$

Secondary reactions will subsequently occur between the primary products and the silane and lead to material deposition on the hot substrate surface.

Silicon Dioxide Deposition. The VUV induced reaction scheme between SiH_4 and N_2O has been studied by several groups [17, 18]. In our experiments, gas mixtures containing between 0.5 to 10% silane in nitrous oxide were introduced into the chamber at a constant flow of 50 sccm, and with a total gas pressure of 100 mbar, the absorption by N_2O of the 172 nm radiation entering the processing chamber was about 50%. Following photodissociation of N_2O, the oxygen radicals formed (7) will react with silane (9-11). Branching reactions (12-14) subsequently lead to SiO_2 formation (15). A more complete model for this scheme has been detailed by Petitjean et al. [20].

$$SiH_4 + O \dashrightarrow SiH_3 + OH \tag{9}$$
$$SiH_4 + OH \dashrightarrow SiH_3 + H_2O \tag{10}$$
$$SiH_4 + O \dashrightarrow SiH_2 + H_2O \tag{11}$$
$$SiH_2 + O \dashrightarrow SiH_2O \tag{12}$$
$$SiH_3 + O \dashrightarrow SiH_2O + H \tag{13}$$
$$SiH_2O + O_2 \dashrightarrow SiH_2O_2 + O \tag{14}$$
$$SiH_2O_2 + O_2 \dashrightarrow SiO_2 + H_2O + O \tag{15}$$

For exposure times up to 30 min, film thicknesses of 50 nm were grown at temperatures below 300°C. At gas pressures exceeding 150 mbar, powder was formed as a consequence of gas phase reactions being stimulated. In contrast, the deposition rate was very low below 100 mbar. At intermediate pressures and between 200 and 300°C, the films formed were scratch resistant and adherent. Refractive indices of 1.458 were obtained on optimised samples, which exhibited full width half maximum (FWHM) values for the 1065 cm-1 IR mode of 70 cm-1, representative of good structural quality SiO_2. Details of the deposition optimisation are published elsewhere [17].

Silicon Nitride Deposition. When NH_3 is mixed with SiH_4 instead of N_2O, SiN layers can be deposited by the 172 nm radiation. In addition to (8) above, the following also occur:

$$NH_3 + hv \rightarrow NH + H_2 \tag{16}$$
$$NH_3 + hv \rightarrow NH + 2H \tag{17}$$

although together these contribute less than 5% of the total dissociation reaction [19]. Following from (8), hydrazine is formed, but is subsequently also dissociated (18-21).

$$NH_2 + SiH_4 \rightarrow NH_3 + SiH_3 \tag{18}$$
$$NH_2 + NH_2 + M \rightarrow N_2H_4 + M \tag{19}$$
$$N_2H_4 + hv \rightarrow H + N_2H_3 \tag{20}$$
$$N_2H_4 + hv \rightarrow NH_2 + NH_2 \tag{21}$$

Secondary reactions of H and NH_2 lead to the formation of silylamine (SiH_3NH_2), which, together with NH_2 and SiH_3, contribute to surface reactions that result in the growth of SiN (22-26).

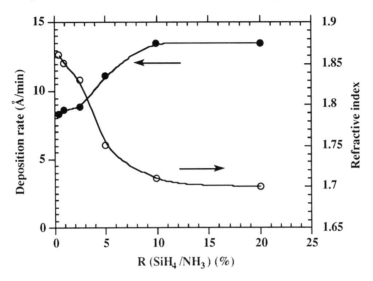

Figure 4. Deposition rate & refractive index of photodeposited SiN for different SiH₄/NH₃ ratios

$$H + SiH_4 \rightarrow H_2 + SiH_3 \tag{22}$$
$$H + SiH_3 \rightarrow SiH_2 + H_2 \tag{23}$$
$$N_2H_3 + H \rightarrow N_2 + 2H_2 \tag{24}$$
$$SiH_3 + NH_2 \rightarrow SiH_3NH_2 \tag{25}$$
$$N_2H_3 + N_2H_3 \rightarrow 2NH_3 + N_2 \tag{26}$$

A total gas pressure of 5-10 mbar was used, with flow rates of the SiH_4 and NH_3 being 2/20 sccm, respectively. Initial experiments revealed that this process requires the use of extremely high purity gases, and the presence of only a few 100ppm of O-containing species (such as O_2, H_2O or CO_2) was found to encourage significant oxygen incorporation into the films. The effect of R, the ratio of SiH_4 to NH_3, on the deposition rate and the refractive index of the films grown is shown in Fig. 4 for an NH_3 flow of 20 sccm, a total gas pressure of 10 mbar, and substrate temperature of 300°C.

The growth rates were more strongly affected by light intensity than substrate temperature in the 200°C to 300°C range, and the values shown here could in principle be increased by using a stronger lamp source. The lack of any absorption in the 2175 cm⁻¹ region of the infrared spectra provided an upper limit of a few percent to the hydrogen content of the films.

The refractive index clearly decreases strongly as a function of SiH_4 concentration. The infrared absorption at 840 cm⁻¹ due to Si-N bonds also decreased markedly for increasing values of R at a fixed NH_3 flow. Apparently, high levels of SiH_4 cannot be efficiently dissociated if the number of NH and NH_2 species is limited, making Si-N bond formation difficult. Any trace presence of O will thus be more likely to become incorporated in the films grown under these conditions.

SIMS analysis was performed on samples within which the infrared measurements indicated that the H and O contents to be low. Figure 5 shows a SIMS profile for an as-deposited 60 nm SiN layer grown on Si. Clearly, the oxygen content is constant with depth, indicating negligible contamination, while the hydrogen content is also very small. The figure also indicates that the film stoichiometry is constant with depth (the tail being due to recoil mixing). All the films produced were scratch resistant, adherent and stable to the atmosphere. Etch rates were 3 nm/s in buffered HF (1:30), while the average dielectric breakdown fields were measured to be close to 10 MV/cm.

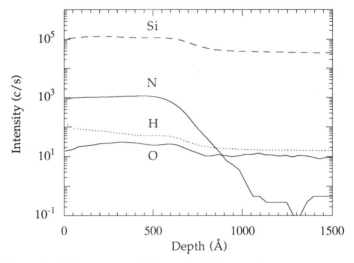

Figure 5. SIMS spectrum of 60 nm thick SiN indicating very low O presence

The conditions outlined above for SiO_2 and SiN film deposition can be applied consecutively within the same reaction chamber to produce multilayers. As shown in Fig. 6, one such possibility is a simple O-N-O structure. It is also possible to deposit SiO_xN_y layers by mixing N_2O and NH_3 in the reaction cell. This opens up the possibility of producing a variety of dielectric film structures with a number of potential applications in optical technology as both active and passive coatings.

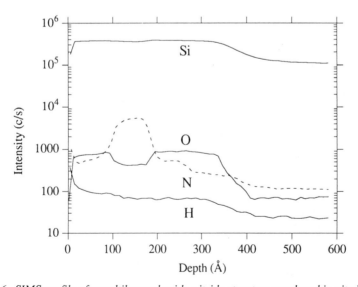

Figure 6. SIMS profile of a multilayered oxide-nitride structure produced in-situ by photo-CVD

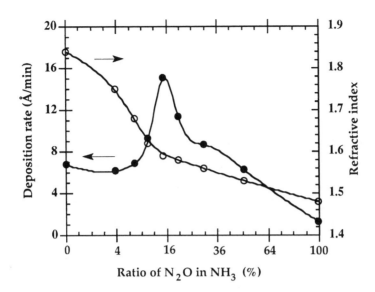

Figure 7: Relationship of the SiO$_x$N$_y$ growth rate and refractive index with N$_2$O/NH$_3$ mixture.

Silicon Oxynitride Deposition. The combination of the two processes also promises to enable the growth of silicon oxynitride with fine tuning of the refractive index being controlled by varying the ratio of the precursor mixtures. In fact, smooth progression from SiN to SiO$_2$ and thus a good control of the SiO$_x$N$_y$ stoichiometry can be achieved during deposition[21]. As shown in Fig. 7, at a constant pressure of 10 mbar, the growth rate of the films is maximum for precursor concentration ratios N$_2$O/NH$_3$ close to 15%. In fact, at low N$_2$O concentrations, mainly ammonia excited species are present in the gas phase, leading to a N rich SiO$_x$N$_y$ film. On the other hand, at much higher N$_2$O/NH$_3$ ratios, the radiation entering the chamber is less absorbed by the gas phase (lower σ) and lower deposition rates result. In the range around 15%, the film obtained is oxygen rich, and the high growth rate is due to the combination of efficient photo-excitation of ammonia species together with a high degree of reactivity of the oxygen species towards the reaction with other species. There is therefore a transferred photo-excitation, or sensitisation, of the N$_2$O precursor by the excited NH$_3$, which leads to O rich SiO$_x$N$_y$ film, as indicated by the decrease in the refractive index value with small increases in the nitrous oxide concentration in the gas mixture.

To evaluate further the structure of the deposited films, infrared spectroscopy measurements were carried out. Table 1 summarises the properties of these films by showing the correlation between the refractive index and the position of the peak resulting from the convolution of the Si-O and Si-N main stretching vibration modes at 1065 and 835 cm^{-1} respectively [22,23]. This ability to selectively predetermine the reactive precursor composition that will ultimately react with the silane gives this photo-CVD process an added advantage over conventional plasma deposition processes for optical coatings and optical fibre material manufacturing.

A promising perspective is also the possibility of laying down silicon nitride and silicon oxide multilayers in the same processing chamber without exposing the sample to air between depositions. Potential benefits provided by such an approach over conventional dielectric film deposition methods include reduced defect density, higher lifetime of the device, lower leakage current, and higher breakdown fields [22, 24, 25].

Table 1: Infrared absorption characteristics of photo-CVD silicon oxynitride thin films

SiH$_4$:NH$_3$:N$_2$O	N$_2$O/NH$_3$ (%)	n	Growth rate (Å/min)	Main Stretching Vibrational Mode (cm^{-1})	FWHM (cm^{-1})
0.5 : 50 : 0	0	1.84	6.74	845 ± 2 cm^{-1}	125 ± 5 cm^{-1}
0.5 : 50 : 2	3.8	1.75	6.16	910 ± 2 cm^{-1}	175 ± 5 cm^{-1}
0.5 : 50 : 4	7.4	1.68	6.88	940 ± 2 cm^{-1}	175 ± 5 cm^{-1}
0.5 : 50 : 6	10.7	1.62	9.25	975 ± 3 cm^{-1}	150 ± 5 cm^{-1}
0.5 : 42.5 : 7.5	15	1.59	15.03	1000 ± 2 cm^{-1}	125 ± 5 cm^{-1}
0.5 : 40 : 10	20	1.58	11.31	1001 ± 2 cm^{-1}	125 ± 5 cm^{-1}
0.5 : 35 : 15	30	1.56	8.68	1015 ± 2 cm^{-1}	110 ± 5 cm^{-1}
0.5 : 25 : 25	50	1.53	6.22	1025 ± 2 cm^{-1}	100 ± 5 cm^{-1}
0.5 : 0 : 50	100	1.47	1.25	1050 ± 1 cm^{-1}	65 ± 3 cm^{-1}

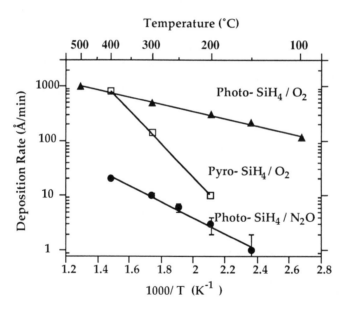

Figure 8. Arrhenius plot of thermal and photo-CVD of SiH$_4$/O$_2$ and photo CVD of SiH$_4$/N$_2$O

Rapid Silicon Dioxide Deposition. The reaction of silane with oxygen is also clearly possible using photo-CVD and as will be shown below, much higher deposition rates in the range of 100 to 500 Å/min are possible [26]. Generally, films deposited by conventional methods using such gas phase mixtures lead to porous oxides, with a large degree of hydrogen incorporation. This is due to the very high degree of reactivity of the oxygen species towards the reaction with silane,

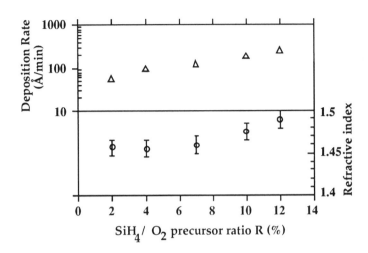

Figure 9. Growth rate and refractive index of oxides grown at 200 °C and 7mbar, for ratios of silane /oxygen.

causing gas phase nucleation and therefore SiO_xN_y film deposition. In our case, however, by using a photo-enhanced chemistry, it appears from a close study of the deposition parameter influences that gas phase nuclei formation can be limited and layers whose properties are very close to thermally grown stoichiometric silicon oxide can be obtained.

In this work, Ar, SiH_4, and O_2 were used in the ratio ranges 55:0.2-1.2:10 at pressures ranging from 5-10 mbar, very far from the known explosion conditions [27]. Figure 8 compares the growth rates of SiO_2 layers produced by conventional CVD with those obtained previously using SiH_4/N_2O mixtures and also with SiH_4/O_2 mixtures used here. As can be seen, a strong decrease in the growth rate is observed at temperatures below 400°C for pyrolytic CVD, which exhibits an activation energy E_a of 0.6 eV. By comparison, the value of E_a obtained for photo-CVD of SiH_4/O_2 was 0.13 eV.

Figure 9 shows the actual deposition rates and refractive index values obtained for SiH_4/O_2 ratios in the range 2-12% at a pressure of 7 mbar and temperature of 200°C. As can be seen, very small differences in the SiH_4/O_2 strongly affect the deposition rate. The refractive index increases with an increase in R, and we have found this to be related to the hydrogen content of the films by monitoring an increase in the 880 cm-1 Si-H absorption band using FTIR [28]. Thus, unless the presence of H presents no problem, a value of R=10% represents an upper limit for obtaining low temperature films whose refractive index values are close to those usually obtained for conventional SiO_2 films. Under these conditions, deposition rates of some 25 nm/min can be achieved.

Figure 10 indicates the level of H incorporated in such layers at 200°C, showing the FTIR spectrum of a 2000 Å film obtained at 200°C in 5 min, and annealed in wet N_2 at 300°C for 3 hours. Sharp peaks appear at 1065, 800 and 460 cm-1 at the stretching, bending and rocking Si-O-Si vibration modes respectively, and near 880 cm-1 indicating the presence of SiH [27]. The reduction in this peak after annealing reveals a level of H of only 0.5%. This is indeed a very low value for such a processing temperature. Typically these films exhibited refractive indices of 1.46, etch rates of 20 Å/min in diluted 1:25 buffered HF:H_2O, and electrical breakdown fields of 5 to 8 MV/cm.

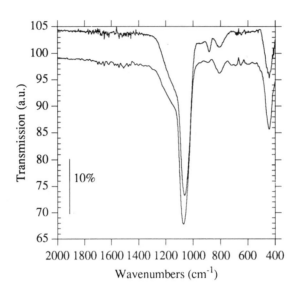

Figure 10. FTIR spectrum of 200 ℃ film prior to, and after annealing for 3 hours at 300 ℃.

Films have also been grown at higher temperatures. Figure 11 shows the deposition rate and refractive index of the silicon dioxide layers grown at temperatures up to 500°C at pressures of 5 and 10 mbar for a fixed silane to oxygen ratio of 7%. For both pressures, as the temperature is increased, higher deposition rates are obtained, and refractive index values approach those usually

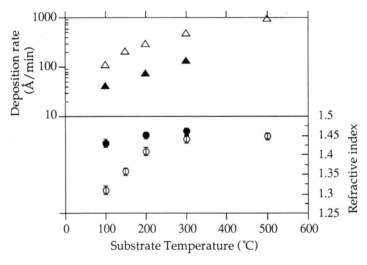

Figure 11. Temperature effect on growth rate and n of SiO_2 for 5(filled) & 10 mbar, (R=7%).

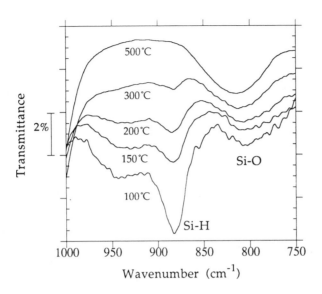

Figure 12. Comparison of the 880 cm-1 Si-H bands for SiO2 films grown between 100 and 500 ℃

found in stoichiometric material. At 10 mbar, although faster growth rates are obtained, generally low values of refractive index are found and hence more porous films are produced. This is confirmed in Fig. 12, which shows the effect of the substrate temperature on the normalised Si-H vibrational mode near 880 cm-1 for films deposited at 10 mbar. Very low substrate temperatures (100°C-200°C) clearly possess high concentrations of hydrogen due to the limitation of the nuclei dissociation kinetics on the relatively cool sample surface. Above this temperature range, the necessary dissociation kinetics are able to operate. Hence, it is clear that while high quality thin films of SiO_2 can be produced at quite low temperatures by this photo-CVD technique, by using low pressures, this can be achieved at the loss of the otherwise very attractive deposition rate.

Conclusion

In summary, the possibility of depositing a variety of high quality silicon dielectric films using a VUV lamp-initiated photochemical technique has been demonstrated. This process is based on the 172 nm wavelength from the excimer continuum of Xe excited within a dielectric barrier discharge lamp. Both SiO_2 and Si_3N_4 as well as any predetermined SiO_xN_y stoichiometry can be grown and the technique can clearly be applied to a variety of substrates. The excellent control offered by this method potentially opens up new areas for the application of such lamps, especially since deposition at high rates and over large areas can readily be achieved by scaling up the geometry of the apparatus used. Future work will also concentrate on the use of a shorter wavelength excimer lamp, using the Ar continuum radiating at 126 nm, to initiate direct photodissociation of silane for amorphous silicon thin film deposition.

Acknowledgements

This project was funded in part by the Science & Engineering Research Council, UK (Awards GR/F02229 and GR/J47750) and by ABB (Switzerland). Many people have contributed to the success of this work, and my sincere gratitude goes to Drs Ulrich Kogelschatz and Philippe Bergonzo, without whom this project would never have been completed, Dr Robin Thompson (Hughes Microelectronics Ltd) for providing the substrates, and Dr Steve Best and Ms Elizabeth Lloyd for the use of the FTIR spectrometers , Prof Mino Green for the ellipsometry, and Ms Christiane Dubois for the SIMS analyses.

References

1. I.W. Boyd: *Laser Processing of Thin Films and Microstructures* (Springer, New York 1987).
2. J.G. Eden: *Photochemical Vapour Deposition*, Chemical Analysis Vol. 122 (J. Wiley & Sons Inc., Canada 1992)
3. T. Szorényi, P. González, M.D. Fernández, J. Pou, B. León, M. Pérez-Amor, Thin Solid Films **193/194**, 619 (1990).
4. D.A. Buchanan, Appl. Phys. Lett. **56**, 1037 (1990).
5. R.A.B. Devine, J.M. Francou, J. Appl. Phys. **66**, 5654 (1989).
6. P.K. Boyer, G.A. Roche, W.H. Ritchie, G.J. Collins, Appl. Phys. Lett. **40**, 716 (1982).
7. S. Nishino, H. Honda, H. Matsunami, Jpn. J. Appl. Phys. **25**, L87 (1986).
8. P. Patel, I.W. Boyd, Appl. Surf. Sci. **46**, 352 (1990).
9. S.D. Baker, W.I. Milne, P.A. Robertson, Appl. Phys. **A46**, 243 (1988).
10. B. Gellert, U. Kogelschatz, Appl. Phys. **B52**, 14 (1991).
11. F. Kessler, G.H. Bauer, Appl. Surf. Sci., **54**, 430 (1992), and references therein..
12. B. Eliasson, U. Kogelschatz, IEEE Trans. Plasma Sci., **19** 309 (1991).
13. Ch. K. Rhodes: *Excimer Lasers*, Topics in Appl. Phys., Vol. 30 (Springer, Berlin, 1984).
14. A. Gilbert, J. Baggot: *Essential of Molecular Photochemistry* (Blackwell Scientific, Oxford, (1991).
15. U. Kogelschatz, Appl. Surf. Sci. **54**, 410 (1992).
16. P. Bergonzo, P. Patel, I.W. Boyd, U. Kogelschatz, Appl. Surf. Sci. **54**, 424 (1992).
17. P. Bergonzo, I.W. Boyd, U. Kogelschatz, Appl. Surf. Sci. **69**, 393 (1993).
18. P. Bergonzo, I.W. Boyd, Appl. Phys. Lett. **63**, 1757 (1993).
19. J.G. Calvert, J.N. Pitts: *Photochemistry* (J. Wiley & Sons, New York 1966).
20. M. Petitjean, N. Proust, J.F. Chapeaublanc, Appl. Surf. Sci. **46**, 189 (1990).
21. P. Bergonzo, I.W. Boyd, in *Proc. E-MRS Spring Meeting*, ed. by Y. Nissim, (Elsevier, Amsterdam, 1994), to be published.
22. I.W. Boyd, J.I.B. Wilson, J. Appl. Phys. **53**, 4166 (1982).
23. M. Berti, M. Meliga, G. Rovai, S. Stano, S. Tamagno, Thin Solid Films **165**, 279 (1988).
23. W. Ting, J. Ahn, D.L. Kwong, J. Appl. Phys. **70**, 3934 (1991).
24. A.B. Joshi, D.L. Kwong, J. Lee, Appl. Phys. Lett. **69**, 1489 (1992).
25. H. Hwang, W. Ting, D.L. Kwong, J. Lee, IEEE Elec. Dev. Lett. **9**, 495 (1991).
26. P. Bergonzo, I.W. Boyd, Electronics Letters (to be published, 1994).
27. J.R. Hartman, J. Famil-Ghiriha, M.A. Ring, H.E. O'Neal, Combustion & Flame **68**, 43 (1987).
28. P. Bergonzo, I.W. Boyd , Electronics Letters (to be published, 1994).

Materials Science Forum Vols. 173-174 (1995) pp. 93-98
© 1995 Trans Tech Publications, Switzerland

CONDITIONS FOR HETEROPHASE JUNCTION FORMATION IN LASER DEPOSITED PbTe FILMS

V. Gaydarova

Faculty of Physics, Sofia University, BG-1126 Sofia, Bulgaria

Keywords: Laser Deposition, Deposition Rate, PbTe Films, Narrow-Gap Semiconductors, Polymorphism, Heterophase Junctions

Abstract. Optical, Far Infra Red and X-ray measurements of laser deposited PbTe films confirmed the existence of strained sublayer with polycrystal GeS - type and/or monocrystal CsCl - type structure. The relative content of the two high pressure PbTe phases and the thickness of the sublayers depended on the technological parameters: pulse power density, target - substrate distance, substrate temperature, as well as on the target and substrate materials. Correlation between growth conditions and film properties was studied by two control parameters: 1)deposition rate and 2)visibility of the interference patterns in transmittance spectra of the films. The fact that PbTe is able to crystallize in three different pressure dependent modifications and the existence of well defined boundary between the strained and the relaxed sublayers of laser deposited films, offered an opportunity heterophase junctions to be created in the process of deposition.

Introduction

Laser assisted deposition (LAD) has been successfully applied to the narrow gap semiconductors, lead salts in particular, - materials mainly used as infrared detectors in the second, 3-5 μm, and in the third, 8-14 μm, windows of transparency of the atmosphere.

The interference transmittance spectra of laser deposited thick ($d > 2$ μm) PbTe films were usually modulated. This was related to existence of sublayer with different optical constants. The X-ray measurements displayed additional reflections besides those of f.c.c. O_h^5 (NaCl - type) structure with (100) direction of crystallization. These additional reflections were attributed to high pressure (HP) PbTe phases with b.c.c. O_{2h}^1 (CsCl -type) and orthorhombic D_{2n}^{16} (GeS - type) structures. Thus the strained sublayer in laser deposited PbTe films was recognized to be with polycrystal GeS - type (lattice constants $a = 4.7$ Å, $b = 11.4$ Å, $c = 4.3$ Å) and/or monocrystal (111) CsCl -type structure (lattice constant $a = 6.19$ Å) [1].

The relative content of HP phases and the thickness of the strained sublayers depended on the conditions of deposition. The influence of the technological parameters on growth of different PbTe modifications was studied by the deposition rate and its dependence on laser power density, target - substrate distance, substrate temperature, as well as on the target and substrate materials.

Experiment

Laser assisted deposition was performed by means of a set-up schematically presented in Fig. 1. The base pressure in the vacuum chamber (VC) was in the range of 10^{-6} - 10^{-7} Torr. The target (T) -

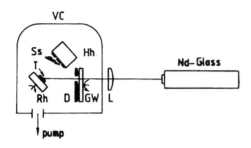

Fig. 1. *Experimental set-up for laser assisted deposition;*
VC - vacuum chamber, Nd-Glass laser, L - lens, T - target,
Rh - rotating holder, Ss - substrates, Hh - heated holder, GW
- rotating glass window, D - diaphragm

a tablet obtained through cold pressing at 5000 kg/cm^2 from rubbed to powder bulk material, was mounted on rotating holder (Rh) to minimize the overlap of damaged areas.

Pulsed Nd-Glass laser radiation (1.06μm) was focused onto the surface at incident angle of about 45°. The laser was free generating a multimode pulse train with a rise time of about 100 ns and combined duration of 400 μs with repetition rate 0.1 Hz. The spatial intensity distribution was rather flat in the far field with an approximate diameter 0.8 cm. Pulse power density in the zone of laser action was adjusted by changing pulse energy between 1 J and 8 J. The substrates (Ss) were placed on a stainless steel heated holder (Hh). The temperature on the substrate surface was measured with accuracy within 5°C. A rotating glass window (GW) with a diaphragm (D) protected the deposition chamber from staining with evaporated material which reduced power density.

The targets were prepared from PbTe, PbTe doped with 0.025wt% and 0.3wt% Cr and PbTe doped with 0.02wt% Co. Two types of substrates were used: amorphous - glass, and crystalline - freshly cleaved KCl and KBr (100) oriented crystals with lattice constants 6.288 Å and 6.599 Å respectively, close to that of f.c.c. PbTe ($a = 6.456$ Å).

Technological parameters varied were:
- number of the pulses, N_m, from 1 to 500;
- target-substrate distance, L_{ts};= 2, 3, 4 and 4.5 cm;
- pulse energy $E_p = 1 \div 6$ J, which resulted in pulse power density $P_s = 10 \div 60$ kW/cm^2 onto target surface;
- substrate temperature, T_s, maintained at various temperatures in the range of $50 \div 300°C$.

Investigations. The following investigations were undertaken:
- Scanning Electron Microscopy (SEM) was used to measure film thickness and to investigate surface morphology;
- X-ray and electron diffraction and also high resolution Transmission Electron Microscopy (TEM) were performed to study the crystal structure;
- film properties were controlled by optical measurements; transmittance and reflectance interference spectra were taken by double beam spectrometer in the range of $400 \div 4000$ cm^{-1}.

The correlation between growth conditions and film properties was studied by means of two control parameters:

1) the visibility of the interference patterns in transmittance spectra of the films $V = \dfrac{T_{max} - T_{min}}{T_{max} + T_{min}}$,

where T_{max} and T_{min} are the values of transmittance extrema. Visibility was used as a measure of the absorption in the films: in the case of low-loss film on transparent substrate the absorption $\alpha \approx -\dfrac{\ln V}{d}$. The optical constants (refractive and absorption indexes n and k), the optical band gap E_g and the electrotransport properties (free-carrier concentration N, Hall mobility μ) and their temperature behaviour, were evaluated from contactless and nondestructive optical measurements only [2];

2) the deposition rate $v_d = \dfrac{d}{\tau N_m}$, where τ = 400 μs is pulse duration, d is film thickness, N_m - number of pulses.

A detailed study of the dependences v_d (P_s, L_{ts}, T_s) showed that the deposition rate is most sensitive to the laser power density and to the target-substrate distance as can be seen in Fig. 2a. The dashed lines present the expected dependence $v_d \approx P_s^2$. The experimental results follow quite well this dependence at P_s > 25 kW/cm^2 at each distance target-substrate L_{ts}. At L_{ts} = 2 cm the deposition rate strongly deviates from the dependence $v_d \approx P_s^2$ - it increases with decrease of P_s bellow 25 kW/cm^2. This unexpected behaviour can be attributed to an ablation character of the process at low power densities, when species, besides atoms and molecules, reach the substrate. The decrease of the deposition rate at longer L_{ts} is obviously due to plasma expansion, while the decrease of the slope correlates with the anisotropic character of the expansion.

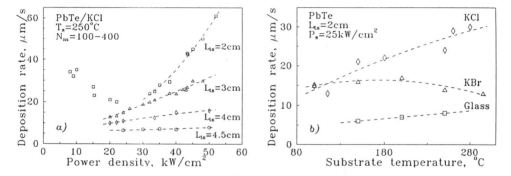

Fig. 2. Dependence of the deposition rate on: a) pulse power density at four target-substrate distances and b) substarte temperature for three different substarte materials (glass, KCl, KBr).

The dependence of the deposition rate on substrate temperature is not clearly pronounced at target-substrate distances L_{ts} > 3 cm: v_d = 15 μm/s +5 μm/s when T_s changes from 50°C to 250°C. The influence of substrate temperature is noticeable at L_{ts} = 2 cm as is seen from the dependences shown in Fig. 2b for three different substrate materials. The saturation of the dependence $v_d(T_s)$ at high temperature in the case of KCl substrate and the decrease of v_d when KBr substrate (with higher thermal conductivity) is used, means that process of re-evaporation limits further increase of v_d.

Results and discussion

The optical band gap E_g of the high pressure PbTe phases was evaluated from the dispersion curves $\alpha(E) \approx (\dfrac{\ln V}{d})(E)$ and $n(E)$, calculated from the extrema positions in transmittance spectra of thick films and from both the reflectance and transmittance spectra of thin films [1, 2]. The band gap of b.c.c. PbTe phase was determined to be E_g^{bcc} = 0.23 eV, and the band gap of the orthorhombic PbTe - E_g^{orth} = 0.42 eV. The heterophase junction formed between the sublayer with orthorhombic structure and the upper f.c.c. layer is schematically presented in Fig. 3. The transitions $E_p^{orth} \Rightarrow E_p^{fcc}$, $E_p^{fcc} \Rightarrow E_n^{orth}$ across the junction appear in the refractive index dispersion $n(E)$ as maxima at 0.16 eV and 0.26 eV respectively [1].

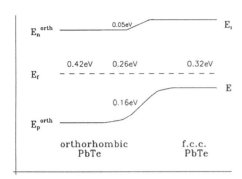

Fig. 3 Heterophase junction, orthorhombic - f.c.c. phase, in laser deposited PbTe film.

The relative content of HP phases and the strained sublayer thickness depended on the technological conditions through the deposition rate v_d used as a measure of the activation degree.

Films with high crystal perfection could be formed according to Petrosian et al. [3] in two limiting cases:

1) at low activation of the impinging species when the film grows in size from few nuclei only;

2) at high activation far from thermodynamical equilibrium, as is in the case of LAD, as a result of integration of great number interacting nuclei.

Laser deposition is in fact condensation of dense high-temperature plasma fluxes with very high energetic activation of the condensing atoms. The requirement for the nuclei number is also satisfied: at supersaturation $\delta\mu \approx kT.\ln\dfrac{P}{P_o}$ (T is plasma temperature, P is the actual vapours pressure, P_o is their equilibrium value) of about 10^5 cal/mol, as is at millisecond pulse duration, even a single atom becomes a growth nucleus. This concept is confirmed by TEM, SEM and electron diffraction studies of the initial stages of growth (Fig. 4) - dense, 15 ÷ 40 nm thick PbTe films with f.c.c. structure were deposited on KCL substrates with 1-2 pulses at L_{ts} = 3 cm, T_s = 200 °C, P_s = 50 ÷ 60 kW/cm^2 , v_d= 50 ÷ 60 µm/s.

 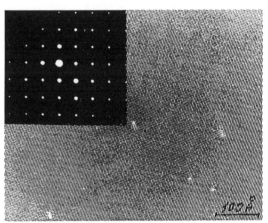

Fig. 4: a)TEM and electron diffraction of PbTe film deposited with 1 laser pulse;   *b)high resolution TEM and electron diffraction of PbTe film deposited with 2 pulses.*

The strong and nonlinear dependence of the deposition rate v_d on laser power density and target - substrate distance (see Fig. 2) makes the f.c.c. PbTe growth conditions very crucial. Even a slight change causes deterioration of the crystal perfection and PbTe film begins to grow in polycrystal orthorhombic (GeS-type) modification. The transition f.c.c. ⇒ orthorhombic structure is sharp with temperature (and/or pressure); an abrupt boundary [4], seen very well on the high resolution TEM picture in Fig. 5, is formed when conditions for growth of f.c.c. PbTe are reached. Presence of GeS phase was registered also by X-ray measurements and Raman scattering investigations [5, 6] in almost all of the samples.

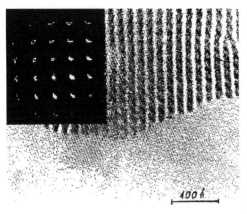

Fig. 5. Electron diffraction and high resolution TEM of two phase boundary in laser deposited PbTe film.

The thickness of the strained sublayer with orthorhombic structure was about 0.7 ÷ 0.9 μm, i.e. growth of f.c.c. PbTe was suppressed, at the following technological parameters: L_{ts} = 2 cm, P_s = 25÷30 kW/cm², T_s = 150 ÷ 250°C. The deposition rate, as defined above, was over 20 μm/s. However films with thickness of about 0.7 ÷ 0.9 μm are deposited during the first 80 pulses when v_d is about 10 μm/s at L_{ts} = 2 cm, which is thought to be due to the initial isothermal plasma formation and expansion [7]. At distances L_{ts} > 3 cm and power density P_s = 20÷30 kW/cm² the deposition rate is 40 μm/s in average for the first 50 - 80 pulses; conditions for f.c.c. PbTe growth are easily reached. This high value of the deposition rate at L_{ts} > 3 cm, N_m < 100 pulses is partially compensated from the adiabatic plasma expansion, which takes place at pulse duration and power density used. So a sublayer with orthorhombic structure grows as thick as 0.3÷0.4 μm at L_{ts} = 3 cm. Longer target - substrate distances cause deviation in stoichiometry of film surface - inversion of the conductivity from *p* to *n* type was proved by thermal-probe measurements.

Films enriched with lead were deposited also at substrate temperatures T_s > 280°C and T_s < 100°C as is seen from carrier concentration-versus-temperature plot in Fig. 6.

Conductivity type is inverted also in films grown at low substrate temperature (not shown in the figure) when the velocity of crystallization is very low for the high cooling rate of the condensate. The unbound atoms of more volatile Te re-evaporate when the next vapour flux reaches the substrate. Lead segregation was proved by electron diffraction of films deposited at 50°C and 100°C.

Growth conditions for laser deposition of b.c.c. PbTe (usually mixed with the orthorhombic phase) are not so clear yet. This phase was dominant in:

- sample grown at L_{ts} = 2 cm, P_s = 25 kW/cm², T_s = 270°C with N_m = 80 pulses;

- samples deposited on KBr substrates with low (< 10 μm/s) deposition rate at T_s > 200°C

- films deposited from Cr-doped targets.

The favourable (111) direction of b.c.c. PbTe growth is easily understandable in case of b.c.c. crystal, in which the unbound valences on [111] surface are twice more then those on [100] surface.

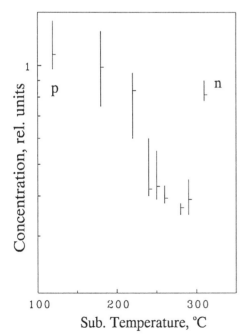

Fig. 6. Dependence of carrier concentration in PbTe/KCl on substrate temperature during the deposition.

Conclusion

A strict control of the deposition rate during laser assisted deposition of polymorphous materials could determine proper conditions for crystallization of one or another phase and is an opportunity for "band-gap engineering" without the use of polycomponent compounds, in which the problem with stoichiometry is hardly to be overcome.

References

[1] M. Baleva, V. Gaidarova, M.Momtchilova: in *Proc. of First Balkan Conf. of Physicist,* ed. by I.Kushev, Sept 1991, Thessaloniki, Greece.
[2] M. Baleva, M. Sendova, V. Gaidarova: in *Proc. of Conference on Optical Characterization of Semiconductors,* July 1990, Sofia, Bulgaria.
[3] W. I. Petrosian, I. Dagman, *Problemi epitaksii poluprovodnikovih plenok,* (Nauka, Moskva 1972).
[4] T. Chattopadhyay, A. Werner, H. G. von Schnering, *Proc. Matt. Res. Soc.,* vol. II, 93, (1984)
[5] M. Baleva, L. Bozukov, M. Momtchilova, J. Phys.: Condens. Matt. 4, 4633 (1992)
[6] M. Baleva, I. Ivanov, M. Momtchilova, J. Phys.: Condens. Matt. 4, 4645 (1992)
[7] Rajiv K. Singh, J. Narayan, Phys. Rev. B 41, 8843 (1990)

Materials Science Forum Vols. 173-174 (1995) pp. 99-104

HIGH-QUALITY NARROW-GAP THIN FILMS PRODUCED BY QUASI-CONTINUOUS-WAVE IR-LASER-ASSISTED EVAPORATION

S.V. Plyatsko and Yu.S. Gromovoj

Institute for Physics of Semiconductors, Ukrainian Academy of Sciences,
Pr. Nauki, 45, Kiev, 252650, Ukraine

Keywords: Laser Deposition, Epitaxy, Semiconductors, Thin Films, Multiple Quantum Wells

Abstract. Thin films of narrow-gap IV-VI doped and undoped semiconductors were grown by laser ($h\omega < E_g$) quasi-continuous evaporation and epitaxy. It was shown that crystalline quality of the layers and physical properties very much depend on the types of substrates, laser power density incident on the target and slightly depends on the substrate temperature. High quality layers can be obtained even at substrate temperature below 100C. Multiple quantum well structures with rather high carrier mobility (10^4 cm^2/V s) were obtained and very pronounced quasi-two-dimensional behaviour was proved. The homogeneous impurity distribution in IV-VI layers obtained by laser-assisted evaporation was confirmed by investigations of electrical characteristics and electron paramagnetic resonance of doped films.

INTRODUCTION

Narrow-gap IV-VI semiconductors are widely used for IR optoelectronics. At present high quality IV-VI layers and structures are mainly produced by hot wall epitaxy (HWE) and molecular beam epitaxy (MBE) methods [1-4]. The properties of individual layers of IV-VI semiconductors grown by either MBE or HWE are similar. However, up to now the lowest substrate temperatures used to obtain monocrystalline layers and quantum size structures have been about 520K [1-3]. But even at this substrate temperature inter diffusion depths are about 30-50 Å [1]. At lower temperature the crystalline quality and electrical properties are not is good as required. To obtain the sharp metallurgical boundary needed for an abrupt interface and for increasing the efficiency of devices of these materials, for example in heterojunctions, superlattices and quantum-well structures, low-temperature deposition methods should be actively developed and should be used.

The aplication of laser beams for evaporation of solid targets of IV-VI semiconductor compounds permit the above mentioned problem to be solved. The method of laser-assisted evaporation and epitaxy is not very well known for producing high-quality IV-VI thin single-crystal layers including quantum-well structures. In the last years very interesting and sometimes unique results have been obtained due to the possibility of varying over a wide range the process parameters. This paper presents the results of the studies of electrical, optical properties, structural characteristics and electron paramagnetic resonance (EPR) of PbTe, PbTe:Mn(Eu) epilayers obtained by the laser-assisted deposition (LAD) technique.

RESULTS AND DISCUSSION

In order to diminish the number of high energy particles in the vapour plume the quasi-continuos regime of target evaporation by the CO_2 laser was used (($10 \leq f \leq 100$ Hz), where f is the modulation frequency of the laser beam). The exposuring period was shorter than the shading one. The ratio of period of the exposure to period of the shade was 1:8. The band gap E_g of target samples used for laser sputtering were greater than energy quantum of laser irradiation ($E_g = 0,2$-$0,3$ eV $> h\omega = 0,118$ eV). The laser-assisted apparatus consists of two parts: the vacuum chamber and laser with assorted optics. The laser power density W was varied by attenuation of laser irradiation from 10^3 to 10^5 W/cm^2. The distance between the target and the substrate was varied from 3,5 to 8,5 cm. The layers were deposited on KCl, KBr, NaF, BaF_2, dielectric substrates in a vacuum chamber with residual vapour pressure of 10^{-6} torr. The temperature of the substrates was in the range $T = 293$-473 K. The thickness of the grown layers $h = 3*10^{-3}$ - 10^1 µm was uniform over the substrate area and surface of the film was mirror-like.

Structural characteristics

The structural perfection of the layers was studied using electron diffraction and X-ray Bragg reflection methods. High quality layers of PbSnTe may be produced by laser deposition even at substrate temperature about 300K. Examination of the Bragg reflection spectra show that the X-ray rocking curves of the (400) reflection with full width at half-maximum (FWHM) of PbSnTe thin layers on (100) KCl, KBr depended upon the layer thickness and did not exceed 2 arc min for the thickest films (10 µm). In the case of PbSnTe films on (111) BaF_2 , the situation is quite different. The clearly resolved peaks are due to (111) and (333) reflection (WFHM = 50arc sec) of the thin films (h≤3000Å) indicate a perfect monocrystal structure [5]. With an increase of the film thickness, the substrate influence becomes insignificant and more favourable thermodynamic direction (100) takes place.

Electrical characteristics

It was found that the electrical properties of layers deposited onto substrates mentioned above depend very much on the type of substrate and its orientation, and also on the laser power densities on the target and slightly depends on the substrate temperature (373K≤T≤473K). Figure 1 shows the dependencies of the carrier concentration and carrier mobility on the laser power densities on the target. The best electrical characteristics of PbTe/KBr layers were obtained with laser power densities in the range $W = 2*10^4$ -$8*10^4$ W/cm^2. One can see that there are regions where films with good electrical characteristics can be controllably manufactured. In this power density range the electrical properties of the films are comparable with those on the best single crystals ($\mu = 4*10^4$ cm^2/V s) at T = 77K and also comparable with electrical characteristics of the best thin films manufactured by MBE and HWE ($\mu = (1,4$-$4,9)x10^4$ cm^2/V s) at T = 77K [6,7]. The electrophysical properties of the PbTe/KCl(KBr) and PbTe/BaF_2 layers are strongly different. The temperature dependence of the Hall coefficient presented in Figure 2 for different samples PbTe/KBr (KCl) and PbTe/BaF_2 grown on equal conditions. The PbTe/BaF_2 undoped films have only p-type conductivity with an exponential decrease of conductivity and similar increase of Hall coefficient (decrease carrier concentrations) with decreasing temperature ($P_{77} = 10^{11}$ - 10^{12} cm-3, $\mu_{77} = 6*10^2$ - $8*10^3$ cm^2/Vs). An extremely high photo conductivity of these films from 77K to 150K was observed. ($\lambda_{max} = 6$ µm). The life time of the photo excited carriers ($\tau = 1.5*10^{-6}$ s) were measured by photo conductivity decay techniques. It was found that by changing the laser power density on the target the concentration values and conductivity type of the films can be changed. To demonstrate the quasi-two-dimensional (2D) conductivity of carriers in multiple quantum wells (MQW) obtained by CO_2 -laser-induced evaporation, weak field magneto resistance (WFMR) experiments were performed.

In Figure 3 the angular dependence of the WFMR for PbTe/PbSnTe $(x=0.2)$ MQW are shown for two configurations labelled B and C. As for these two configurations the experimental data are very similar, only then data for configuration B are presented. Configuration B involves the rotation of the magnetic field in the plane perpendicular to the current direction and configuration C involves rotation of the magnetic field from the transverse direction to the longitudinal direction. For PbSnTe films the angular dependence of WFMR was measured for configuration B. As one can see from Figure 3 the angular dependencies of the WFMR in the PbTe/PbSnTe MQW on a KCl,KBr substrates exhibit quasi-2D behaviour.

Optical properties

The optical absorption and photosensitivity measurements were carried out at T = 80-350 K. The position of the inter band absorption edge in thin films (h≤1 μm) depends strongly on the temperature of the substrate and the mismatch between the lattice constants of the materials deposited and those of the substrates. As the substrate temperature decreases, an increasing shift in the inter band transition edge to a shorter wavelength band is observed. For example for PbSnTe layers a shift $\Delta E=24$ and 110meV in the case of KCl and LiF substrates respectively that was observed at T=300K. For thick layers (h≥1.5 μm) deposited on substrates at T= 420K there was almost no band edge shift. In thin (d≤400Å) PbSnTe $(x=0.2)$ layers on BaF_2 substrates, a dimensional quantization of the optical absorption spectra was observed, due to transition between the quantized states [8].

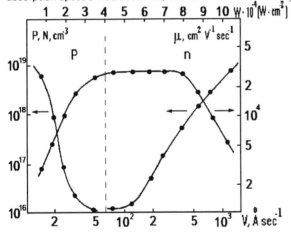

But clear step-like absorption (T=300K) corresponding to electron transition between valence and conduction subbands are not observed for PbTe/PbSnTe $(x=0.2)$ MQW on the KCl(KBr) substrate with equal thickness of well and barrier. The inter band absorption spectra only shift to the short wave region with a decrease in thickness of PbSnTe wells.

Electron paramagnetic resonance spectra

Fig.1 Dependencies of carrier concentrations and mobility T=77K) on power density and deposition ate of PbTe/KBr layers ($T_s = 423K$).

The high crystalline quality of the layers obtained was also confirmed by studying by electron paramagnetic resonance (EPR) spectra of the films grown from the targets doped with manganese and europium. Figure 4 shows the EPR spectrum of the PbTe:Mn single crystal used as the target and the spectrum of the layer. Six isotropic lines of the hyper-fine *structure (HFS) interaction with half-width $\Delta H = 20$ Oe were observed in the PbTe target. This rather broad half-width is caused by inhomogeneous distribution of manganese ions in the PbTe lattice. This EPR spectrum is characterised by the spin hamiltonian (SH) parameters g = 2.0033±0.0005 and A = $(59.9\pm0.2)/10^{-4}$ cm^{-1}, where g and A are the g factor and HFS constant respectively. In the films deposited from the PbTe:Mn target exhibited (in addition to six mach narrow

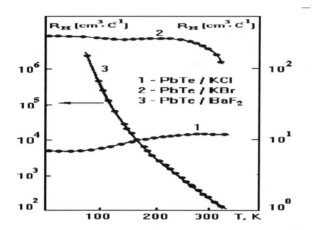

($\Delta H = 4.5$ Oe) HFS lines) satellite lines near each HFS, caused by the super-hyper-fine interaction (SHFI) of Mn^{2+} $3d^5$ electrons with nuclear moments of ^{125}Te and ^{123}Te isotopes in the first co-ordination sphere. In this case the SH parameters are $g = 1.9975 \pm 0.0005$, $A = (59.9 \pm 0.2) \times 10^{-4}$ cm^{-1}, and SHFI constant a_{Te} $=$ $(15.80 \pm 0.2) \times 10^{-4}$ cm^{-1}. The intensity relations between SHFI lines and HFS lines in the best films was about 1:6:1

Fig.2 Temperature dependencies of the Hall coefficient for PbTe films . 1 - n-PbTe/KCl ; 2 - n-PbTe/KBr; 3 - p-PbTe/BaF2 T_s = 423K, W = $5*10^4$ W/cm^2, h = 4µm.

which is very close to the ratio 1:4:1 of the ideal case when all the Mn^{2+} ions are in the metal sub lattice positions. Figure 5 shows the part of the EPR spectra near the central $+1/2 <--> -1/2$ electron transition of the PbTe:Eu films on a KBr (KCl) (001) substrate. For the first time super hyperfine interaction (SHFI) of Eu^{2+} ions with ^{123}Te and ^{125}Te isotopes of the first coordination sphere was investigated.

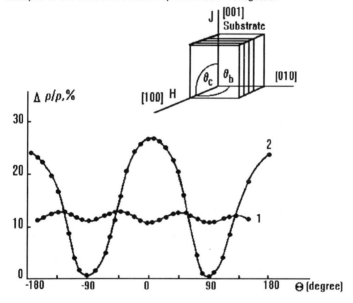

Fig. 3 The WFMR dependencies in a p-type Pb $_{08}Sn_{02}Te$ film (curve 1) and a PbTe/Pb$_{08}Sn_{02}$ Te MQW (d(PbSnTe) = d(PbTe) = 220 Å: number of periods n = 100) on KCl substrates. For the epitaxial films, d = 2,6 µm, P77 = 1.17*10^{17} cm^{-3}, µ77 = 8.5*10^3 cm^2/Vs. For the MQW P77 = 2.91*10^{17} cm^{-3}, µ77 = 9.8*10^3 cm^2/Vs.

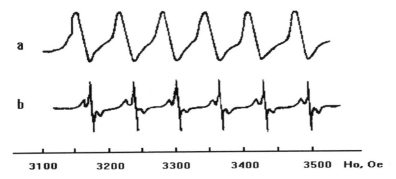

3100   3200   3300   3400   3500   Ho, Oe

Fig 4. The EPR spectra of Mn^{2+} ions of the PbTe target ($N_{Mn} = 0.0002$ wt %) (spectrum a) and a PbTe:Mn film (d = 5.5 μm, T = 20K) (spectrum b).

The higher resolution of the EPR spectrum of Eu^{2+} in the films in comparison with the single crystals allowed the SH constant to be refined and SHFI constant to be determined. The EPR spectrum at 20K is characterised by the following parameters: $g = 1.9975 \pm 0.00055$, $b_4 = (40.1 \pm 0.45) \times 10^{-4}$ cm^{-1}, $b_6 = (-0.63 \pm 0.45) \times 10^{-4}$ cm^{-1}, $A = (27.0 \pm 0.45) \times 10^{-4}$ cm^{-1}, $a_{Te} = (11.8 \pm 0.45) \times 10^{-4}$ cm^{-1} for the ^{151}Eu isotope, $A = (11.9 \pm 0.45) \times 10^{-4}$ cm^{-1} for ^{153}Eu isotope. For comparison, the EPR spectrum on the PbTe:Eu single crystal used as target is also shown in Figure 5.

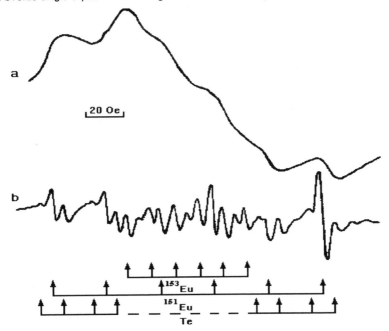

Fig 5. EPR spectra of Eu^{2+} in a PbTe:Eu single crystal (target) (spectrum a) and a PbTe:Eu film on a KBr substrate (spectrum b) near the central $+1/2 <\text{--}> -1/2$ electron transition (H// <001>, T = 20K).

Here b_4 and b_6 are the cubic field splitting parameters. Due to the heavy overlap of the EPR lines, the value of the SHFI constant for the ^{153}Eu isotope could not be determined.

CONCLUSION

The electrical, optical and structural characteristics of PbSnTe films obtained by laser assisted evaporation using the quasi-continuous regime of a CO_2 laser proves the high quality of these doped and undoped films which can be used to obtain multiple layered structures including superlattices and MQW.

ACKNOWLEDGEMENTS

Thanks are due to F.F.Sizov for many helpful discussions, to V.P.Klad'ko for measurements of the structural characteristics and to M.Yu.Gusev for his experimental assistance.

REFERENCES

1   F.F.Sizov, Acta Phys.Polon., **29**, 83 (1991).
2   S.Shimomura,I.Urakava,S.Takaoka,K.Murase,A.Ishida and H.Fujiyasu, Superlattices Microstruct., **7**, 5   (1990)
3   M.Kriechbaum,K.E.Ambrosh,E.J.Fantner,H.Clemens and G.Bauer, Phys.Rev.B, **30**, 3394 (1984).
4   D.L.Partin, IEEE J.Quantum Electron., **24**, 1716 (1988).
5   S.V. Plyatsko, Yu.S. Gromovoj, G.E. Kostyunin, F.F. Sizov and V.P.Klad'ko, Thin Solid Films, **221**, 127 (1992).
6   D.L.Partin, J.Electron.Mater., **10**, 313 (1981).
7   A.Lopez-Otero,Thin Solid Films, **49**, 3 (1978).
8   S.V.Plyatsko,Yu.S.Gromovoj and F.F.Sizov, Infrared Phys, **3**, 173, (1991).

Materials Science Forum Vols. 173-174 (1995) pp. 105-108

LASER-INDUCED ETCHING OF Si SURFACES;
THE EFFECT OF WEAK BACKGROUND LIGHT

H. Grebel and T. Gayen

The Optical Waveguide Laboratory, Electrical and Computer Engineering Department, The New Jersey Institute of Technology, Newark, New Jersey 07102, USA

Keywords: Laser Induced Etching, Etching of Si Surfaces, Etching with Two Laser Frequencies

Abstract. Two different lasers were used to etch patterns on Si surfaces employing a thin film cell configuration. A strong, pulsed, 20 W KrF excimer laser was used for etching. A weak, CW, 5 mW HeNe laser was serving as background light which by itself was incapable to etch the Si surface. A substantial enhancement of the laser etching process with light background was observed. Etching process using many pulses was compared to etching by use of only one laser pulse.

Introduction

Laser-induced ablation of semiconductors [1-2] and other non-organic [3] surfaces gained interest recently owing to the simplicity of the reaction system and the possibility for a true, three dimensional, controllable patterns. The patterning process leads to very effective material removal. In a thin film reaction cell configuration, a thin electrolyte layer is held by capillary action between the semiconductor wafer and a cover glass [4] as seen in Fig. 1. The laser beam transverses the cell and induces a reaction between the electrolyte and the semiconductor in addition to locally ablate the Si surface.

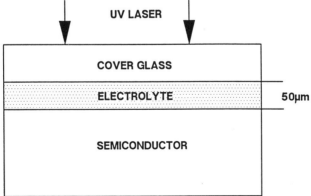

Figure 1. A cross section of a thin film cell configuration.

Experiment

A thin film configuration was employed in these experiments as shown in Fig. 1. The cover glass made of polycrystalline quartz material is held by the capillary action of a 50 μm thick layer of an etchant. Two laser beams, one from each laser were used in these experiments. A KrF excimer laser (λ=0.248 μm, 0.2 J/pulse) and a HeNe laser (λ=0.63 μm, I=5 mW) were used as light sources. The experimental configuration is shown in Fig. 2. The excimer laser beam was first collimated and then focused into a 15 μm x 5 cm line by use of a cylindrical lens. The HeNe laser beam was transversing a variable attenuator before impinging on the sample. The beam of the HeNe laser was approximately 2 mm in diameter. The sample was put on an X-Y-Z translational stage. Successive lines were etched by moving the stage one step each time. One laser-pulse experiments were performed by letting the motorized stage move in a constant pace and tune the excimer laser pulse rate appropriately. The periodicity of the etched features was 50 μm and the shape of an individual feature resembles that of a guassian. Large grooves were analyzed by use of SEM micrographs of cleaved surfaces. Small grooves were analyzed by light scattering [4].

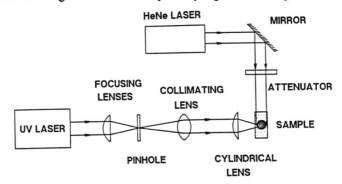

Figure 2. The experimental setup.

A typical background light illumination enhancement is shown in Fig. 3.

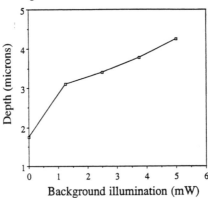

Figure 3. The effect of background lighting.

Shown is the depth of a line etched with 30 pulses of KrF laser and various background illumination levels of <100> p-Si with 10^{16}/cm^3 dopant. The etchant was 5% KOH:H$_2$O solution. An enhancement of 2.5 times is achieved with a background of 5 mW HeNe compared to no background lighting.

A similar trend was detected with only one pulse. The enhancement factor with a 5 mW HeNe laser lighting is also about 2.5 time the value for no background lighting at all. However, the average depth per pulse for a multi pulse experiment with 5 mW HeNe laser background lighting was 0.15 μm. This was larger then the depth per pulse for one pulse experiment which was 0.045 μm. The difference may be explained by the presence of a thin layer of oxide on fresh wafers.

The enhancement of laser etching in the presence of solutions may be explained by photoelectrochemical process that takes place in addition to the ablation process by the KrF laser. Owing to the light dampening effect of the solution and the relatively long life time of the photo generated carriers, one may postulate that the light intensity at the Si surface is low enough to direct the HeNe photo carriers into the small feature region ablated by the KrF beam. Small feature etching enhancement has been observed before for two laser etching in InP [5].

Asymmetric Schottky barrier solar cells have been fabricated using KrF laser in the presence of 5% KOH solution. The pattern is shown in Fig. 4. The solar cell light collection efficiency was 20% larger than its flat surface counterpart with a penalty of 0.04 eV barrier height reduction.

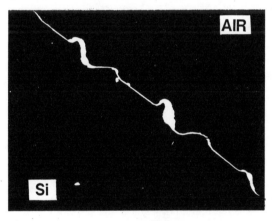

⊢—⊣ 10μm

Figure 4. A cross section of an asymmetric pattern on Si.

References

[1]   R. Kullmer and D. Bäuerle, App. Phys. A **47**, 377 (1988).
[2]   G. B. Shinn, F. Steigerwald, H. Stiegler, R. Sauerbrey, F. K. Tittel and W. L. Wilson Jr., J. Vac. Sci. Tech. B **4**, 1273 (1986).
[3]   D. J. Ehrlich and J. Y. Tsao, *Laser Microfabrication*, Academic, NY 1989.
[4]   H. Grebel, B. Iskandar and K. Sheppard, App. Phys. Lett. **55**, 2655 (1989).
[5]   H. Grebel and P. Pien, J. App. Phys. **71**, 2428 (1992).

Materials Science Forum Vols. 173-174 (1995) pp. 109-116

GAS-DYNAMIC EFFECTS IN LASER PULSE SPUTTERING OF ALUMINIUM

A. Peterlongo, A. Miotello and R. Kelly

Dipartimento di Fisica and Consorzio Interuniversitario Nazionale Fisica della Materia,
Unversità di Trento, I-38050 Povo (Trento), Italy

Keywords: Aluminium, Gas-Dynamics, Laser Ablation, Laser Sputtering

Abstract. We have developed a numerical method to describe pulsed laser sputtering of Al in a thermal regime. Heat transport is described by.the heat equation with boundary conditions for both evaporation and/or boiling. When the particle emission rate exceeds about 1 monolayer in 20 ns, the motion of the particles leaving the Knudsen layer can be described in the gas-dynamic approximation. Numerical results concerning the efficiency of laser sputtering in producing craters in irradiated Al, as well as some gas-dynamic aspects of the emitted particles are illustrated.

Introduction

Pulsed laser irradiations of metals or semiconductors involve many physical processes including reflection-transmission of the radiation itself, photoelectric effect, electron excitation which induces the formation of electron-hole pairs, ionization, atom or cluster emission due to the breakage of interatomic bonds and so on [1]. Moreover, when the radiation intensity is high enough, energy transfer from excited electrons to the lattice causes appreciable heating of the solid. As a consequence phase changes, evaporation, boiling and even phase explosion may occur. In addition, the emitted particles may interact both with the terminal part of the laser pulse and with the target surface, reflection or recondensation effects being possible in the latter case. Due to the complexity of the involved processes, approximations are necessary to describe laser-solid interaction [2].

However, to develop a model to describe *laser ablation* (or *laser sputtering*) some approximations may be easily justified on the basis that many of the above mentioned processes are threshold-processes with respect to laser pulse specifications such as wavelength, duration and energy density. Moreover some effects become dominant with respect to others depending on the phase (solid, liquid) of the irradiated sample, on the evaporation rate, or on the value of T_{tc}, the thermodynamic critical temperature.

In this paper we describe the main features of the thermal regime model that we have developed to account for laser sputtering of Al. We consider both heat transport within the target as well as gas-dynamic processes occurring with the emitted particles.

Heat transport and phase transitions

It is well known that laser light absorption may be described according to the relation:

$$I = (1 - R)I_0\, e^{-\mu x}, \tag{1}$$

where I_0 is the intensity of the laser pulse, I is the intensity at distance x beneath the surface which is located at $x = 0$, μ is the absorption coefficient ($5.0 \times 10^5\, cm^{-1}$ for Al), and R is the reflection coefficient (0.79 for Al at $\lambda = 600\, nm$).

To describe heat transport we make use of the heat flow equation in one dimensional form, as is appropriate to many experimental conditions [3]:

$$c(T)\,\rho\frac{\partial T}{\partial t} = \nabla(K(x,T)\nabla T) + S(x,t), \tag{2}$$

where T, $c(T)$, ρ, K are respectively temperature, specific heat ($0.94\, J/g\, K$), mass density ($2.7\, g/cm^3$), and thermal conductivity ($2.25\, J/s\, cm\, K$ for the solid phase and $1\, J/s\, cm\, K$ for the liquid phase). $S(x,t)$ is the heat source term, which is derivable from Eq. 1. We have adopted a triangular shape for the laser power pulse with a duration $\tau_e = 30\, ns$.

To consider solid-liquid or liquid-solid transitions, two boundary conditions are required at the interface where phase transition occurs, namely [4]:

$$\rho\,\Delta H(T_i)\,v_{int} = K_{sol}\left.\frac{\partial T}{\partial x}\right|_{x_{int}^+} - K_{liq}\left.\frac{\partial T}{\partial x}\right|_{x_{int}^-}, \tag{3}$$

where T_i is the temperature at which the transition takes place, $\Delta H(T_i)$ is the heat of fusion at $T = T_i$ ($\Delta H(T_m) = 395.6\, J/g$) and v_{int} is the solid-liquid interface velocity. Eq. 3 is essentially an energy balance at the solid-liquid interface. Moreover, the thermodynamics of crystal growth requires that v_{int} be a function of $T_i - T_m$:

$$v_{int} = f(T_i - T_m) \tag{4}$$

where T_m is the equilibrium melting temperature ($933.52\, K$) and f is determined by crystal growth thermodynamics.

During evaporation the surface recedes with velocity v_r. It is, however, still possible to label its position with $x = 0$ if one chooses a reference frame moving with the receding surface. Then, neglecting mass accumulation, Eq. 2 becomes:

$$c(T)\,\rho\frac{\partial T}{\partial t} = \nabla(K(x,T)\nabla T) + c(T)\,\rho\, v_r\, \nabla T + S(x,t). \tag{5}$$

A simple way to compute v_r as function of temperature is to consider what happens when a liquid is in thermal equilibrium with its saturated vapor. In this case the number N_a of particles evaporating per unit time and area, is:

$$N_a = \frac{p}{(2\pi k_B T m)^{\frac{1}{2}}} C_S \tag{6}$$

where p is the gas pressure, k_B is the Boltzmann constant, m is the mass particle, and C_S is the sticking coefficient ($C_S \simeq 1$). One may assume that Eq. 6 also holds in a non-equilibrium situation as when particles are emitted into vacuum thus yielding:

$$v_e = \frac{p}{\rho\,(2\pi k_B T/m)^{\frac{1}{2}}} C_S \tag{7}$$

where v_e is the evaporation velocity.

The relation between the equilibrium vapor pressure and the temperature may be computed

Table 1: Main results obtained by irradiating an Al sample with a single $30\,ns$ laser pulse.

HEATING PHASE CHARACTERISTIC					
WITHOUT BOILING					
Energy density (J/cm^2)	*Start of liquid phase* (ns)	*Duration of liquid phase* (ns)	*Maximum T* (K)	*Maximum* v_r (cm/s)	*Crater depth* (nm)
3.00	8.3	97.1	2588	0.7	4.9×10^{-2}
3.10	8.0	103.5	2679	1.0	7.7×10^{-2}
3.20	7.9	118.0	2769	1.5	1.2×10^{-1}
3.30	7.7	124.5	2859	2.2	1.7×10^{-1}
WITH BOILING					
Energy density (J/cm^2)	*Start of boiling* (ns)	*Duration of boiling* (ns)	*Maximum T* (K)	*Maximum* v_r (cm/s)	*Crater depth* (nm)
3.00	-	-	2588	0.7	4.9×10^{-2}
3.10	-	-	2679	1.0	7.7×10^{-2}
3.20	18.5	2.5	2740	44.9	8.2×10^{-1}
3.30	17.2	4.5	2740	127.7	4.1

from the Clausius-Clapeyron equation in the limit $V_{liq} \ll V_{gas}$, V_{liq} and V_{gas} being the molar volume of liquid and gas respectively.

When the surface temperature reaches the boiling temperature, T_b (for Al T_b is $2740\,K$ while $\Delta H(T_b)$ is $10536\,J/g$) then the vapor pressure equals the external pressure p_b. As a consequence bubble nucleation may occur in the bulk, bubbles themselves may move toward the surface,

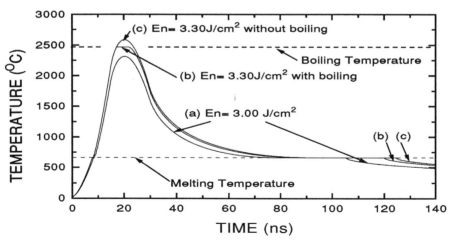

Figure 1: Surface temperature of an Al sample irradiated with a single laser pulse with energy density equal to $3.00\,J/cm^2$ for (a) and to $3.30\,J/cm^2$ for (b) and (c). In (b) it has been assumed that boiling is *possible* on the relevant time scale, $30\,ns$, while in (c) it has been assumed *not possible*.

and the evaporation rate may be enhanced. At the same time the temperature of the sample will tend to assume a constant value. The energy, E_{min}, needed for bubble formation is given by [5]:

$$E_{min} = -(\mu_l - \mu_v)\frac{4}{3}\pi r^3 n + 4\pi r^2 \sigma_s, \tag{8}$$

where μ_l, μ_v are the chemical potentials of the liquid and gas phase respectively, r is the bubble radius, n is the number density of the gas, and σ_s is the surface tension of the liquid.

As a matter of fact if the rate at which heat is deposited on the sample by the laser pulse is higher than the rate at which the bubbles "absorb" energy, then a significant superheating may occur, with the temperature finally reaching T_{tc}. Alternatively if the velocity at which the surface recedes is high enough then nucleation cannot occur and again bubbles will not form.

Since the transition from simple evaporation to boiling needs a characteristic time, τ_b, of about $10^{-9} - 10^{-8}\,s$ for a liquid dielectric and $10^{-13} - 10^{-11}\,s$ for metals [5] then we do not expect boiling in the picosecond regime while in the nanosecond regime the question is still open to a definitive answer. This is why we present results for both situations, boiling and superheating: see Fig. 1 and Table 1.

Gas dynamics

Considering a thermal mechanism for particle emission, then the velocity distribution of the emitted particles must be given by a Maxwell distribution restricted to v_x positive [6, 7]:

$$f_M^+ = \begin{cases} n_S \left(\dfrac{m}{2\pi k_B T_S}\right)^{\frac{3}{2}} exp\{-\dfrac{1}{k_B T_S}[\dfrac{1}{2}m\,(v_x^2 + v_y^2 + v_z^2) + E_i]\} \\ \\ v_x \geq 0;\; -\infty < v_y, v_x < +\infty \end{cases} \tag{9}$$

where E_i is the internal energy of the gas and T_S, n_S are respectively temperature and number density of the gas at the surface. Beyond the surface the particles will, for sufficiently low density, go into free flight but otherwise form a Knudsen layer (KL). This is a region having a width of a few mean free paths in which the particles come to equilibrium.

If the density of the gas at the boundary of the KL is high enough the particles enter into a so-called *unsteady adiabatic expansion* (UAE) well described by the gas-dynamic equations [2]. The threshold for this regime corresponds roughly to an emission rate of about 1 monolayer in $20\,ns$, that is $v_t = 0.5\,cm/s$ for the recession velocity of the surface [6].

If the evaporation rate is below the above mentioned threshold then gas-dynamics effects do not occur at all, except possibly for formation of the KL, and we neglect them. On the other hand as soon as the evaporation rate reaches v_t at time t_i we make full use of the gas-dynamic equations in one dimension [6]:

1. continuity equation:

$$\frac{\partial \rho_g}{\partial t} + \frac{\partial}{\partial x}(\rho_g\,u) = 0, \tag{10}$$

2. Euler equation

$$\frac{\partial}{\partial t}(\rho_g\,u) + \frac{\partial}{\partial x}(\rho_g\,u^2 + p) = 0, \tag{11}$$

3. energy conservation equation:

$$\frac{\partial}{\partial t}\left(\rho_g\left(\frac{1}{2}u^2 + \mathcal{E}\right)\right) + \frac{\partial}{\partial x}\left(\rho_g\, u\left(\frac{1}{2}u^2 + \mathcal{E} + \frac{p}{\rho_g}\right)\right) = \frac{\partial\Phi}{\partial x}. \tag{12}$$

Here u is the flow velocity, p, ρ_g, $\mathcal{E}$ are respectively gas pressure, mass density and internal

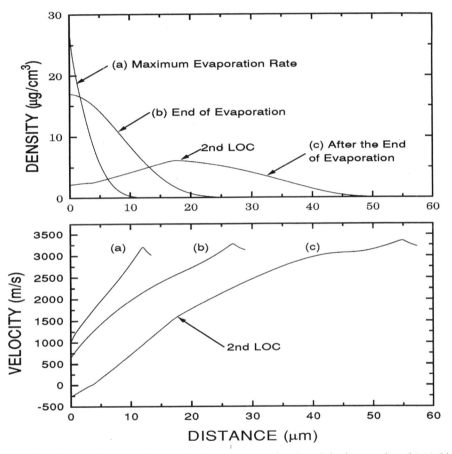

Figure 2: Density and flow velocity profiles for an energy density of the laser pulse of $3.10\, J/cm^2$ and three values of t. The emission is found to begin at $t_i = 16.5\ ns$ with $T = 2497K$. Curve (a) corresponds to the time when the surface temperature is at its maximum, namely $t = 20.3\,ns$. Curve (b) shows the time at the end of the "emission", namely $t_f = 24.8\,ns$ when we have $v_e = v_t$. Curve (c) shows the time when a period equal to the emission duration (i.e. $t_f - t_i \simeq 8\,ns$) has passed since the end of the emission, namely $t = 33.1\,ns$. A recondensation boundary condition is assumed.

energy per unit mass, and Φ is the laser heat input into the escaping particles. In addition, we note that for any adiabatic ($\partial\Phi/\partial x = 0$), reversible, isentropic expansion the following is true (*Poisson equation*):

$$\frac{p}{p_o} = \left(\frac{\rho_g}{\rho_{go}}\right)^\gamma, \tag{13}$$

where p_0, ρ_{go} are respectively pressure and density of the gas at some arbitrary position and time, and $\gamma = C_p/C_V$ is the specific heat ratio.

The gas-dynamic equations were integrated numerically by following the Godunov scheme [8] with the KL jump conditions and $M = 1$ (M=Mach number) as boundary conditions. Moreover, since the crater depth is relatively small compared to the scale distance of gas-dynamics, we neglect the motion of the receding surface (v_r in Eq. 5) in connection with the gas-dynamics.

Concerning Eq. 13, appropriate choices for p_o and ρ_{go} are obviously the values of pressure and density at the boundary of the KL, i.e. p_K and ρ_K respectively. Nevertheless these values are not connected to each other by a linear relation and so we have to face the problem that the gas flow is not actually barotropic as the proportionality factor, α_p, between p and ρ^γ is a monotonically decreasing function of the surface temperature. We have therefore assumed that during the emission phase α_p assumes, for each value of t, a unique value for the whole gas equal to that computed at the boundary of the KL. When the evaporation rate goes below the threshold for KL formation at time t_f we assume that α_p retains a constant value.

In Fig. 2 we report the results obtained using a recondensation boundary condition. Three different aspects are illustrated.

(a) Consider the gas density and flow velocity when the surface temperature reaches its maximum at $t = 20.3\,ns$. The gas density has its maximum at the target surface ($\rho_K = 28\,\mu g/cm^3$) and drops to 0 well away from the surface. The flow velocity of the gas at the KL boundary is $u_K = 959\,m/s$ but increases with distance. According to previous analytical work, u at the expansion front should be a maximum, given by $\hat{u} = 4u_K$ for atoms with $\gamma = 5/3$ [2]. The observed discrepancy, that is that the flow velocity reaches a maximum slightly before the expansion front, is due to the fact that near the front the density is so low that the numerical method and rounding error play a heavy role. More accurate results may be obtained by decreasing the time and spatial step lengths, a procedure which however implies a heavy increase in the CPU expense.

(b) At the end of the emission phase, when we have $v_e = v_t$ and $t_f = 24.8\,ns$, no important differences with respect to the previous case are observed even if the flow velocity ($u_K = 654\,m/s$) and density of the gas ($\rho_K = 17\,\mu g/cm^3$) at the KL boundary are a bit lower.

(c) Finally at $t = 33.1\,ns$ the density and flow velocity display a line of contact (LOC), which is a remembrance of the density drop at the end of the emission. For short enough times this LOC is located just at the absolute density maximum. Near the surface the velocity exhibits negative values due to the recondensation process. These results are very similar to those obtained by Sibold and Urbassek [9] in solutions for one-dimensional expansions obtained with the Boltzmann equation. At this moment some 4.5% of the emitted gas has already recondensed.

Conclusions

We have described a laser ablation process as would occur in a thermal regime when both a primary mechanism (due to laser absorption) and a secondary mechanism (gas-dynamics) are taken into account.

In spite of there being no possibility of direct comparison with experimental results, due both to the lack of published data (crater depth for instance) and to the difficulty in imaging the gas cloud, qualitative agreement with previous theoretical work was found [2, 9]. The latter, however, involved rather heavy restrictions in the choice of boundary conditions which made the flow velocity assume a rather simple structure given by a single straight line during the evaporation and by two nearly straight lines after the end of the evaporation. On the contrary, the boundary conditions used in our model are less restrictive, the flow velocity and gas density (Fig. 2) being free to follow the temperature changes of the sample surface (Fig. 1). This implies a more structured flow velocity function which departs from having straight or nearly straight components.

The scheme adopted here for describing the laser ablation process may be improved by taking into account the temperature and phase dependence of density, reflection and absorption coefficients, specific heat, conductivity, heat of evaporation . Also one would ideally consider the energy absorbed into the gas cloud, i.e. the term $\partial\Phi/\partial x$ in Eq 12. At the same time the model is quite general: the laser ablation of other materials may be described simply by providing alternative parameters.

References

[1] N. Bloembergen, H. Kurz, J.M. Liu and R. Yen in "*Laser and Electron-Beam Interactions with Solids*", ed. by B.R. Appleton and G.K. Celler (Elsevier Science Publishing, 1982), p. 3.

[2] R. Kelly, A. Miotello, B. Braren, A. Gupta and K. Casey, Nucl. Instr. Meth. **B65**, 187 (1992).

[3] H.S. Carslaw and J.C. Jaeger, *Conduction of Heat in Solids*, (Oxford University Press, Oxford, 1959), p. 50.

[4] L.F. Donà dalle Rose and A. Miotello in "*European Scientific Laser Workshop on Mathematical Simulation,* " ed. by H.W. Bergmann, (Sprechsaal Pub. Group, Coburg, 1989), p. 242.

[5] F.V. Bunkin and M.I. Tribel'skiĭ, Sov. Phys. Usp. **23**, 105 (1980).

[6] R. Kelly, J. Chem. Phys. **92**, 5047 (1990).

[7] R. Kelly and R.W. Dreyfus, Surf. Sci. **198**, 263 (1988).

[8] S.K. Godunov, A.V. Zabrodin and G.P. Prokopov, U.S.S.R. Comput. Math. Math. Phys. **4**, 1187 (1962).

[9] D. Sibold and H.M. Urbassek, Phys. Rev. A **43**, 6722 (1991).

Materials Science Forum Vols. 173-174 (1995) pp. 117-122

LASER DEPOSITED PHTHALOCYANINE- AND FULLERENE-BASED PHOTOVOLTAIC CELLS

A.I. Shevaleevskii, V.P. Poponin and L.L. Larina

N.N. Semenov Institute of Chemical Physics, Russian Acad. of Sciences,
Kosigin str.4, 117977 Moscow, Russia

Keywords: Thin Films, Solar Cells, Molecular Semiconductors, Fullerenes, Laser Deposition

Abstract. Laser evaporation technique based on infrared CO_2 laser emission was applied for obtaining thin films of molecular materials for photovoltaic cells constructing. Thin zinc phthalocyanine and fullerene single layers were sublimated. Photovoltaic behavior of TiO_2 monocristal phthalocyanine based cells was demonstrated. Photoeffects in the single layers and phthalocyanine based heterojunctions were investigated in wavelengths dependence measurements.

Introduction

Thin films of different molecular semiconductors are of great interest according to the possibility of constructing on its base new electronic devices including solar photovoltaic cells [1]. Many investigations have already been performed on thin films of metallophthalocyanines (MePc) based organic heterojunctions [2]. It is also of interest the succesfull experimental results of photosensitive semiconductor polymer-fullerene heterojunction diode fabrication [3].

Vacuum evaporation technique can be easily applied for obtaining thin layers with controlled thickness. Meanwhile a problem arises to avoid uncontrolled doping process of organics during deposition that come from the material of electric heater and other heated parts of the vacuum chamber. One of the possibilities to improve the experimental conditions can be fulfilled in applying the laser beam instead of resistive heater. Previously we reported the construction and application of a laser evaporation method for depositing thin CdS films for solar cells [4]. It was shown that film photoconductivity possessed more reproducible improved parameters when compared with that obtained by standard evaporation technique. In this work we present the experimental data descibing the parameters of photovoltaic devices and single layers that were constructed according to laser deposition technology.

Experimental

The detailed description of the deposition system was reported earlier [4]. The output of an infrared CO_2 laser beam operating at 10,6 micron wavelength was used in the present work. The 80 watt beam of about 1,2 mm in diameter was transited into the chamber through a special port made of NaCl monocrystal. Inside the chamber the incoming beam was focused at the target by the sphere mirror that was incident at an angle of 30^0 to the target. We varied the density of beam energy by changing the laser sport size on the target the variering the mirror to target distance. The pressure was always around $4 \cdot 10^{-6}$ mbar. The controlling system gave us possibility to obtain the films at the constant evaporation rate at values of 1-5 nm/sec for organics and of 0,5-2 nm/sec for fullerene layers. Zink phthalocyanine (ZnPc) was used as one of the components for cell constructing. ZnPc powder was predominantly purified at 500 K in argon flow. Fullerene powder contained around 85% of C_{60} molecules with 15% of C_{70} and small amounts of C_{80} . Single layers of ZnPc and fullerenes were deposited in a sandwich structure between resistyvely evaporated semitransparent gold and ITO layers of around 40 nm thick. ZnPc layers were also sublimated at the surface of TiO_2 niobium doped monocrystal with the following evaporation of Au contacts 30 nm thick.

Volt-ampere photocurrent characteristics were obtained with the help of measuring unit SMU Keithley 236. All spectral measurements were carried out at constant foton flux for all wavelengthes between 350 and 850 nm. For photoresponse measurements 200 Watt xenon lamp and single-grating monochromator were used.

Single deposited layers

To get information about the absorption properties of the materials we deposited single layers of ZnPc and C_{60} - C_{70} films. Figure 1 shows the absorption spectra of 360 nm thick ZnPc layer and 80 nm C_{60} - C_{70} film deposited on thin glass supports. The same figure presents the absorption spectra of niobium doped TiO_2 monocrystal.

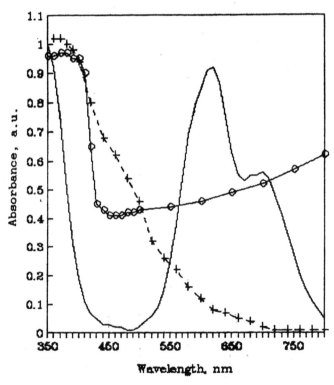

Fig. 1.   Solid   state absorption   characteristics of (ZnPc) dye (solid line) and C_{60} - C_{70} (crosses) films. Circles - absorption of niobium doped titanium dioxide crystal.

A number of experiments have been performed to find out the proper ohmic contacts for conductivity measurements of photoelectric properties of the constructed cells. For this purpose single layers of the materials mentioned were prepared in sandwich structure of different configurations. To insure that these contacts were really ohmic current voltage characteristics were analyzed. The proportional volt ampere behavior and the lack of rectifying effect was observed for the following configurations: ITO/ZnPc/Au and for the second material - Al/C_{60}-C_{70}/Al. Here we must note that asymmetric configuration for ZnPc layer with upper Au electrode was chosen according to the data obtained in [5]. The presented investigation does not discuss the problems that appear when electric field higher than 10^5 V cm^{-1} was applied to the sample as long as in our measurements we did not exceed the values of about 10^4 V cm^{-1} .

P/N cells

U/I -characteristics. Fig. 2 shows photocurrent density - voltage characteristic of In/TiO$_2$/ZnPc (320 nm)/Au cell. The voltage is applied to the Au electrode. Under illumination of 70 mV/cm^2 white light from ZnPc side through Au electrode the TiO$_2$ side of the cell shows negative photovoltage of about 0,38 mV and the observable value of short circuted photocurrent about 0,65 μA cm^{-2} . The calculated fill factor of 0,27 indicates that this cell is influenced by a series resistance as often found for cells made of organic materials [2]. On the same figer one can see the behavior of the cell when illuinated through titanium dioxide side. Indium electrodes were formed at the back side of TiO$_2$ apart the illuminated region that was possible according to high conductivity of doped crystal. For this configuration the efficiency of the cell is much larger than for the previous one. Short circuited current arises to the value of 1.84 μA cm^{-2} and the total efficiency of the cell is about 0.005 %.

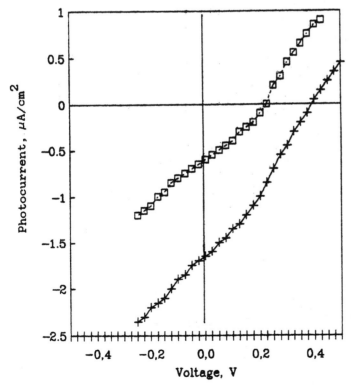

Fig.2. Photocurrent-voltage characteristics of TiO$_2$/ZnPc cell illuminated from ZnPc side (solid curve) and from titaniun dioxide side (dashed curve).

Photocurrent spectra. Fig.3 presents short circuited photocurrent spectra for $TiO_2/ZnPc$ cell when illuminated from ZnPc side (i.e. through 36 nm thick Au contact) as well as from TiO_2 side. As long as the depth of the dye layer is thick when compared with penetration depth of the light, the obtained spectra should be due to the absorption characteristics of TiO_2. On the other hand when illuminated from TiO_2 side, the resulted curve repeats the absorption spectra of ZnPc, that is cut at its left part (at wavelength around 420 nm) according to transparency of TiO_2 monocrystal (see Fig. 1). It may be concluded that short circuited current generating region is in the vicinity of the p/n interface and mostly situated in the dye layer.

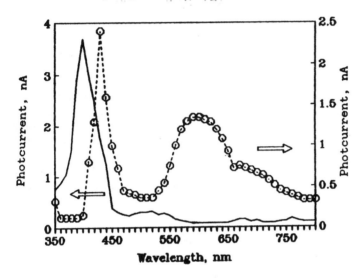

Fig. 3. Short-circuited current spectrum of a In/TiO_2 /ZnPc/Au sample illuminated through Au electrode (solid curve) and TiO_2-side (circles).

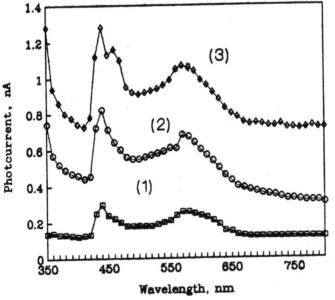

Fig.4. Short circuited current spectra for $TiO_2/ZnPc$ cell (1) and photocurrent under bias voltage of 0,2 (2) and 0,4 (3) volt applied.

Photocurrent under bias voltage. The strong effect of photoconduction is demonstrated on the same cell by photocurrent spectra obtained for different bias voltages applied. Fig. 4 shows the comparative intensities of photo currents for 0,2 and 0,4 volts applied under illumination from titanium dioxide side. Note that the effect of increased values between 350 ad 420 nm when bias voltage applied was due to relaxation properties and could be diminished by increasing the time between spectra points measurements.

P/N junctions with fullerene layers. To investigate the photocurrent properties of C_{60}-C_{70} deposited films we tried different structure organization between mentioned material and ZnPc and also tried different metal electrodes. We faced some technological problems when depositing C_{60}-C_{70} /ZnPc combination without exposing to the air. At last we prepared laser deposited C_{60}-C_{70} layers on aluminum foil and after that covered it with resistively evaporated ZnPc layers. The Au contact was evaporated at the upper surface of the cell. The thickness the deposited fullerene layer was about 80 nm and ZnPc - 300 nm. Here we also faced the situation when photocurrent parameters of the cell gave significant variety with time. So we observed the broad variation of photocurrent values in different samples. One possible explanation may be connected with uncontrolled doping process of dye layer by carbon clusters that was already mentioned for organics in [6]. In Fig. 4 we present the photocurrent behavior of one of the samples that was measured just after the deposition. The spectral dependence of short circuited current is shown here for Al/C_{60}-C_{70} /ZnPc/Au cell.

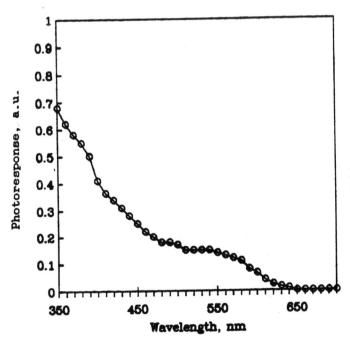

Fig. 5. Spectral response of the photocurrent in C_{60}-C_{70} /ZnPc junction.

The observed spectra may be treated as the combination of the absorption spectra of carbon clusters film (see Fig. 1) and the one of ZnPc after the light absorption in rather thick dye layer. The small intensity peak at about 580 nm also shows the influence of dye layer and correlates with the one obtained for TiO/ZnPc cell when illuminating from the dye side.

Discussion

The presented paper demonstrates the rectifying behavior and photoresponse of the heterojunction between doped titanium dioxide and zinc phthalocyanine as well as phthalocyanine and fullerene layers. For the first structure we observed high values of open circuit voltage and showed the spectral response to be partially adopted to solar spectra. The spectral curves under bias voltage show that efficiency of the whole cell is limited by high series resistance of the dye layer. So the technology that may form very thin ZnPc without pinholes and other defects might give possibility to increse dramatically the total efficiency of the cell. We feel that the investigated structure after its optimization may be applied for solar energy conversion devices [7].

It is interesting to compare our results with the investigations that were carried out with photoelectrochemical cell based on TiO_2 nanoparticles and organic dyes [8]. Our TiO_2 based cell and the one described in [8] both applied high conductivity titanium dioxide as one of the p/n junction parts supports. We applied doped crystals (with conductivity 50 Ω^{-1} cm^{-1}) where's nanoparticles are conductive according to the high concentration of surface defect states. Meantime the absorption curve of doped TiO_2 monocrystal show the steady increase in 400 - 800 nm region to compare with nearly full transparency for undoped particles. This fact may play role in the efficiency of the constructed cell. From the other hand the molecular semiconductor (organic dye) applied in [8] have broad absorption region in visible spectra and have great advantage when compared with narrow absorption peaks in ZnPc. We suppose that the future possibilities for increasing the efficiency of organic-based cells can be find in connecting the advantages of doped TiO_2 crystals (or small particles) with the proper organic dyes.

One of the authors (O.I.Sh.) is grateful to Dr. D.Meissner and Prof. R.Memming from The Institute for Solar Energy Research, Hannover for helpful discussions and constant interest to the problem.

References

[1] J.Simon, J.J.Andre: *Molecular Semiconductors. Photoelectrical properties of solar cells.* (Springer-Verlag, Berlin 1985).
[2] D. Wohrle and D.Meissner, Adv. Mater. **3**, 129 (1991).
[3] N.S. Sariciftci, D.Braun, C.Zhang, V.I.Srdanov, A.J.Heeger, G.Stucky and F.Wudl. Appl.Phys.Lett. **62**, 585 (1993).
[4] O.K.Karjagina, A.A.Kharlamov and O.I.Shevaleevskii, Dokladi Ak. Nauk SSSR, **292**, 365 (1986).
[5] S.Siebentritt, S.Gunster and D.Meissner, Synth. Metals, **41-43**, 1173 (1991).
[6] G.Gustafsson,Y. Cao, G.M.Treacy, F.Klavetter, N.Colanery and A.J.Heeger, Nature **357**, 477 (1992).
[7] O.I.Shevaleevskii, Dokladi Ak. Nauk Russia (to be published).
[8] B. O'Regan and M.Gratzel, Nature. **353**, 737 (1991).

Materials Science Forum Vols. 173-174 (1995) pp. 123-128
© *1995 Trans Tech Publications, Switzerland*

SYNTHESIS OF β-FeSi₂-FILMS BY PULSED LASER DEPOSITION

M. Panzner[1], H. Mai[1], B. Schöneich[1], B. Selle[2], H. Lange[2], W. Henrion[2],
Th. Wittke[1], A. Lenk[1], A. Teresiak[3] and R. Grötzschel[4]

[1] Fraunhofer- Institut für Werkstoffphysik und Schichttechnologie, Helmholtzstrasse 20,
D-01069 Dresden, Germany
[2] Abt. Photovoltaik, Hahn Meitner Institut Berlin GmBH, Rudower Chaussee 5,
D-12489 Berlin, Germany
[3] Institut für Festkörper- und Werkstofforschung, Helmholtzstrasse 20,
D-01069 Dresden, Germany
[4] Institut für Ionenstrahlphysik und Materialforschung, Forschungszentrum Rossendorf,
Bautzner Landstrasse 128, D-01474 Rossendorf, Germany

Keywords: Pulsed Laser Deposition, Laser Ablation, Thin Film Synthesis, Deposition, Iron Disilicide, FeSi₂-Films

Abstract. Thin Films of ß-FeSi₂ were synthesized by PLD (Pulsed Laser Deposition) using Nd:YAG and excimer lasers. The characteristics of the ablation process are studied by short time photography and investigation of the visual light emission. The particle flux distribution in the ablation plume is derived from ellipsometric measurements of the film thickness profile at the substrate. In the case of excimer laser PLD the stoichiometry of the deposited films differs only slightly from that of the target. By contrast deviations from FeSi₂-stoichiometry are observed when λ = 532 nm (the second harmonics of Nd:YAG laser radiation) is used for material ablation. In addition to the ß-FeSi₂ the presence of α-FeSi₂ and FeSi was confirmed by X-ray diffraction. A gap energy of 0.85 eV is derived from reflectivity measurements.

Introduction

The investigation of transition metal silicides has found growing interest because of its suitability for photovoltaic applications. Ecological benefits and the widely available iron and silicon based substances are substantial arguments for a broad technical use. For thin film absorbers in solar cells ß-FeSi₂ could offer certain advantages due to its direct band transition with a gap energy of 0.85 eV [1] and its high absorption coefficient (10^5 cm^{-1} at 1064 nm) [2]. Lefki and coworkers [3] measured the photovoltage of a mesa structure metal/ß-FeSi₂/n-Si for zero bias voltage at room temperature as a function of photon energy. However, no ß-FeSi₂-based solar cells have still been developed until now. A high density of lattice defects of the synthesized films in combination with a high recombination rate seems to be the reason for the failure of previous efforts. The synthesis of epitaxial almost single crystalline films is necessary to minimize the defects. A wide range of deposition methods is used for the development of suitable films. Komabayashi and coworkers [4] prepared ß-FeSi₂-films on glass substrates by rf-sputtering. Si discs combined with Fe plates were used as targets. A variation of the Fe and Si cross sections allowed the control of film composition. Another promising effort is the deposition of iron films on oxide free silicon surfaces with subsequent annealing to induce solid state reactions. Single phase polycrystalline ß-FeSi₂-films were for the first time synthesized by Bost and Mahan [2]. Epitaxially grown ß-FeSi₂-films were reported by Cherief and coworkers [5] using the same method under UHV-conditions. Gerthsen and coworkers [6] obtained epitaxial growth of ß-FeSi₂-films on silicon substrate by implantation of 200 keV Fe$^+$ ions into Si(111) wafers at elevated temperatures. The PLD is an alternative method to be tested for the development of thin film ß-FeSi₂-based solar cells. Results about PLD-synthesized films were published by Karpenko and coworkers [7]. The films deposited by KrF-excimer laser ablation from FeSi₂-compound target show prevailing ß-phase. Deviations from stoichio-metry were compensated using oxide free silicon substrates providing a source for Si-diffusion from the substrate to the growing film.

    The aim of our contribution is to investigate the influence of the PLD parameters on film synthesis and the upscaling of the method for large area coating for technical application.

Sample Preparation

The experimental setup used for thin film deposition consists of a high vacuum chamber with a basic pressure of 5×10^{-6} Torr and the laser. The laser beam is focused on the target surface by a quartz lens with a focal length of 545 mm at 532 nm wavelength. The lens represents also the laser port window of the chamber. By variation of the distance between lens and target, power densities of 10^9 to 10^{10} Wcm^{-2} were realized at the surface. The substrate located at a distance of 80 mm from the target can be heated up to 800°C. Sintered pellets with a nominal composition of Fe:Si = 1:2 were used as targets. The density of the targets is 2.4 gcm^{-3}, which is about 50 % of the X-ray density.

The Nd:YAG-laser (Continuum, NY82S-10) is working at 10 Hz with 1,8 J pulse energy and 8 ns pulse duration. 900 mJ per pulse are available by the second harmonic generation (λ = 532 nm). From the pulse energy (2 J) of the XeCl excimer laser (308 nm, 40 ns, KWU Siemens, XP 2020) only 600 mJ the almost homogeneous part of the beam was focused on the target surface.

During the ablation process the chronological development of the material effusion from the target is analyzed by short time photography. Through a chamber window the ablation plume with its following stages was observed by a high speed framing camera (PCO). In the same way the optical spectra, integrated over the life time of the plasma, were recorded.

Results and Discussion

<u>Ablation Process</u> The analysis of the evaporation process caused by a laser shot (10^9 Wcm^{-2}, 532 nm, 8 ns) demonstrates the following chronological scale. The spatial expansion of the light emitting plasma cloud increases up to 80 ns after the laser pulse. A short time photograph with 240 ns delay shows only a vanishing plasma glow. The spreading direction of the plasma seems to be perpendicular to the target surface if the laser beam (incident angle 45°) irradiates the target surface for the first time. Subsequently, droplets are emitted (figure 1). After a time delay of 300 μs to 500 μs no visible material emission is detected. A droplet velocity of 10 ms^{-1} to 100 ms^{-1} can be derived from the exposure time of the photograph and the length of the shining trace.

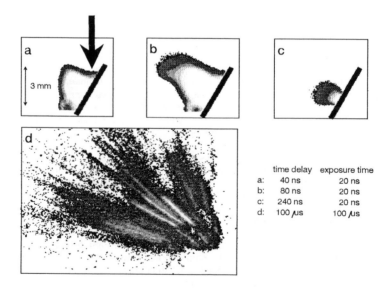

	time delay	exposure time
a:	40 ns	20 ns
b:	80 ns	20 ns
c:	240 ns	20 ns
d:	100 μs	100 μs

Figure 1   Short time photographs of the ablation process caused by a laser shot (532 nm, 10^9 Wcm^{-2}, 8 ns) on a sintered FeSi$_2$ target

These values correspond to those reported by Lubben and coworkers [9]. Many proposals are made in literature as to avoid droplet emission from the target [10]. Barr [11] was the first who tried to use the lower velocity of the particles to produce droplet free films by a mechanical filter. We can deduce from our experiments that a shutter would have to be open only for some microseconds immediately after every laser pulse.

The spatial distribution of the ejected material during ablation can be characterized by $M(\alpha) \sim \cos^n \alpha$ where $M(\alpha)$ is the material flux distribution for a spatial direction α. We deduced for n values between 3 and 10 from ellipsometric measurements of the thickness distribution of a film prepared on 4" wafer. The power n varies countercurrently to the irradiated target area.

Also the stoichiometry of the material flux varies slightly with direction α. A Decrease of 8 % in the Si:Fe ratio detected at the deposition centre of the film and 30 mm outside is found by Rutherford backscattering analysis (RBS).

Thin Film Analysis An important prerequisite for the synthesis of single phase ß-FeSi$_2$-films is to get the required 2:1 Fe:Si stoichiometry. The laser parameters as well as the temperature dependent sticking coefficient can change the film stoichiometry. Films at room temperature were prepared in order to investigate the influence of the laser parameters. We found a composition transfer of Si/Fe $\geq$ 2/1 for the XeCl-excimer laser and Si/Fe $<$ 2/1 for the second harmonics of the Nd:YAG-laser by RBS investigations. The 2:1 stoichiometry of the target was checked by RBS too. Film stoichiometry variations were also found by changing the substrate temperature at constant laser parameters and duration of deposition. At higher substrate temperatures a Si deficit of the films seems to be compensated by diffusion from the silicon substrate into the film. This is supported by measurements of the concentration profiles of silicon and iron by Auger electron spectroscopy (AES). A widening of the transition region between FeSi$_x$-film and Si-substrate is observed at higher substrate temperatures because of the diffusion processes. In our case the natural oxide at the substrate surface represents a diffusion barrier. It can be concluded from the experiments that silicon losses might be completely compensated by using clean surfaces.

The substrate temperature is an important parameter also for the crystallisation process of the ß-phase. No crystallisation of the film is observed for substrate temperatures below 500°C. α -FeSi$_2$ and FeSi beside the prevailing ß-FeSi$_2$ were found by X-ray diffraction (XRD) at substrate temperatures above 700°C (figure 2).

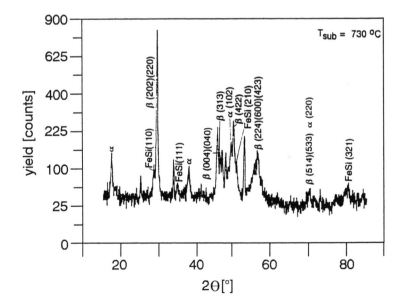

Figure 2   XRD spectra of a FeSi$_x$ film grown at a substrate temperature of 730°C

Scanning electron micrographs (SEM) of the film surface show crystallites of a few hundred nanometres in diameter. A decrease of the substrate temperature to values below 600°C broadens the XRD-peaks of ß-FeSi$_2$ and removes those of the α-FeSi$_2$ and FeSi. It is known from the Fe-Si-phase diagram [8] that the metallic α-phase exists stably only above 915°C. The existence of α-FeSi$_2$ in our films grown at substrate temperatures far below 900°C cannot be explained by a transfer of α-phase containing target particles because of the XRD results at room temperatures showing no crystalline peaks. This films are amorphous. The non equilibrium process during condensation of the highly dense particle beam (ion energies up to 1000 eV) at the substrate plays the important role for the development of the α-phase. The formation of the FeSi phase seems to be a reaction on the shortage of silicon in the films.

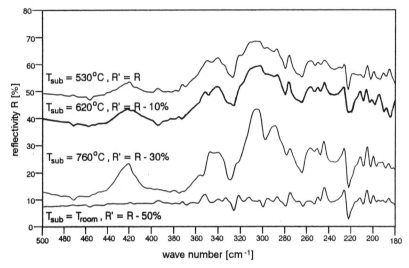

Figure 3   FTIR spectra of FeSi$_x$-films grown at several substrate temperatures

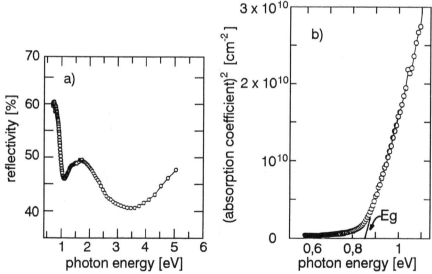

Figure 4   Reflectivity spectrum (a) and absorption edge (b) of a PLD synthesized FeSi$_x$-film
                  (x $\approx$ 1,7) measured by a spectrometer with integrating sphere to detect also
                  the scattered light

The ß-phase was also identified by IR reflectivity measurements. Figure 3 shows the reflectivity of films deposited at several substrate temperatures. The well-known five characteristic peaks [3] caused by IR active lattice vibrations arise with increasing substrate temperature.

A gap energy of 0,85 eV could be deduced from calculations of the absorption coefficient on the basis of reflectivity and transmission measurements in the corresponding spectral range (figure 4).

Conclusion

$FeSi_x$ films were grown by PLD. $ß-FeSi_2$ could be identified as main phase by XRD and reflectivity measurements. The detected silicon loss in the films synthesized by Nd:YAG laser can be avoided by using UV-laser and silicon substrates with oxide free surfaces. Substrate temperatures above 500°C have to be used for the synthesis of crystallized films by in situ PLD. Substrate temperatures below 700°C and an optimization of the laser parameters are necessary to suppress the undesired phases.

References

[1]   C. A. Dimitriadis, J. H. Werner, S. Logothetidis, M. Stutzman, J. Weber and R. Nesper,
        J. Appl. Phys. **68**, 1726 (1990).

[2]   M. C. Bost and J. E. Mahan,
        J. Appl. Phys. **58**, 2696 (1985).

[3]   K. Lefki, P. Muret, N. Cherief and R. C. Cinti,
        Sensors and Actuators A (Physical) **A33**, 81 (992).

[4]   M. Komabayashi, H.I. Hijikata and S. Ido;
        Japanese J. of Appl. Phys. **30**, 563 (1991).

[5]   N. Cherief, C. D' Anterroches, R.C. Cinti, T.A. Nguen Tan and J. Derrien
        Appl. Phys. Lett. **55**, 1671 (1989).

[6]   D. Gerthsen, K. Rademacher, Ch. Dieker and S. Manti,
        J. Appl. Phys. **71**, 3788 (1992).

[7]   O. P. Karpenko, C. H. Olk, G. L. Doll, J. F. Mansfleid and S. M. Yalisovs,
        MRS Fall Meeting, Boston, December 1993, to be puplished.

[8]   I. N. Strukov and P. V. Geld,
        Fiz. Metallov i Metallovedenie **4**, 190 (1957).

[9]   D. Lubben et. al.,
        J. Vac. Sci. Techn. **B3**, 968 (1985).

[10]   H. Sankur and J. T. Cheung,
        Appl. Phys. **A47**, 271 (1985).

[11]   W. P. Barr,
        J. Phys. **E2**, 1064 (1969).

Materials Science Forum Vols. 173-174 (1995) pp. 129-134
© 1995 Trans Tech Publications, Switzerland

COMPLEX SEMICONDUCTOR COMPOUND TARGETS FOR PULSED LASER DEPOSITION

H. Dittrich[1], M. Klose[2], M. Brieger[2], R. Schäffler[3] and H.W. Schock[3]

[1] Zentrum für Sonnenenergie- und Wasserstoff-Forschung,
Hessbrühlstr. 21c, D-70565 Stuttgart, Germany

[2] Institut für Technische Physik, DLR,
Pfaffenwaldring 38-40, D-70569 Stuttgart, Germany

[3] Institut für Physikalische Elektronik, Universität Stuttgart,
Pfaffenwaldring 47, D-70569 Stuttgart, Germany

Keywords: $CuInSe_2$, Target Preparation, Pulsed Laser Deposition, Uniaxial Hot Pressing, X-ray Characterization, Laser - Target Interaction

Abstract. $CuInSe_2$ targets for pulsed laser deposition were prepared by uniaxial hot pressing in a fully automized commercial hot pressing system. Also few experiments were carried out with an isostatical cold pressing system. Several starting materials were used under different pressing parameters like temperature, pressure, holding time etc. Targets were characterized by their density, composition and structural aspects. Results indicate, that temperatures of more than 550°C for reactive hot pressing from binary compounds and 600°C for pure $CuInSe_2$ (CIS) powder is necessary for the formation of dense and homogeneous targets. The optimum pressing temperature lies inbetween 630 and 650°C. Highest densities of 96,1% of the theoretical density is observed using CIS powder with some excess of In_2Se_3 resulting in off-stoichiometric targets as needed for the deposition of photovoltaic active thin films. Targets were exposed to YAG:Nd laser light with quadrupeled frequency resulting in a wavelength of 266nm. Specific erosion structures are observable: top shielded cones with different cone angles depending on target material and light intensity were observed similar to erosion structures from sputtering. With a special target movement the formation of the cones could be reduced.

Introduction

The availability of suitable targets is the basis for the deposition of complex semiconductor thin films by pulsed laser deposition (PLD). In preceeding publications informations about CIS target preparation is extremely scarce [1,2]. Only single crystal targets and targets from melt synthesis are mentioned. Already commercially available targets suffer from being ideal dense or homogeneous with respect to their phase composition. A further restriction is the non-availability of defined off-stoichiometric targets, which is very important for the deposition of (CIS) thin films with appropriate photovoltaic properties.

For this purpose preparation methods have to be investigated to optimize target properties. Uniaxial pressing and isostatical pressing are considered as being suitable and experiments are carried out to optimize preparation parameters. Within the framework of a project supported by the state of Baden-Württemberg, these targets are transferred to the PLD for solar cell preparation and device characterization [3].

Being a new and unconventional deposition method in the field of compound semiconductor deposition for photovoltaic applications the knowledge about laser - target interaction is scarce and of great interest for an optimized deposition process. Therefore first observations are presented and discussed.

Target Preparation

As already mentioned in the introduction uniaxial hot pressing and isostatical cold pressing are taken into consideration for the target preparation. Special attention is directed to the uniaxial pressing and only few additional experiments are carried out by isostatical cold pressing at the Max-Planck-Institut für Pulvermetallurgie, Stuttgart, Germany.

The diversification of investigated target preparation possibilities is summarized in Fig. 1. For the uniaxial hot pressing four different starting materials are considered:
- mixture of binary powders from $CuSe$ and In_2Se_3
- mixture of binary powders from Cu_2Se and In_2Se_3
- presynthesized stoichiometric $CuInSe_2$
- mixture of $CuInSe_2$ and In_2Se_3

(All materials from CERAC, purity classification 5N)

Two different materials for the hot pressing tools are tested with respect to their chemical inertness: quartz glass and Al_2O_3 ceramic. Both materials are sufficiently inert so that the direct contact to the hot pressed CIS does not lead to measurable contaminations. The advantage of using quartz glass lies in the IR transparency, which accelerates the dynamic of heating processes, whereas the advantage of Al_2O_3 ceramic lies in a better stability under high pressures.

Prepared powders and homogenized mixtures are filled into the pressing tools and situated inbetween the punches of a fully automized commercial hot pressing system (KCE corp.). Vacuum condition for the experiments lies in the medium high vacuum region about 1 Pa. The free parameters for the hot pressing process are: temperature, pressure, temperature and pressure ramps, and temperature and pressure holding times.

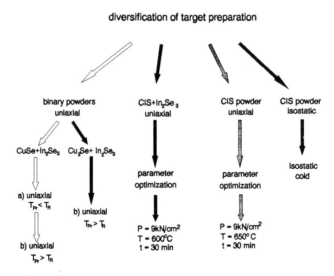

Fig. 1: Diversification of target preparation

In Fig. 2 a process protocol of a standard uniaxial hot pressing process is given. Two distinct compression activities are observed:

    1. during the setting of the process pressure a compression is observable due to the mechanical compression of the powder particles

    2. during the setting of the process temperature a compression is observable due to the reaction of binary compounds and recrystallization processes. This second step is more effective and therefore more important than the first one.

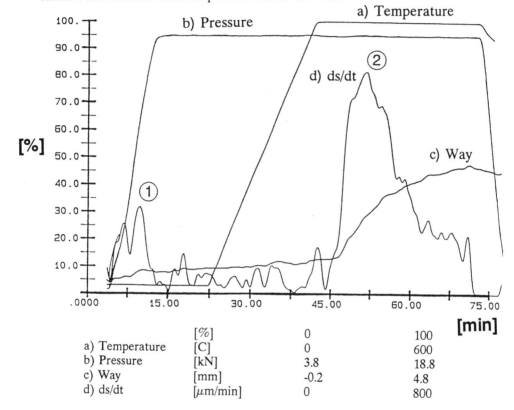

	[%]	0	100
a) Temperature	[C]	0	600
b) Pressure	[kN]	3.8	18.8
c) Way	[mm]	-0.2	4.8
d) ds/dt	[μm/min]	0	800

Fig. 2: Process protocol of a standard uniaxial hot pressing experiment
    ①: mechanical compression
    ②: recrystallization compression

Target Characterization

Target Density. Uniaxial hot pressed targets have a disk shape with diameters of 25mm and thicknesses from 3 - 6mm. Due to the precisely treated pressing tools, the sharp geometrical shape was used for the density estimation by measuring the volume and weighting the target. Highest densities observed reach 96.1% of the theoretical density of CIS.

    These densities are observed in experiments using CIS powder and amounts of In_2Se_3 in addition to obtain off-stoichiometric targets on the In-rich side. From X-ray characterization these targets show single phase chalcopyrite or thiogallate structure (see X-ray characterization this chapter) without resting In_2Se_3 phases. It is therefore to be mentioned that this In_2Se_3

surplus acts as a catalyst for the recrystallization compression.

Target Composition. In spite of the macroscopic flatness of the target shapes, the surfaces show a very rough morphology on a microscopic scale. For this reason the composition measurements done by EDS analysis are very critical. A distinct statistical difference between weighted ingot composition and measured compositions by EDS is observed. This is also caused by differences observed for bulk and surface compositions. Being a semi-open system, compositional changes at the surface of the target are possible considering the decomposition reaction:

$$2CuInSe_2 \rightarrow Cu_2Se + In_2Se\uparrow + Se_2\uparrow$$

resulting in two mobile gas phase reaction products. As a result from this observation, virgin targets should be first cleaned by dummy PLD experiments.

X-ray Phase Analysis. X-ray phase analysis is carried out on a Siemens D-501 powder diffraction system using an iron X-ray tube (FeKα_1: $\lambda = 0.193604$nm). A graphite monochromator on the diffracted beam side enhances the system resolution.

Targets from mixed and homogenized powders of binary compounds can be transformed to single phase CIS with perfect chalcopyrite structure at temperatures above the observed reaction temperature of about 450°C. Fig. 3 shows the X-ray patterns of the starting CuSe and In$_2$Se$_3$ powders compared with the reacted CIS target and a calculated standard CIS pattern. The calculation was done by the POWD10 software program based on kinematical scattering theory taken into consideration a perfect chalcopyrite structure. The correspondence of measured pattern and simulated pattern is obvious. Similar results can be obtained from a Cu$_2$Se/In$_2$Se$_3$ powder mix, so that the following reactions take place during the uniaxial hot

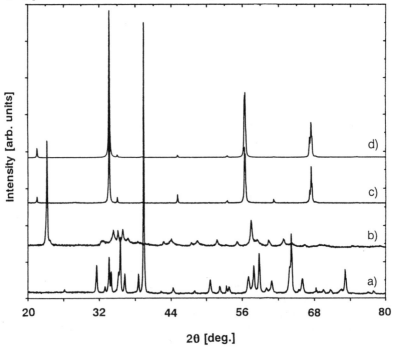

Fig. 3: XRD patterns for a) CuSe powder, b) In$_2$Se$_3$ powder, c) hot pressed target, and d) calculated CIS standard

pressing experiments:

a) $Cu_2Se + In_2Se_3 \rightarrow 2CuInSe_2$ and   b) $4CuSe + 2In_2Se_3 \rightarrow 4CuInSe_2 + Se_2$

In case b) intrinsic Se-rich targets are observed. The advantages of this approach are a free Cu/In ratio according to the molecularity line in the pseodobinary Cu_2Se - In_2Se_3 phase diagram and the possibility of obtaining homogeneous and perfect crystallized CIS. Unfortunately the density of these targets is lower than that of the optimum ones observed.

Targets from presynthesized $CuInSe_2$ powders uniaxially hot pressed at temperatures above 600°C show the same perfect chalcopyrite structure with a higher density. As the Cu/In ratio is a fixed value, off-stoichiometry processed during the transfer from target to substrate and during thin film growth cannot be corrected.

Targets from $CuInSe_2$ powders with additional In_2Se_3 surplus have measurable amounts of a thiogallate phase (spacegroup: I4) indicating a composition of $CuIn_2Se_4$. The thiogallate CIS structure is very similar to the chalcopyrite CIS structure but contains ordered Cu vacancies. The proof of the thiogallate structure was given by the coincidence of the measured X-ray powder diffraction (XRD) pattern (see Fig. 4) with a simulated one (see bars in Fig. 4). These targets are characterized by the highest densities observed. As photovoltaic active CIS layers need to have a Cu/In ratio <1 these targets are considered to be best suited.

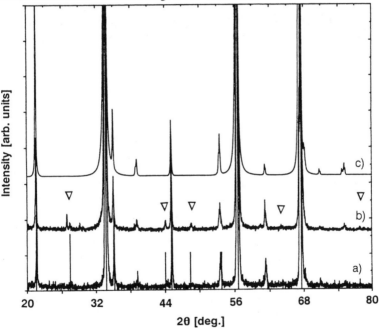

Fig. 4: XRD patterns of a) $Cu_{1.0}In_{1.1}Se_{2.15}$ and b) $Cu_{1.0}In_{1.2}Se_{2.30}$ uniaxial hot pressed compared with c) the calculated chalcopyrite CIS pattern. The symbol ∇ indicates new peaks belonging to the CIS thiogallate phase

Laser - Target Interaction

Targets are exposed to pulsed YAG:Nd laser light to carry out laser ablation. After the deposition experiments we observe material specific erosion structures dependent on target compositions and light intensities. Generally cone-shaped structures with tops shielded by dust particles and liquid or solid secondary phase segregations are directed towards the incident light. Combined with the occurrence of these surface structures at the target is a high density of μm size particles at the grown thin film, strongly influencing the photovoltaic activity of the solar cell. By introducing a specific target movement, the tendency of the cone formation can be reduced drastically. A comparison of surfaces of simply rotated targets and targets with the specific movement is given in Fig. 5.

a) 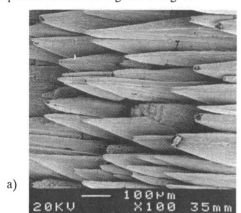 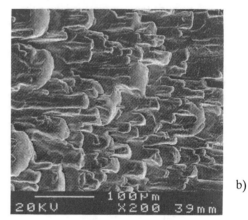 b)

Fig. 5: Surface structures of target with a) simple rotation and b) specific movement

Conclusion

CIS targets for PLD of CIS thin film solar cells are prepared by uniaxial hot pressing and isostatic cold pressing. Best results are obtained in uniaxial hot pressing experiments using $CuInSe_2$ plus In_2Se_3 powders for target preparation. The optimized hot pressing paramaters for these experiments are: pressure P = 9kN/cm^2 and temperature T_{pr} = 630°C. Targets are very dense and show up to 96.1% of the theoretical density. Stoichiometric targets have well crystallized, single phase chalcopyrite structure as demonstrated in XRD patterns. Indium-rich targets show amounts of a CIS thiogallate structure, an ordered vacancy compound with a composition of $CuIn_2Se_4$. During laser - target interaction cone shaped surface morphologies occure similar to sputtered CIS surfaces, limiting solar cell efficiencies by particle transfer to the growing thin film. Occurring of cones and therfore particle transfer can be reduced by a specific target movement.

Acknowledgement: SEM investigations of P. Allenspacher is gratefully acknowledged. This work was supported by the Ministery of Economics, Baden-Württemberg, Contr.-No.: IV 4332.62-D 721.46

References

[1] J. Levoska, A.E. Hill, S. Leppävuori, O. Kusmartseva, R.D. Tomlinson, R.D. Pilkington, Jpn. J. Appl. Phys., Vol. 32 (1993), Suppl. 32-3, p. 43
[2] V.F. Gremenok, E.P. Zaretskaya, I.V. Bodnar, I.A. Victorov, Jpn. J. Appl. Phys., Vol. 32 (1993), Suppl. 32-3, p. 90
[3] R. Schäffler, H. W. Schock, M. Brieger, M. Klose, and H. Dittrich, this conference

Materials Science Forum Vols. 173-174 (1995) pp. 135-140
© 1995 Trans Tech Publications, Switzerland

PULSED LASER DEPOSITION AND CHARACTERIZATION OF CuInSe$_2$ THIN FILMS FOR SOLAR CELL APPLICATIONS

R. Schäffler[1], M. Klose[2], M. Brieger[2], H. Dittrich[3] and H.W. Schock[3]

[1] Institut für Physikalische Elektronik, Pfaffenwaldring 47,
D-70569 Stuttgart, Germany

[2] Institut für Technische Physik, DLR, Pfaffenwaldring 38-40, D-70569 Stuttgart, Germany

[3] Zentrum für Sonnenenergie- und Wasserstoff-Forschung, Hessbrühlstr. 21c,
D-70565 Stuttgart, Germany

Keywords: Pulsed Laser Deposition, CuInSe$_2$, Solar Cell, Electrical Characterization, Thermopower

Abstract. Thin films of CuInSe$_2$ (CIS) were deposited on soda-lime float glass with and without molybdenum back contact. The substrate temperature was 520..580°C. Pulsed Laser Deposition (PLD) of the CIS absorber layer was obtained under high vacuum conditions at 10^{-4} Pa. The beam of an excimer laser or a quadrupled Nd:YAG laser was directed onto a rotating target. The resulting films have been characterized by SEM, EDX and XRD. It will be shown that by choosing adequate deposition parameters, homogeneous films of single phase CIS can be prepared. Films prepared without additional selenium partial pressure during deposition do not result in active photovoltaic devices. Electrical conductivity and thermopower measurements were carried out on a variety of films. They show a correlation between selenium partial pressure and the electrical properties, as expected from the defect chemistry model of CIS. Depending on selenium partial pressure, the electrical characteristics (for In-rich CIS) range from p-type (metallic) through n-type to high resistive p-type behaviour. CuInSe$_2$/CdS/ZnO heterojunctions prepared from the deposited layers reach conversion efficiencies of 8.5%.

Introduction

Solar cells based on CuInSe$_2$ (CIS) and related compounds recently have shown highest efficiencies [1], and therefore are a very promising candidate for large scale production. The excellent performance of the method of Pulsed Laser Deposition (PLD) in the preparation of high-temperature superconductors [2] was the motivation to look into the feasibility of thin film deposition of CuInSe$_2$ (CIS) for solar cell applications. PLD implies not a steady state growth, but a cyclic process of a short pulse of incoming high flux on the surface and an intermittent period, in which suface mobilies of adatoms and recrystallization processes are of importance. This characteristic feature of PLD could result in specific structures and their physical properties, which have to be analysed.

Pulsed laser deposition (PLD) - Experimental

Presynthesized CIS powder (5N) with an admixture of In$_2$Se$_3$ was uniaxially hot pressed in order to obtain a slightly In-rich target material. Details on this subject are described elsewhere [3].

Fig. 1 shows the scheme for the PLD set-up. Thin film deposition was obtained under high vacuum at about 10^{-4} Pa. The beam of an excimer laser ($\lambda=308$nm, pulse width 30..50nsec) or a quadrupled Nd:YAG laser ($\lambda=266$nm, pulse width 6nsec) was directed onto an under 45° - with respect to surface normal - rotating target. For both lasers the power density in the spot was in the range of $10^7..10^8$ Wcm^{-2}. The distance between target and substrate varied from 65 to 120 mm and the substrate temperature was kept at 520..580 °C.

Different approaches have been followed in preparing CIS by PLD. Due to the high substrate temperature during deposition, an additional Se supply is necessary. Using the Nd:YAG laser, this has been provided either during a post deposition selenization step (approach 1) or by an additional selenium partial pressure (by means of an a selenium source for thermal evaporation) during deposition (approach 2). In order to further increase the selenium partial pressure, the target-substrate space was encapsulated and an effusive selenium source incorporated for thermal evaporation (approach 3). Using the excimer laser (approach 4), so far only some few experiments where done without additional selenium.

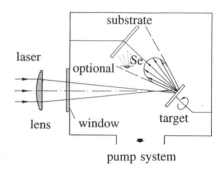

Fig. 1   schematic experimental setup for PLD

Surface morphology and phase analysis

No obvious difference in surface morphology is observed when comparing approach one and two. The typical morphology of a slightly In-rich film prepared by approach one or two is shown in Fig.2a. Structural analysis with a Siemens D501 X-ray diffractometer (FeKα_1, λ=0.193604nm),

Fig.2   typical film morphologies obtained for the different approaches - a: Nd:YAG open system - b: Nd:YAG, encapsulated system - c: excimer

shows single phase material for sufficient excess of selenium. When working with Nd:YAG laser and without any additional selenium, secondary phases appear [4]. Subtle distinctions in the XRD 204/220 peak, representing the crystal quality (peak splitting in case of high crystal quality), are displayed in Fig.3. Fig.3a shows a weak 204/220 peak splitting for a film prepared under Se-excess.

Compared to films prepared in the open system (approach one and two) an increased grain size is found (Fig.2b) when indroducing the encapsulated target substrate space (approach three). On

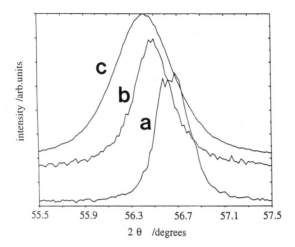

Fig. 3   Tetragonal splitting of the 204/220 peak due to different "crystallinities" of the CIS layers

the other hand the crystal quality is reduced (missing peak splitting of the 204/220 peak - Fig.3b).

Films prepared by the excimer laser (Fig.2c) are very dense with a totally different morphology compared to the films deposited by the Nd:YAG laser radiation (at comparable laser power density). XRD measurements show single phase material, an extreme 204/220 fiber texture, but poor crystal quality (Fig.3c)

Electrical Characterization

According to the previous paragraph single phase (within the detection limits of XRD, about one vol. %) $CuInSe_2$ can be observed for various deposition conditions. In addition, suitable electronic properties are required for device applications. It is well known that the defect chemistry and free carrier concentration in $CuInSe_2$ varies drastically in dependence of substrate material (soda-lime float glass or borosilicate glass)[1] and processing conditions (e.g.[5]).

Therefore, electrical properties were determined by means of thermopower and conductivity measurements as a matter of routine. Thermopower measurements were done with a temperature difference of ten Kelvin at a temperature level slightly above room temperature. Since neither the process of scattering nor the level of compensation nor the homogeneity of the film are known, the raw data (i.e. thermopower α in mV/K, positive for p-type - , negative for n-type material) are presented instead of calculating the carrier concentration.

In the case of low effective doping the minority carrier concentration also has to be taken into account. For the case of a homogeneous (single phase) film - using Boltzmann statistics - the thermopower is given by

$$\alpha = -\frac{k_B}{\sigma}[\mu_n \, n \, (s+\frac{5}{2}-\ln\frac{n}{N_C}) - \mu_p \, p \, (s+\frac{5}{2}-\ln\frac{p}{N_V})] \tag{1}$$

where μ_n,μ_p: mobility of electrons respectively holes
             σ: conductivity
             s: scattering mechanism
             k_B: Boltzmann constant
             N_C,N_V: effective density of states in conduction band, valence band

With increasing compensational doping, for example caused by selenium deficiency, thermopower theoretically increases to a maximum and then decreases again or even reverses (depending on the ratio of the mobilities and the predominant scattering mechanism). It should be noted that an *effective* thermopower is measured in the case of an inhomogeneous (multiphase) film.

As already mentioned, different approaches have been followed preparing thin films of CIS. The following electrical characterization focusses on three approaches.

First approach (Nd:YAG $\lambda=266$nm, open system, post-PLD selenization treatment):

In this case, the usual PLD - without any additional selenium - is followed by a post deposition selenization step. Tab.1 demonstrates, that the electrical characteristics are far from optimum, when preparing without additional selenium. Using a post deposition selenization step, a decisive improvement could be achieved. As a guideline, electrical data of a high efficiency ($\eta\approx13..14\%$) CIS absorber layer close to stoichiometrie - deposited by conventional thermal evaporation and using standard window technology [6] - are $\sigma\approx0.01..0.05$ $(\Omega\text{cm})^{-1}$ **and** $\alpha_p\approx0.81..0.87$ mV/K.

Tab.1:   electrical characteristics before and after selenization (all films slightly In-rich, i.e. In/(Cu+In)=0.53) - solar grade absorber material marked grey

before Selenization	after Selenization	before Selenization	after Selenization
$\sigma=5$ (Ωcm)-1 $\alpha=0.03$mV/K p-type	$\sigma=0.05$ (Ωcm)-1 $\alpha=0.38$mV/K p-type	$\sigma=0.02$ (Ωcm)-1 $\alpha=-0.53$mV/K n-type	$\sigma=0.04$ (Ωcm)-1 $\alpha=0.57$mV/K p-type
$\sigma=6$ (Ωcm)-1 $\alpha=0.03$mV/K p-type	$\sigma=0.8$ (Ωcm)-1 $\alpha=0.18$mV/K p-type	$\sigma=0.004$ (Ωcm)-1 $\alpha=0.50$mV/K p-type	$\sigma=0.02$ (Ωcm)-1 $\alpha=0.61$mV/K p-type

This approach has led to one of the best results ($V_{OC}=402$mV, $j_{SC}=34.3$mA/cm^2, ff=61%, $\eta=8.4\%$, measured at 100mW/cm^2 [4]).

Second approach (Nd:YAG $\lambda=266$nm, open system, in-situ selenization during PLD):

This approach led to the best device so far ($V_{OC}=381$mV, $j_{SC}=36.9$mA/cm^2, ff=60%, $\eta=8.5\%$, measured at 100mW/cm^2 [4]). Fig.4 displays the electrical characteristics for a variety of films prepared by PLD under different conditions (variable in-situ selenium partial pressure, laser power density). Measurements of thermopower and resistivity of the same sample are marked by a unique symbol. The cross hatched area represents device relevant CIS absorber material. The extension towards the In-rich region is strongly process dependant and therefore marked with a dotted line. It has to be emphasized, that both thermopower **and** resistivity have to be situated in the cross hatched region in order to end up

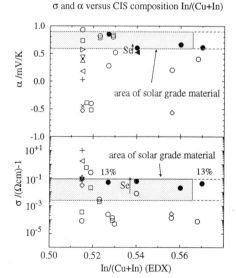

Fig. 4   observed values for α and σ. Device relevant films are marked with filled circles

with solar grade CIS absorber material. It can be seen that films of the same composition in the In-rich region electrically can behave totally different. In addition some films deposited in moderate selenium excess are situated far from the solar grade region, despite the fact that they were of single phase material according to XRD. Furthermore the electrical characteristics can be influenced by either post deposition selenization (see Tab.1) or variation of the in-situ selenium excess (keeping other process parameters constant - marked with Se-arrow). The results are sumarized in Tab.2

Tab.2:   Model of electrical behavior of CIS close to stoichiometry depending on Se-offer during deposition
increasing selenium deficiency during deposition (from left to right)

	p-conductivity	p-conductivity	n-conductivity	p-conductivity
conductivity	"high" conductivity $\sigma \approx 5E-2..5E-3$ $(\Omega cm)^{-1}$	low conductivity $\sigma \approx 5E-3..5E-6$ $(\Omega cm)^{-1}$	ranging from low to high conductivity	metallic behaviour $\sigma > \approx 1E-1$ $(\Omega cm)^{-1}$
thermopower	$\alpha_p \approx 0.6..0,85$ mV/K	reduced thermopower (see eq. 1)	α_n ranging from high to low values	metallic behaviour $\alpha_p \approx 30 \mu V/K$
possible origin	"high" efficiency absorber material $\eta \geq 10\%$	compensational effects (Se-vacancies ?)	compensational effects **and** beginning segregation of Cu-In intermetallics	segregation of Cu-In intermetallics $\rightarrow$ predominant effect

These observations are very similar to those found for other processes preparing CIS under selenium deficiency [7],[8].

The beginning segregation of intermetallic phases can also be recognized in solar cell devices when looking at the open circuit voltages V_{OC}. Tab. 3 represents electrical characterisitcs and open circuit voltage data.

Tab.3:   Correlation of electrical characteristics of the absorber layer with cell device characteristics

Ident	σ $1/\Omega cm$	α /mV/K	V_{OC} /mV	comment
2906	2E-3	0,78...0,82	346	no shunt
2806	1E-4	0,7	285	
1802	4E-5	0,52	236	slightly shunted
1702	8E-5	0,4	-	totally shunted

Third approach (Nd:YAG, encapsulated substrate chamber with the selenium source incorporated): Despite the fact of a higher selenium partial pressure, which generally is supposed to improve electrical characteristics, the films still remain highly resistive. Remembering the XRD results and this high resistivity, it is quite obvious why conversion efficiencies are very low.

Conclusions

CIS films prepared by PLD show various morphologies, chemical compositions and electronic properties depending on preparation conditions. Films deposited without any additional selenium show metallic behaviour. The variation of the electronic properties of the absorber layer depending on the excess of selenium is strongly correlated to device properties and can be explained in terms of the CIS defect chemistry model.

Slighty In-rich targets, additional selenium supply, and a homogeneous laser spot on the target with laser power densities of about $5*10^7$ Wcm^{-2} are suitable for the deposition of device relevant absorber material. To our knowledge it is the first time, that a photovoltaic conversion efficiency of 8.5% has been achieved by PLD using a CIS absorber layer. Therefore the principal feasability of PLD in preparing CIS absorber layers has been demonstrated.

Acknowledgement

The authors wish to acknowledge the skilled contributions of D. Hariskos in preparing CdS by chemical bath deposition and P. Allenspacher in SEM investigations. This work was supported by the Ministery of Economics, Baden-Württemberg, Contr. No.: IV 4332.62-D 721.46

References

[1] J. Hedström, H. Olsén, M. Bodegård, A. Kylner, L. Stolt, D. Hariskos, M. Ruckh and H.W. Schock, in *Proc. 23rd IEEE Photov. Spec. Conf.*, (IEEE, New York, 1993), p. 364
[2] T. Venkatesan, X.D. Wu, A. Inam and J.B. Wachtmann, Appl. Phys. Lett. **52**, 1193 (1988).
[3] H. Dittrich, M. Klose, M. Brieger, R. Schäffler and H.W. Schock, this conference
[4] H. Dittrich, M. Klose, M. Brieger, R. Schäffler and H.W. Schock, in *Proc. 23rd IEEE Photov. Spec. Conf.* (IEEE, New York, 1993), p. 617-620.
[5] H.R. Moutinho, D.J. Dunlavy, L.L. Kazmerski, R.K. Ahrenkiel, and F.A. Abou-Elfotouh, in *Proc. 23rd IEEE Photov. Spec. Conf.* (IEEE, New York, 1993), p. 572.
[6] J. Kessler, K.O. Velthaus, M. Ruckh, R. Laichinger, H.W. Schock, D. Lincot, R. Ortega and J. Vedel, in *Proc. 6th Int. Photov. Sci. Eng. Conf. (PVSEC-6)*, (Kluwer, Dordrecht, 1992), p. 1005.
[7] B. Dimmler, D. Schmid and H.W. Schock, in *Proc. 6th Int. Photov. Sci. Eng. Conf. (PVSEC-6)*, (Kluwer, Dordrecht, 1992), p. 103.
[8] J. Kessler, D. Schmid, S. Zweigart, H. Dittrich and H.W. Schock, presented at 12th EC PVSEC, Amsterdam, 11-15 April 1994, in press.

Materials Science Forum Vols. 173-174 (1995) pp. 141-152
© 1995 Trans Tech Publications, Switzerland

LASER CHARACTERIZATION OF SEMICONDUCTORS

M. Stutzmann

Walter Schottky Institut, Technische Universität München,
Am Coulombwall, D-85748 Garching, Germany

Keywords: Laser, Raman Scattering, Photothermal Deflection, LBIC, Photoinduced Absorption, Time of Flight, Photocarrier Grating, Moving Grating, Holographic Recrystallization, Silicon, Si-Ge Alloys, GaAs, $FeSi_2$, $CuInSe_2$

Abstract. Various experimental methods for the structural, optical, and electronic characterization of semiconductors based on lasers are briefly reviewed. Special attention is given to techniques which can be used at room temperature and can easily be implemented for materials, devices, and processes used in photovoltaics.

Introduction

Ever since their development in the 1960's, lasers have been used as a very versatile and powerful tool for the characterization and processing of semiconductors. Compared to other light sources commonly employed in semiconductor physics, lasers stand out due to their high intensity, the monochromaticity and coherence of the emitted radiation, their ability to produce short and highly energetic pulses, and the excellent properties of many laser sources for applications requiring non-diverging beams or spatial resolution in the µm range. On the other hand, common problems encountered when using lasers are high costs, lack of stability and tuneability, difficulties with thermal stressing and laser-induced degradation, and the complex handling of more sophisticated laser systems. In any case, the many applications of lasers in modern semiconductor physics prove beyond doubt that their advantages outweigh possible disadvantages. As a consequence, new developments in laser physics quickly result in related progress in semiconductor spectroscopy. The purpose of this article is to provide a brief overview over recent applications of lasers in the characterization of semiconductor materials, with a particular emphasis on problems related to applications in photovoltaics. For the sake of clarity, most examples mentioned in the following will be dealing with silicon as the most common semiconducting material. Similar applications in other photovoltaic systems are quite straightforward, but generally much less advanced. Also, we shall restrict our discussion to characterization techniques which can be used around room temperature, i.e. under usual operating conditions of solar cells.

The Tools

Today many different laser systems are commercially available and are more or less widely used in semiconductor characterization. A schematic overview is given in Fig. 1. Most applications to date rely on high performance gas or ion lasers such as HeCd, Ar^+, Kr^+, HeNe, CO, or CO_2 lasers for cw-radiation, or Excimer, N_2, Cu- or Au-vapor lasers for pulsed radiation. In particular, HeNe, Ar^+ and Kr^+ lasers are used in many laboratories due to their ease of operation combined with high power, small linewidth, good amplitude and mode stability, their low beam divergence and small beam diameter. These lasers are ideally suited for applications such as Raman scattering, luminescence, spatial mapping of optical properties, mirage techniques, and holography. Pulsed laser systems such as N_2-lasers are of importance for time-resolved measurements dealing with the dynamics of photo-excited carriers in semiconductors and devices. The time resolution is generally limited by the length of the exciting laser pulse. Pulse-lengths ranging from 10^{-6} s to 10^{-14}s are currently possible, but electronic processes relevant for photovoltaic applications usually require less extreme time resolutions of $\geq 10^{-9}$s.

More important for the use of lasers in semiconductors is the accessible spectral range. Until recently, the generation of coherent radiation with continuously variable wavelength was mainly the domain of dye lasers. As shown in Fig. 1, the spectral region covered by common dyes is from 300 to 1000 nm, i.e., ideally suited for optical investigations of most semiconductors. A problem, however, is the limited lifetime of dyes and the need to change dyes when large spectral scans are desired. Here, progress in solid state lasers has opened new opportunities. Ti-sapphire lasers are now established light sources for optical spectroscopy in the red and near-infrared part of the spectrum (700 - 950 nm). An even broader range is provided by optical parametric oscillators (OPO) [1], which are becoming commercially available. Thus, Fig. 1 shows the output characteristics of an OPO pumped by the third harmonic of a Nd-YAG laser and providing continuous coverage of wavelengths between 400 and 2000 nm. Wavelengths further into the UV can be generated by additional frequency doubling.

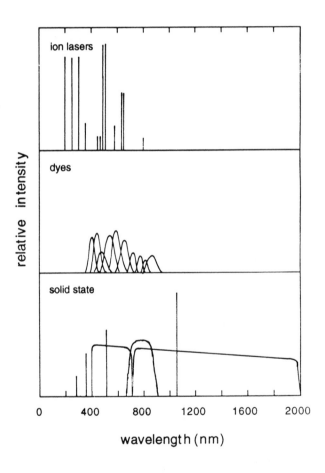

Fig. 1:
Schematic comparison of the spectral characteristics of commercial laser systems for semiconductor characterization: ion and gas lasers (top), dye lasers (middle), and solid state lasers (bottom).

Examples and Applications

Vibrational Spectroscopy. One of the first applications of lasers in solid state and semiconductor physics was as an intense and monochromatic source for Raman and luminescence spectroscopy. Raman scattering as an alternative to IR-absorption for the study of vibrational properties in semiconductors is also of interest for photovoltaic applications. Thus, Raman scattering can be used to probe strain and crystallite sizes in poly- or microcrystalline semiconductors [2]. Another example is the investigation of alloys, where Raman scattering serves as a fast, nondestructive method for the determination of alloy composition or phase separation. In Fig. 2, Raman spectra of crystalline Si-Ge alloys grown by liquid phase epitaxy are shown as a function of Ge content [3]. The intensity, shape, and exact position of the main peaks corresponding to Si-Si, Si-Ge, and Ge-Ge vibrations can be analysed in terms of chemical composition, macroscopic strain, wrong bonding, etc. Similar investigations have also been helpful in amorphous Si-Ge, Si-C, or Si-N alloys [4,5].

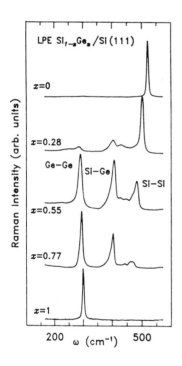

Fig. 2:
First order Raman spectra of crystalline Si-Ge alloys grown by LPE on Si(111) substrates as a function of Ge-content, x [3].

In addition, Raman scattering can be used to investigate bonding properties and diffusion of impurity atoms, e.g. boron or hydrogen in silicon, carbon or silicon in GaAs, etc. In these examples, the light impurity atoms in a host lattice consisting of heavier matrix atoms give rise to characteristic local vibrational modes which can be detected as discrete peaks in the Raman spectra. Fig. 3 demonstrates this for the case of deuterium diffused into a boron-doped crystalline Si sample during a D_2-plasma treatment at 170°C. Under these conditions, atomic hydrogen or deuterium is known to form specific B-H (B-D) complexes which give rise to an almost complete passivation of the boron acceptor levels [6]. In addition, excess H or D is incorporated in various Si-H(D) bonding configurations. Similar passivation steps are also used to increase the conversion efficiency of polycrystalline Si solar cells. The deuterium-related stretch-vibrations appear as a number of Raman peaks with shifts between 1300 and 1700 cm^{-1} from the laser line. The boron-related local vibrations appear at 620 cm^{-1} (^{11}B) and 643 cm^{-1} (^{10}B) close to the first-order Raman peak of crystalline Si. The typical detection limit for such local modes in Si under excitation in the green (e.g. Ar$^+$ laser, 514.5 nm) is a concentration of about 10^{18} cm^{-3} over the probe depth of 500 nm, corresponding to an areal density of impurities of about a tenth of a monolayer. Using different laser wavelengths, one can also vary the Raman probe depth ($d \approx 1/2 \, \alpha$, where α is the absorption coefficient at the laser energy) and thus obtain information about depth profiles of impurities.

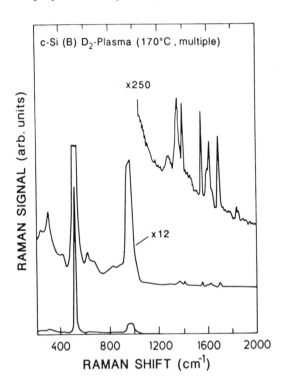

Fig. 3:
Raman spectra of boron-doped crystalline Si after passivation in a deuterium plasma. The lowest trace shows the first order Raman peak of the silicon lattice with a shift of 520 cm^{-1}. Amplification of the spectrum by an order of magnitude reveals the boron local modes at 620 cm^{-1} and 643 cm^{-1}, and in addition features due to scattering by two lattice phonons. The upper trace shows the deuterium stretch vibrations between 1350 cm^{-1} and 1700 cm^{-1} from the laser line.

Photothermal deflection. Photothermal deflection spectroscopy (PDS) is a very versatile technique for the determination of small absorption coefficients due to defects in semiconductors or at surfaces or interfaces [7]. The principle of the technique is outlined in Fig. 4. Monochromatized and chopped light from a strong light source is focused on the sample, which is surrounded by a transparent, non absorbing liquid (deflection medium). At room temperature, most of the light absorbed by the sample is converted into heat via nonradiative recombination. The resulting temperature rise creates a gradient of the refractive index in the deflection medium close to the illuminated sample surface. The focused beam of a HeNe-laser propagating through this refractive index gradient very close and parallel to the sample surface is slightly deflected from its path, and the deflection is monitored by a differential position detector. In the small deflection limit, the output voltage of the position detector is proportional to the amount of heat generated in the sample and, thus, to the absorptance.

Using high power Xe-lamps or tungsten lamps, optimized PDS set-ups can measure absorptances as low as 10^{-5}, which corresponds to an absorption coefficient of 10^{-1} cm^{-1} in a 1μm thick film. The sensitivity limit is mainly determined by the amplitude and pointing stability of the HeNe deflection laser, so that special care has to be taken in the selection of this critical component.

PDS has mainly been used for the investigation of absorption due to midgap defects and band tails in amorphous Si thin films and related materials, but it can also provide useful information about other semiconductors. In Fig. 5, we compare PDS spectra of different materials which are at present used or considered for solar cells.

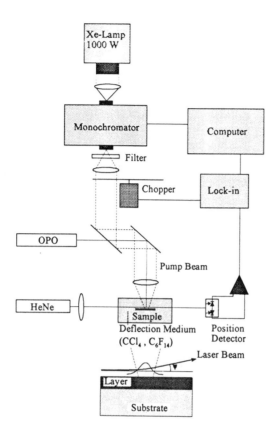

Fig. 4:
Basic outline of a photo-thermal deflection spectro-meter. See text for details.

Typical for PDS spectra in semiconductors with a direct bandgap or in thick, bulk samples is a saturation of the signal when all of the incident pump light is absorbed ($\alpha d \geq 1$). In $CuInSe_2$, GaAs, and $FeSi_2$, this occurs close to the bandgap, followed by a more or less rapid decay of the absorption coefficient in the subgap region. Among the different materials in Fig. 5, $FeSi_2$ shows the largest subgap absorption, probably due to small concentrations of segregated metallic phases [8]. Subgap absorption in bulk, semi-insulating GaAs is due to intrinsic defects such as EL2, but details still need to be investigated. The same is true for subgap absorption in $CuInSe_2$, which depends strongly on the exact preparation conditions. The PDS spectrum of the silicon-on-sapphire film in Fig. 5 corresponds to the weak indirect optical transitions of bulk Si without a noticeable contribution due to defects.

A promising new development in photothermal deflection spectroscopy is the use of an OPO instead of high-power lamps as the pump light source (cf. Fig. 4). OPOs provide much higher intensities and better spectral resolutions than monochromatized lamps, and first measurements show an improvement of the deflection signal by two to three orders of magnitude. This very high sensitivity together with the excellent spectral resolution will certainly open up new applications for PDS in the near future.

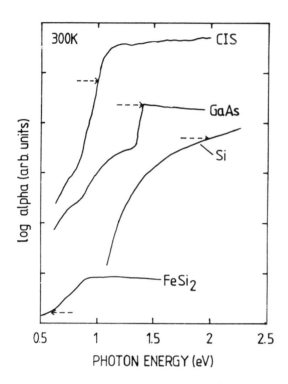

Fig. 5:
Photothermal deflection spectra of CuInSe$_2$ (CIS, d = 1 µm), bulk GaAs (d = 300 µm), silicon on sapphire (d= 0.6µm), and FeSi$_2$ (d= 0.5 µm). The absorption coefficient is shown on a logarithmic scale. Dashed horizontal arrows indicate a value of α = 10^3 cm^{-1}.

Laser scanning techniques. We mention in passing the large class of scanning techniques which also make use of lasers because of their pointing stability and low beam divergence. Thus, scanning force microscopy (SFM) often employs a detection system very similar to that used in PDS (focused laser beam and position detector) to record very small movements of an elastic cantilever. Similarly, lasers are used in near-field optical microscopes (NFOM) or confocal imaging. Closer to applications in photovoltaics are laser-based wafer probing techniques such as laser beam induced current (LBIC), where a laser beam is focused onto a solar cell and then scanned across the entire area of the cell while recording for example the short circuit current of the cell. In this way, lateral homogeneity of cell performance can be tested. An example of LBIC testing of a 10cm x 10cm a-Si:H solar cell is shown in Fig 6. Other variants of this technique are micro-Raman or laterally resolved luminescence spectroscopy. Using pulsed or tuneable lasers, laser scanning techniques in addition can provide spatially resolved information about carrier dynamics (see below) or depth profiles.

Short pulses and carrier dynamics. A major advantage of lasers compared to conventional light sources is the ability to produce short, intense light pulses. Pulsed lasers are widely used to investigate the dynamical properties of charge carriers in semicondcutors via pump and probe experiments. A typical example of such an experiment is photoinduced absorption (PA). Here, carriers in a semiconductor are excited by a single intense pulse, and the resulting change in absorption at the same or a lower photon energy is probed by a time-delayed weaker pulse. By varying the delay time and the energy of the probe pulse, the thermalization of the excited carriers back into the groundstate can be monitored over a large time scale ranging from femtoseconds to quasi-steady state [9]. Also PA experiments have profited a lot from new developments in laser physics, e.g. by the use of optical parametric oscillators [10].

A different technique using pulsed lasers for the investigation of carriers in semiconductors and devices is the time-of-flight method (TOF). The schematic experimental set-up is shown in Fig. 7.

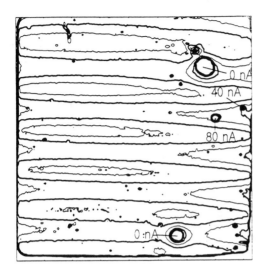

Fig. 6:
Laser beam induced current image of a 10cm x 10cm amorphous silicon solar cell after intentional damaging by a strong reverse bias. Four shunted areas are clearly visible (Courtesy of ISI-PV, Jülich).

The sample under investigation (e.g. a p-i-n diode or a Schottky diode) is put into reverse bias by a suitable voltage pulse. After a short delay, the voltage source also triggers a N_2 laser, which emits a short UV-pulse (337 nm, 3 ns). If desired, the wavelength of the pulse can be changed with the help of a dye laser. The emitted pulse then passes a beam splitter. One part of the pulse is used to trigger a storage oscilloscope via a fast pin-photodetector.The main pulse is directed onto the sample, where electron-hole pairs are created close to the illuminated contact. Depending on the polarity of the applied field, either electrons or holes then drift across the sample and cause a corresponding voltage drop across the 50 Ω resistor. The result of the experiment is a characteristic current transient which is determined by the courier density the drift mobility of the carriers and the total electric field. The arrival of the center of charge at the back electrode causes a drastic drop in the displacement current from which the mobility of carriers can be calculated:

$$\mu_D = \frac{d}{t_T\left(F_{int} + F_{ext}\right)} \tag{1}$$

Here t_T is the transit time of the carriers obtained from the TOF transient, d is the sample thickness, μ_D the electron or hole drift mobility, and F_{int} (F_{ext}) the built-in (externally applied) electric field. For $F_{ext} \gg F_{int}$ the internal field can be neglected and the movement of carriers is entirely dominated by F_{ext}.

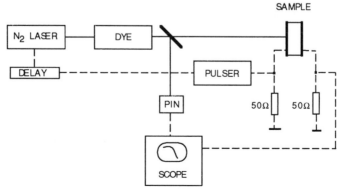

Fig. 7: Schematic set-up of a time-of-flight experiment. Details are described in the text.

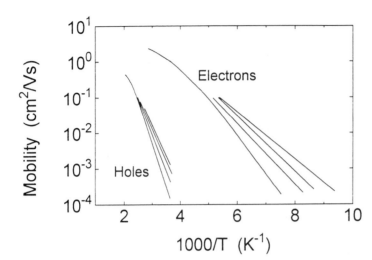

Fig. 8:
Drift mobilities of electrons and holes in undoped a-Si:H as a function of temperature for different applied electric fields.

   TOF has been widely used in the field of amorphous semiconductors, where carrier mobilities are much lower than in crystalline semiconductors. Typical results for amorphous hydrogenated Si are summarized in Fig. 8. The drift mobility is thermally activated at low temperatures due to trapping in band tail states. At a given temperature, the mobility increases more or less pronounced with increasing field. This effect is ascribed to the dispersive nature of transport in a-Si:H [11].
   Holographic techniques. Because of the excellent coherence of laser radiation it is fairly straightforward to create periodic intensity variations on the surface of a sample via two- or four-beam laser interference. As shown in Fig. 9, in the simplest case a laser beam is split into two equal parts which are allowed to interfere under an angle, α, on the sample surface. This gives rise to a periodic modulation of the laser intensity across the sample with the period, p, given by:

$$p = \lambda/\sin(\alpha/2), \qquad (2)$$

where λ is the laser wavelength. Note that according to Eq. (2) the period, p, of the grating can be varied either by changing the laser wavelenght or the interference angle, α. In particular, it is possible to compensate a change in λ by a corresponding change in α, so that the absorption depth of the incoming light and the lateral intensity modulation can be chosen independently from each other, provided that a laser with a sufficiently large tuning range is available.
   Steady state intensity gratings as shown in Fig. 9 have been used successfully for the determination of diffusion lengths of photoexcited carriers in thin semiconducting films (steady state photocarrier grating, SSPG) [12]. To this purpose, a strong bias beam of intensity I_1 interferes with a weaker modulation beam of intensity I_2, thereby creating a small spatially periodic modulation of the total incident intensity on top of a constant bias illumination. Then, the coplanar photoconductivity perpendicular to this intensity grating is measured as a function of the grating period, p, and is compared to the case when both beams are superimposed without interference, e.g. by placing a $\lambda/2$ wave plate into one of the two beams. If the ambipolar diffusion length of photoexcited carriers is larger than the grating period, the diffusion of the carriers will wash out the lateral variation of incident light intensity, and the observed photoconductivities for coherent versus incoherent superposition of the two beams will be equal. On the other hand, when the grating period is larger than the diffusion length, the smaller photoconductivity in the interference minima will result in a smaller coplanar photoconductivity in the presence of an intensity modulation. In this way even the small diffusion lengths in amorphous semiconductors can be determined very accurately.

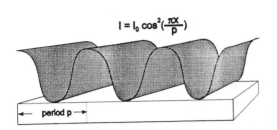

$$I = I_0 \cos^2(\frac{\pi x}{p})$$

Fig. 9:
Coherent superposition of
two laser beams for the
generation of periodic intensity gratings.

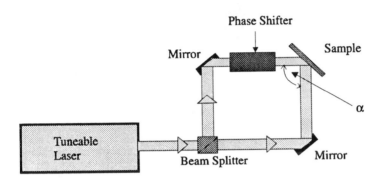

A number of variations of the photocarrier grating method have also been described in the literature. Thus, by adding acousto-optic modulators in one or both beams of the set-up of Fig. 9, it is possible to change the steady state photocarrier grating into a moving grating. The acousto-optic modulator changes the frequency of one beam by a small shift δf with respect to the other beam. As a result, the intensity grating produced by the interfering beams moves across the sample surface with a velocity $v = \delta f \, p$, where p is the grating period given by Eq. (2). In this way, it is possible to obtain grating velocities ranging from μm/s up to several m/s. Using such a moving grating method, it was possible to determine the mobility and the lifetime of electrons or holes in amorphous silicon simultaneously [13]. Other variants include the optical detection of the carrier grating via the diffraction of a third laser beam, or the use of pulsed laser excitation to generate transient gratings, which then can be investigated by a pump and probe technique.

The periodic intensity distribution generated by laser interference is at present also widely used for fast lateral structuring of semiconductors in the sub-μm regime. For example, in holographic lithography a photoresist is exposed to an interference pattern (e.g. of a HeCd laser in the UV), and the resulting line or point grating is then transferred to the semiconductor by conventional lithography. An alternative way, which is of potential interest for applications in thin film solar cells, is the use of a high power pulse laser for lateral structuring via thermal processes. One example of current interest is the creation of crystalline Si wire or dot lattices by interference recrystallization of amorphous silicon. To this end, a thin film of amorphous Si is crystallized by a single shot of two or four interfering beams of a frequency doubled or tripled Nd-YAG laser. The crystalline grating is then developed by preferential etching of the remaining amorphous regions in a hydrogen plasma [14]. A c-Si dot lattice with a period of about 2 μm produced in this manner is depicted in the electron micrograph of Fig. 10. The Si dots have a rather well defined diameter of about 600 nm. For the characterization of such lattices, one can now employ the same laser set-up which was used for the structuring as a convenient excitation source, e.g. for Raman measurements, luminescence, etc. An interesting application for structures such as the one shown in Fig. 10 may be the use as an array of point contacts in thin film solar cells.

Fig. 10:
Electron micrograph of a two-dimensional lattice of crystalline silicon dots produced by laser interference recrystallization of amorphous silicon.

Summary and Outlook

The purpose of this brief review was to provide a qualitative introduction to laser-based characterization methods which can be applied advantageously to materials and devices of current interest for photovoltaic applications. Obviously, it was not possible to mention all of the applications of lasers in modern semiconductor physics, and so I have restricted the discussion to those methods which already have been used successfully for the investigation of solar cells and photovoltaic materials. A general comment is that some of the methods described here are widely used for the study of one material system, but are little known in other areas. Here a lot remains to be done to transfer existing knowledge and expertise to other communities. Finally, it is clear that recent developments in laser technology will open up new opportunities and possibilities for the characterization of semiconductors in the near future.

Acknowledgements

The author would like to thank R. Carius, C. Nebel, W. Rieger, and N. Reinacher for discussions and their help in preparing this manuscript.

References

[1] C. L. Tang, W. R. Rosenberg, T. Ukachi, R. J. Lane, and L. K. Cheng, Proc. IEEE **80**, 365 (1992).

[2] I. H. Campbell, and P. M. Fauchet, Solid State Commun. **58**, 739 (1986).

[3] M. I. Alonso and K. Winer, Phys. Rev. B **39**, 10056 (1989).

[4] M. B. Schubert and G. H. Bauer, Phil. Mag. B **62**, 59 (1990).

[5] A. Morimoto, S. Oozora, M. Kumeda, and T. Shimizu, Solid State Commun. **47**, 773 (1993).

[6] C. P. Herrero, M. Stutzmann, and A. Breitschwerdt, Phys. Rev. B **43**, 1555 (1991).

[7] W. B. Jackson, N. M. Amer, A. C. Boccara, and D. Fournier, Appl. Opt. **20**, 1333 (1981).

[8] C. A. Dimitriadis, J. H. Werner, S. Logothetidis, M. Stutzmann, J. Weber, and R. Nesper, J. Appl. Phys. **68**, 1726 (1990).

[9] J. Tauc, and Z. Vardeny, in *Critical Reviews in Solid State and Material Sciences* **16** (CRC Press, 1990), p. 403.

[10] J. A. Moon and J. Tauc, J. Non-Cryst. Solids **164 - 166**, 885 (1993).

[11] C. E. Nebel and R. A Street, Int. J. Modern Physics B **7**, 1207 (1993).

[12] D. Ritter, E. Zeldov, and K. Weiser, Appl. Phys. Lett. **49**, 791 (1986).

[13] U. Haken, M. Hundhausen, and L. Ley, J. Non-Cryst. Solids **164-166**, 497 (1993).

[14] M. Heintze, P.V. Santos, C.E. Nebel, and M. Stutzmann, J. Appl. Phys., in press (1994).

Materials Science Forum Vols. 173-174 (1995) pp. 153-158
© 1995 Trans Tech Publications, Switzerland

SILICON SURFACE NONLINEAR OPTICS

C. Jordan[1], G. Marowsky[1], R. Buhleier[2], G. Lüpke[2], E.J. Canto-Said[1], Z. Gogolak[2] and J. Kuhl[2]

[1] Laser-Laboratorium Göttingen, Hans-Adolf-Krebs-Weg 1, D-37077 Göttingen, Germany

[2] Max-Planck-Institut für Festkörperforschung, Heisenbergstrasse 1, D-70569 Stuttgart, Germany

Keywords: Nonlinear Optics, Surface Second Harmonic Generation, Degenerate Four Wave Mixing, Silicon Symmetry Analysis

Abstract. By use of second order and third order nonlinear processes the crystalline geometry and topography of silicon surface and bulk were investigated. Wavelength dependence studies of surface second harmonic generation showed phase transitions and resonance effects of the nonlinear response when absorption exists at fundamental or harmonic wavelengths. Symmetries not resolvable by second harmonic generation were investigated by use of a third order nonlinear effect in a degenerate four wave mixing arrangement.

Introduction

Nonlinear optical techniques are widely applicable to the diagnostics of topological and structural properties of surfaces [1]. The tensor characteristic of the nonlinear response offers insight in the geometrical structure of surfaces, interfaces and even bulk material. A strong correlation between the order of nonlinearity and the resolvable order of symmetry of the investigated material limits the applicability of second harmonic generation (SHG). By use of a third order process in a degenerate four wave mixing (DFWM) arrangement we could get around these limitations.

By surface SHG we investigated the crystalline symmetry of (111) oriented silicon (Si(111)) and also the spectral dependence of the nonlinear response in the region of bulk absorption at harmonic wavelength [3]. Due to absorption electronic excitation by high intense laser irradiation was detected in the nonlinear response of the surface [4]. Investigating the nonlinear response by DFWM we observed the symmetry of a (100) oriented silicon surface (Si(100)) [7] which is not resolvable by SHG.

Nonlinear Optics in Surface and Bulk

The dipolar nonlinear polarization P^{NL} responsible for the generation of higher order harmonics can be written as a power expansion in the fundamental electric field amplitude:

$$\vec{P}^{NL}(2\omega, 3\omega, ...) = \vec{\chi}^{(2)} : \vec{E}(\omega)\vec{E}(\omega) + \vec{\chi}^{(3)} : \vec{E}(\omega)\vec{E}(\omega)\vec{E}(\omega) + ... , \qquad (1)$$

where $E(\omega)$ are the electric fields in the polarization sheet and $\chi^{(n)}$ denotes the n-th order nonlinear electric surface/bulk susceptibility tensor displaying the symmetry of the investigated material. An ideal Si(111) surface displays C_{3v} symmetry and a Si(100) surface C_{4v} symmetry, respectively. The properties of the $\chi^{(n)}$ tensors are defined by the corresponding symmetry class of the crystal. Of particular interest for the analysis is the ratio between the anisotropic and isotropic components of the nonlinear response.

In dipole approximation for centrosymmetric media like silicon SHG is allowed only at the surface where the inversion symmetry is broken. However, third order effects like DFWM are allowed both in the bulk and at the surface. Due to the longer interaction length in the bulk third order effects are dominated by the bulk nonlinear response while second order effects are dominated by the surface response.

Surface Nonlinear Optics in the Frequency Domain

The experimental setup used in the wavelength dependence study of the nonlinear anisotropy is shown in Fig. 1.:

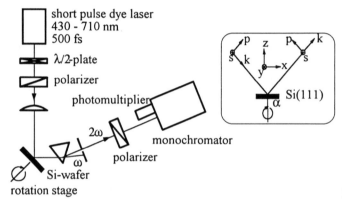

Fig. 1 Experimental setup

The polarized output of a fs dye-laser [2] was incident on the Si(111) sample at 45° by a focusing lens. The sample was rotated about its surface normal and the reflected second harmonic (SH) generated was detected at different polarization states of the fundamental and harmonic signals as a function of the rotation angle α. The polarization states were set in the pks-coordinates defined in the inset of Fig. 1. The SH rotational anisotropy was investigated for different incident wavelengths.

Surface Absorption and Phase Effects

Since the symmetry analysis in the case of no absorption is well established we demonstrate a wavelength dependence study. First we show for comparison the rotational anisotropy of Si(111) as a function of α at two polarization states without absorption at 1064 nm:

$$I_{sp}(\alpha) \qquad I_{ss}(\alpha)$$

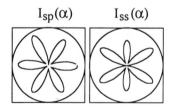

Fig. 2 Rotational anisotropy of the SH signal for different polarization states

The subscripts of the harmonic intensities I displayed in Fig. 2 indicate the polarization state of the fundamental and generated harmonic beams in the pks-coordinates, e.g. sp mean s-polarized excitation and p-polarized detection.

The pure 6-fold intensity rotational pattern (3-fold in the electric field) displays the 3-fold crystal symmetry. In the case of absorption at the harmonic wavelength we obtained the results displayed in Fig. 3 (fundamental wavelength is shown):

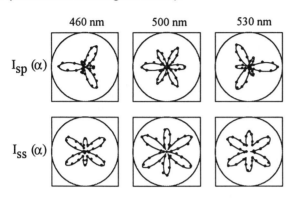

Fig. 3 Rotational anisotropy of the SH at various wavelengths

For analysis the rotation patterns were Fourier analyzed according to:

$$I(\alpha) = \sum_k \left[a_k \cos(k\alpha) + b_k \sin(k\alpha) \right],\qquad(2)$$

where b_k equals 0 from theory. The phase transition in the rotational symmetry of the I_{sp} measurement could be attributed to changes in the ratio of the isotropic and anisotropic nonlinear responses [3] and is demonstrated by the wavelength dependence of the newly defined parameter η:

$$\eta(\lambda) = \frac{a_3(\lambda)}{a_6(\lambda)} = \frac{\text{isotropy}}{\text{anisotropy}} \ .\qquad(3)$$

The experimental data of $\eta(\lambda)$ are shown in Fig. 4 as a function of the fundamental wavelength

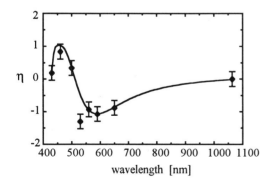

wavelength [nm]

Fig. 4 Ratio of $\eta(\lambda)$ at various wavelengths. Error bars of 20 % are shown

together with a fit function given by the theory. The theory of SHG including absorption effects provides an expression for wavelength dependent η as follows:

$$\eta_{\text{theo}}(\lambda) \ \propto \ \cos(\varphi_{\text{surface}}(\lambda)) ,$$

where $\varphi_{surface}$ equals the phase of a surface susceptibility component that occurs due to the existence of absorption. From comparison with the anharmonic oscillator model the experimental phase transition could be interpreted as resonant behavior [3]. The surface electronic resonance is shown to coincide nearly with the bulk resonance.

Surface Absorption and Electronic Excitation

With p-polarized excitation we measured in p-polarized detection (I_{pp}: pks-coordinates as defined in Fig. 1) a qualitatively different wavelength dependent rotational symmetry of Si(111). In Fig. 5 the rotational anisotropy is plotted in case of no absorption (1064 nm at fundamental) and with increasing absorption towards shorter wavelengths:

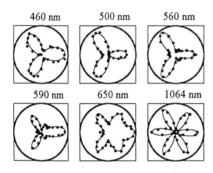

Fig. 5 Rotational nonlinear anisotropy of Si(111) in pp polarization state at various wavelengths

A symmetry transition from pure 6-fold to mostly 3-fold is apparent.

Analyzing the data again by calculating the ratio $\eta(\lambda)$ of isotropic and anisotropic responses we obtain the result displayed in Fig. 6.

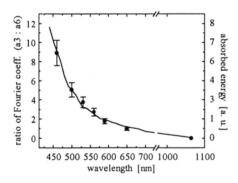

Fig. 6 Experimental wavelength dependence of $\eta(\lambda)$ and absorbed laser energy

An exponential dependence of $\eta(\lambda)$ with wavelength could be fitted to the laser energy absorbed in a thin slab of silicon responsible for surface SHG [4] and was explained by an arising additional isotropic nonlinear response. The absorbed laser energy is related to an electronic excitation from valence to conduction band and is proportional to the generated free electron density.

The measured behavior could be well explained by a jellium model [5] used for describing the nonlinear response of a free electron plasma sheet: the exclusive appearance in the pp-polarization state and its proportionality with the electron density.

Bulk Silicon(100) Nonlinear Optics in the Time Domain

Since the second order nonlinear response provides no structural information of the Si(100) as demonstrated in Fig. 7,

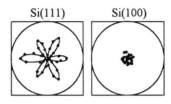

Fig. 7 SH rotational anisotropy
of Si(111) and Si(100)

we investigated the third order nonlinear response with the widely used DFWM technique [6]. The experimental setup is shown in Fig. 8.

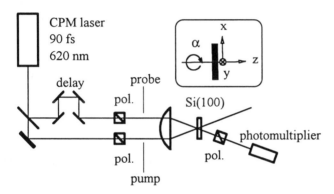

Fig. 8 The experimental setup
of the DFWM arrangement

A short pulse-dye laser was split in a probe and a pump beam of parallel polarization. Both were focused in a 10 μm thick Si(100) sample mounted on a rotation stage and the DFWM signal was detected after a polarizer that was set perpendicular to the fundamental beams polarization.

The DFWM signal was measured in transmission as function of the rotation angle α. In addition the DFWM signal was investigated dependent on deviations from the orthogonal polarization alignment between the fundamental and the DFWM signal and also dependent on a time delay introduced in the probe beam path [7].

In Fig. 9 the DFWM intensity is plotted as a function of the rotation angle α. The 8-fold intensity rotation pattern indicates the 4-fold symmetry of the Si(100) surface.

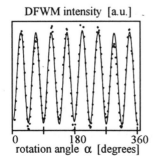

Fig. 9 Rotational anisotropy of the DFWM response

In Fig. 10 the dependence of the DFWM signal as a function of the rotation angle α is demonstrated while the polarizer of the DFWM signal is rotated away from the orthogonal state and also a time delay is introduced between the pump and probe beam.

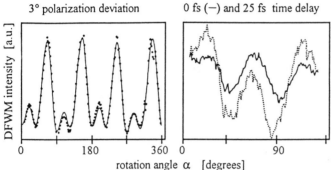

Fig. 10 Rotational symmetry induced by polarization and temporal mismatch

Nearly identical qualitative rotational anisotropy changes from the pure 6-fold symmetry are demonstrated in Fig. 10 but the physical origin is still unknown. Relaxation phenomena changing the ratio of isotropic and anisotropic response on a time scale less than 25 fs provide a reasonable explanation [7].

Summary

We have presented experimental results for the symmetry analysis of silicon crystals using surface SHG in reflection and DFWM in transmission. A wavelength dependence study of surface SHG on Si(111) showed the influence of absorption on the anisotropy of the nonlinear response. Phase transitions in the rotational anisotropy indicated resonant behavior of a surface $\chi^{(2)}$-component near the bulk resonance. Due to absorption free electrons were generated in the silicon surface and were detected by an additional isotropic component in the SH rotational anisotropy.

Since the 4-fold symmetry of Si(100) is not resolvable by surface SHG we analyzed the crystal symmetry using the third order nonlinear effect in a DFWM arrangement. Identical rotational symmetry changes in the nonlinear response caused by polarization misorientation and temporal mismatch were demonstrated. Relaxation phenomena were identified with temporal resolution of less than 25 fs in the rotational anisotropy of silicon.

References

[1] Y. R. Shen: In *New Laser and Optical Techniques*, ed. by R. B. Hall and A. B. Ellis (VCH, Deerfield Beach, FL, 1986), p. 151

[2] S. Szatmári and F. P. Schäfer, Appl. Phys. B46, 305 (1988)

[3] C. Jordan, E. J. Canto-Said and G. Marowsky, Appl. Phys. B58, 111 (1994)

[4] C. Jordan, G. Marowsky, U. Emmerichs and C. Meyer, submitted to Opt. Comm.

[5] J. Rudnick and E. A. Stern, Phys. Rev. B4, 4274 (1971)

[6] D. S. Chemla, D. A. B. Miller, P. W. Smith, A. C. Gossard and W. Wiegmann, IEEE J. Quantum Electron. QE-20, 265 (1984)

[7] R. Buhleier, G. Lüpke, G. Marowsky, Z. Gogolak and J. Kuhl, accepted for publication in Phys. Rev. Lett.

Materials Science Forum Vols. 173-174 (1995) pp. 159-164
© 1995 Trans Tech Publications, Switzerland

MAPPING OF DEFECT-RELATED SILICON PROPERTIES WITH THE ELYMAT TECHNIQUE IN THREE DIMENSIONS

J. Carstensen, W. Lippik, H. Föll

University of Kiel, Faculty of Engineering, Kaiserstr. 2, D-24143 Kiel, Germany

Keywords: Diffusion Length Mapping, Surface Recombination Velocity, Depth Dependence of Diffusion Length

Abstract. The ELYMAT technique uses silicon electrolyte junctions as a Schottky-like contact for measuring photocurrents. It allows to map diffusion lengths as well as surface defects. New modes have been obtained by applying Lasers with different penetration depths and combining these measurements. This allows to extract information about the depth dependent bulk diffusion length as well as the recombination velocities of front and back surface.

Introduction

The ELYMAT (short for Electrolytical Metal Tracer) technique has been developed as an in-line tool for manufacturers of Si wafers *and* for wafer fabs to allow a fully automated mapping of the minority carrier diffusion length (or life time) with a sufficient lateral resolution in a reasonable time [1–3]. The main drive behind this approach was the need to detect a contamination of the wafer with metals (life time killers) at a very low level (ppb - ppt range), and, if possible, to identify the contamination source by its "finger print" in the life time picture. This appraoch was not without success, but competing methods, especially life time mapping with microwave absorbtion [4] or extensions of the surface photo-voltage methods [5] yield similar, albeit less directly interpretable, life time maps and do not require electrochemical knowledge and methods.

The strength of the ELYMAT technique lies in its possible extensions that go well beyond the relatively simple task of mapping diffusion lengths in a convenient range. In the field, more uses than just the visualisation of the life time distribution are emerging, cf. e.g., [6–8]. Monitoring the leakage currents of the Si-electrolyte junction or measuring very small variations of induced photo currents may provide valuable clues to manufacturing processes drifting out of specification [6], whereas measurements of the diffusion length as a function of injection levels helps to identify the chemical nature of the contamination.

New measurement modes are under development, either because new hardware becomes available (e.g. a suitable Laser with wavelength of 1040 nm and a respective penetration depth of 500 μm), or because certain properties of the Si-electrolyte contact are better understood. In addition, the combination of several independent measurements allow to obtain maps of the surface quality (usually displayed as surface recombination velocity S) as well as of the depth dependence of the diffusion length L and thus allows the three dimensional characterization of a Si wafer in unprecedented detail.

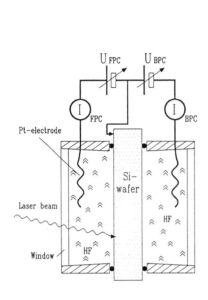

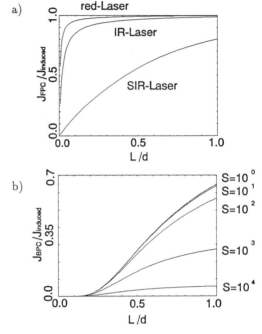

Fig. 1: The electrolytical double cell of the ELYMAT. The two Si-electrolyte junctions can be biased independently; electrolytes other than HF are possible

Fig. 2: a) calculated frontside photocurrent (FPC) for different penetration depth (red Laser: 4 μm, IR-Laser: 13 μm, SIR-Laser: 500 μm); b) calculated backside photocurrent (BPC) for IR-Laser and different values of front surface recombination velocities S [cm/s].

Basic Set-up and Standard Modes

Fig. 1 shows a simplified cross-section through the heart of the ELYMAT, the electrolytical double cell. The Si wafer is in contact with a suitable electrolyte on front- and backside. The wafer as well as the electrolytes are electrically contacted by needles around the perimeter of the wafer, or Pt electrodes, respectively. Both Si-electrolyte junctions can be biased independently, the currents flowing on the frontside or backside of the wafer are the measured quantities in most cases. Since the junction behaves very much like a good Schottky contact if diluted HF is used as an electrolyte, proper biasing essentially allows to use a junction as a solar cell. The peculiarities of the Si electrolyte junction are described in [9] and have to be kept in mind, but are of no particular interest for the standard modes except for the convenient feature of the contact between (clean) Si and HF: Its interface recombination velocity is rather small and can be mostly neglected [10].

If minority carriers are locally generated by a Laser beam, they can be collected at the frontside of the wafer (frontside always defined as the illuminated surface) or at the backside. Front- and backside photo current measurements (FPC and BPC mode) as a function of the position of the Laser beam provide the basic modes of the ELYMAT. If S can be neglected, the current measured in both modes contains only the diffusion length as unknown quantity.

Fig. 2 shows the calculated dependence of the photo currents on the diffusion length (scaled to fractions of the wafer thickness d) with the parameters surface recombination velocity S_F at the frontside for BPC mode and penetration depth of the Laser beam for the FPC mode, respectivly. It can be seen that for a red Laser (wavelength $= 640\ nm$, penetration depth $= 4\ \mu m$) the FPC mode is insensitive to large diffusion lengths whereas the BPC mode fails for small L values. The interesting intermediate region of L may be hard to measure in this case. A Laser with a penetration depth of 13 μm is advantageous in this case, even better is a penetration depth of 500 μm provided by a 1040 nm Laser. The red Laser, however, has some merits in the FPC case, too. The current in this case is particular sensitive to defects in the space charge region and may pick up metal precipitations with extremely high sensitivity [6].

New Measurement Modes and Evaluation Techniques

Using Lasers with markedly different penetration depths allows independent measurements with significantly different depth distributions of minority carriers. The recombination behavior of a minority carrier depends on its depth position if the recombination at the surface can not be neglected or if the depth distribution of recombination centers is not uniform. Thus the currents measured will contain information about the surface recombination velocity on the front- and backside and about the depth dependence of the diffusion length ($L(z)$). Depending on the penetration depth of the light and the mode chosen, the sensitivity of the measured currents to these parameters is quite different. A FPC measurement with a red Laser, e.g., is practically totally insensitive to recombinations at the backside; this is also true for any BPC measurement since the space charge region belonging to the Schottky-like junction on the back-side prevents surface recombination. A FPC measurement with the deeply penetrating Laser (dubbed SIR-Laser), however, is quite sensitive to the recombination velocity on the backside. Since we obtain up to 12 measurements (3 wavelengths, 2 measuring modes, front- and backside interchanged), the data is used to fit a model that allows for finite surface recombination velo-cities and two different diffusion length within the wafer. For sake of mathematical simplicity, $L(z)$ is approximated by a step function (which may not be a good approximation in some cases); these values as well as the depth of the assumed step are extracted from the raw data by a least square fitting procedure.

Using the regular modes, the data obtained contains some noise as, e.g., fluctuating leakage currents or unknown or fluctuating Laser intensities which adversely influence the computations. This problem is reduced in a new mode unique to the ELYMAT technique: Using the SIR-Laser and biasing *both* sides of the wafer as a minority carrier collecting junction (i.e. a FPC and BPC measurement is taken simultaneously), easily measurable currents are obtained, the relation of which can be shown to be almost independent of the intensity of the Laser and of certain noise contributions. Used as an additional input into the numerical calculation, the results are smoothed to a considerable extent and relatively small variations of S and L can be detected.

Another unique feature of the ELYMAT is the socalled "restricted photo current"-mode (RPC). In contrast to the above discussed modes the frontside current is measured applying a bias which is well below the value for photocurrent saturation. This RPC-photocurrent shows a strong dependence on the diffusion length L and surface recombination velocity S, which is not seen in the standard FPC-mode, but requires extensive theory for interpretation. In addition to the parameters L and S it is a function of parameters like concentration of the electrolyte and doping of the wafer and, making things worse, the relative influence of these parameters on the photocurrent depends strongly on the applied bias.

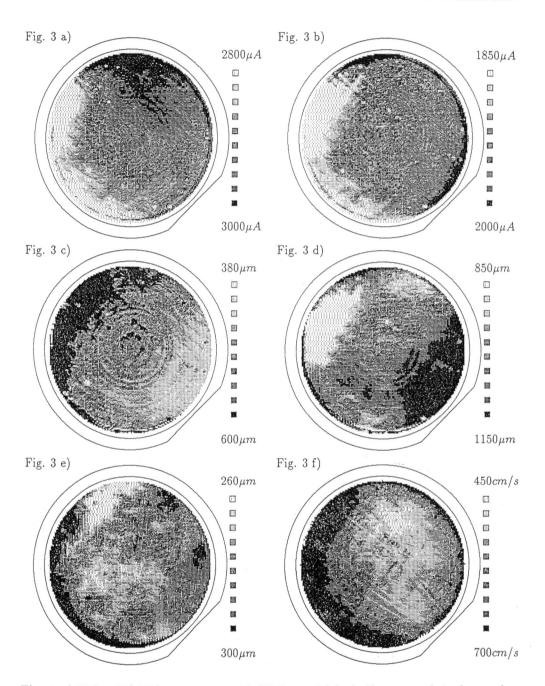

Fig. 3: a) FPC and b) BPC measurement with SIR-Laser with both sides measured simultaneously; c) calculated diffusion length of the front half of the wafer; d) calculated diffusion length of the back half of the wafer; e) calculated depht of the step; f) calculated recombination velocity of the back surface

Fig. 4 a)   Fig. 4 b)

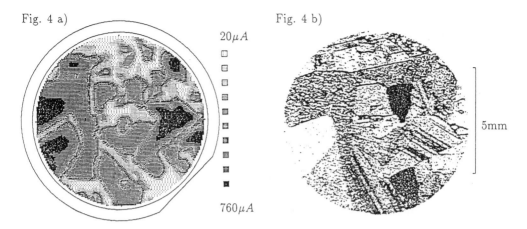

Fig. 4: a) ELYMAT measurement of multicrystalline Si with higher lateral resolusion; b) corresponding structure of the specime obtained by a digital scanner.

A particularily attractive feature of the ELYMAT technique is its extendability to multi-crystalline Si in use for solar cell applications. The minority carrier diffusion length is the prime factor responsible for solar cell yields (and for costs of the material). The ELYMAT not only allows extensive characterisation of L (and S) but offers the possibility of high resolution measurements in lateral directions because the Laser beam can be focussed easily and the electrolyte is fully transparent. Since the measured quantity is a photocurrent, which is also the quantity of prime interest in solar cells, the results are directly and easily interpretable in terms of solar cell performance.

Experimental results

Fig. 3 a) and b) show the photocurrent maps obtained in the new mode where both FPC and BPC currents are measured simultaneously. Surface recombination can be safely neglected in this case; the visible features are predominately bulk. It should be noted that a rare artifact is also observed: The system of several concentric rings in the top half of the wafer of Fig. 3 a) is due to optical reflections. The other measurements which are used for the computations are, for lake of space, not shown.

Fig 3 c) and d) show computed maps of the diffusion length in the front or back region of the wafer; Fig 3 e) the calculated depth of the step where the change-over occures. In this particular case the diffusion length is considerably smaller in the region close to the polished surface, an observation which has been made repeatedly and which may indicate that chemical-mechanical polishing adversely effects diffusion length.

Fig. 3 f) finaly shows the surface recombination velocity map; it clearly denotes defects, probably scratches. The rather large values of S in this case may be due to the fact that this particular wafer was not cleaned and may well carry some (e.g. organic) contamination producing large values of S which are not compensated by the HF.

Fig. 4 a) shows first results of ELYMAT measurements with multicrystalline Si. All modes mentioned so far can be used for characterising this material. It is, however, more difficult to contact the thin ($\approx 300\mu m$) and mechanically "weak" multicrystalline slices with needles.

For lake of space only the possibility of obtaining reasonable high lateral resolution without any special optical means is demonstrated. Fig. 4 b) shows the corresponding structure of the specime simply obtained by a digital scanner. The correlations between electronic surface and bulk properties and crystal defects are clearly visible.

Summary and Conclusion

The ELYMAT principle allows to obtain three-dimensional information about the bulk and surface perfection of a Si wafer in several different modes. It was shown that maps of interesting properties as, e.g. surface recombination velocities or depth dependence of L, can be obtained by relating independent measurements of the well understood standard mode with the help of a complete theory.

An extension of the ELYMAT technique to multicrystalline Si is possible despite of some problems concerning the contacting of the wafer by needles.

Acknowledgements

The authors appreciate help and encouragement from Dr. P. Eichinger and Dr. W. Koch.

References

[1] V. Lehmann and H. Föll. *J. Electrochem. Soc.* **135**, 2831 (1988).

[2] H. Föll, V. Lehmann, G. Zoth, F. Gelsdorf, and B. Göttinger. In *Proc. Symp. Analytical Techniques for Semiconductor Mat. and Process Characterisation*, ed. by B.O. Kolbesen, D.C. McLaughan, and W. Wandervorts, 90-11, p. 44 (Electrocem. Soc., New York 1990).

[3] H. Föll, V. Lehmann, and W. Lippik. *"Characterisation of Single and Polycrystalline Silicon by Extension of the Elymat Technique"*. In *Proceedings of ECS Conference of Crystalline Defects and Contamination*, p. 252, Grenoble (1993).

[4] F. Shimura, T. Okui, and T. Kusama. *J. Appl. Phys.* **67**, 7168 (1990).

[5] L. Jastrzebski. In *Semiconductor Silicon*, ed. by H.R. Huff, K.G. Barraclough, and J.-i. Chikawa, Electrochem. Soc., p. 614 (1990).

[6] W. Bergholz, D. Landsmann, P. Schauenberger, and B. Schoepperl. *"Contamination Monitoring and Control in Device Fabrication"*. In *Proceedings of ECS Conference of Crystalline Defects and Contamination*, p. 69, Grenoble (1993).

[7] R. Falster. *"Low Level Metal Contamination and Wafer Performance"*. In *Productronica Munich, Technical Program*, (1991).

[8] E.E. Fisch. *"Metal Contamination Monitoring in Semiconductor Manufacturing Environment"*. In *Proceedings of ECS Conference of Crystalline Defects and Contamination*, New Orleans (1993).

[9] H. Föll. *Appl. Phys.* **A 53**, 8 (1991).

[10] E. Yablonovitch, D.L Allara, C.C Chang, T. Gmitter, and T.B. Bright. *Phys. Rev. Lett.* **57**, 249 (1986).

Materials Science Forum Vols. 173-174 (1995) pp. 165-170
© 1995 Trans Tech Publications, Switzerland

LASER EXAMINED AlGaAs-SOLAR CELLS WITH WIDE-GAP TUNNELING-THIN CAP LAYERS

V.M. Andreev, V.P. Khvostikov, V.R. Larionov and V.D. Rumyantsev

A.F. Ioffe Physico-Technical Institute, Polytechnicheskaya 26, 194021 St. Petersburg, Russia

Keywords: Solar Cells, Heterostructures, Gallium Arsenide

Abstract. This paper presents the results of a photoresponse (PR) and photoluminescent (PL) investigations of p-$Al_xGa_{1-x}As$/p-GaAs/n-GaAs solar cells with variable ultra-thin (W=2–50 nm) $Al_xGa_{1-x}As$ (x=0.7–0.8) cap (window) layers. He-Ne and Ar lasers were used for a PL and generation. PL and PR measurements give an information about planarity and tunneling properties of wide-gap $Al_xGa_{1-x}As$ layers. A study of the dependence of a PR signal induced by He-Ne laser (hv=1.96 eV) on the thickness W of the wide-gap layer showed, that the internal collection efficiency of photogenerated carriers (Q) remains constant ($Q > 0.95$) down to $W = 6$–7 nm. At $W < 6$ nm Q values decrease rapidly due to a tunneling ejection of photogenerated carriers from the photoactive p-GaAs region to the surface of the cap AlGaAs layer. These investigation allowed us to manufacture solar cells having a concentrator AMO efficiencies of 24–24.7% at 20–100 suns (confirmed at NASA LeRC).

Introduction

Due to development of the low-temperature (400–600 °C) liquid phase epitaxy of multilayer AlGaAs-heterostructures solar cells (SC's) with ultrathin wide gap "windows" have been fabricated [1–4]. The maximum short-wavelength spectral sensitivity and consequently high efficiency of the SC's were achieved due to a decrease in the absorption of the short wavelength radiation in the wide gap layer and a decrease in the surface recombination losses. Investigations of photoresponse (PR) and photoluminescent (PL) excited by laser beam on these structures have been carried out.

Experimental

We have prepared and investigated solar cells based on two types of heterostructures.

Structures of the n$^+$-GaAs-n-$Al_{0.1}Ga_{0.9}As$- n-GaAs-p-GaAs-p-$Al_xGa_{1-x}As$ (x=0.85–0.9) -p$^+$-GaAs type were grown on n$^+$-type GaAs substrates by low-temperature LPE (Fig.1,a). The wide-gap window layer was crystallized at the temperature about 600 °C when the cooling rate was 0.5 °C×min^{-1}. As a dopant we have used Te to obtain n-type layers and Mg and Ge to obtain p-type layers. The densities of free carriers in the p-n junction region were $n=10^{17}$cm^{-3} and $p=(3$–$8)\times10^{18}$cm^{-3}; in the $Al_xGa_{1-x}As$ layer the free carrier concentration was about 10^{18}cm^{-3} and in the cap heavily doped p$^+$-GaAs layer it was $P^+=10^{19}$cm^{-3}. The distribution of the composition across the thickness of the structures was determined and it was found that if x=0.85–0.9, the p-$Al_xGa_{1-x}As$ and p$^+$-GaAs layers were separated by a transition layer 2–4 nm thick with the AlAs content 20–40 mol.%, which was due to special features of the crystallization process.

The other type of investigated heterostructures (Fig.1,b) consists of n-GaAs-p-GaAs-p-$Al_xGa_{1-x}As$ (x=0.5–0.9). Initially an epitaxial n-type GaAs layer of thickness 5–10μm was grown on an n$^+$-type

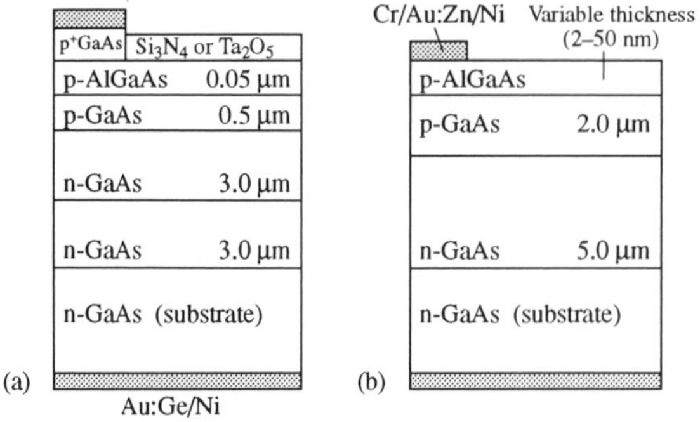

Fig. 1 AlGaAs/GaAs solar cell heterostructures.

GaAs substrate and this was followed by p-type GaAs layer of 2 μm thickness doped with germanium and a p-$Al_xGa_{1-x}As$ layer of variable (2–50 nm) thickness doped with magnesium. In the present study the use of structures (Fig.1,b) with relatively thick (d_p=2μm) p-GaAs layer made it possible to demonstrate more clearly the process of tunnel release of photocarriers from photoactive p-GaAs to the surface of the structure on reduction in the thickness and the band gap energy of the solid solution layer.

The properties of epitaxial structures were investigated using the method of ellipsometry, x-ray Auger photoelectron spectroscopy and precision anodic oxidation as well as Raman scattering method [5–7]. The thicknesses of the tunneling-thin layers were estimated also by a method of stepwise anodic oxidation accompanied by measurements of the photoresponse (PR) and photoluminescence (PL), excited by He-Ne and Ar+lasers. We first established the ratio of the thickness of $Al_xGa_{1-x}As$ which was etched away by anodic oxidation, which was 1–1.5 nm/V in our experiments. In determining this ratio we used several procedures for measuring the thickness of ultrathin layers, including Auger profilometry and Raman scattering method.

The photoluminescense map of samples was induced by a defocused Ar+laser beam (hv=2.4 eV) and was observed via an infrared camera. The PL map had a detectable difference in the p-GaAs PL intensity for a GaAs, covered by as thin as 2–3 nm AlGaAs layer. Therefore the substrate treatment and the epitaxial regime can be chosen from the point of view of a AlGaAs layer planarity.

Results

Fig.2 shows the photoresponse spectra of structures with the wide-gap layers of thicknesses 20 nm (Fig.2,a) and 5 nm (Fig.2,b) in which AlAs content ranged from $x = 0.9$ to $x = 0.5$. Attainment of the value $Q_{i\,max}$ >0.95 in the spectral region 0.5–0.8 μm (Fig.2,a, curves 1,2) demonstrated a practically complete collection of carriers in a structure with wide gap $Al_xGa_{1-x}As$ ($x = 0.7–0.9$) layer of thickness 20 nm. It is clear from curves 3 and 4 (Fig.2a and 2b) that reduction in AlAs contents in $Al_xGa_{1-x}As$ layers in the range $x < 0.5$ resulted in reduction in the value of Q_i both in the transparency range of the wide gap window and in the short-wavelength part of the spectrum.

A more detailed analysis of the photoresponse spectra was made by plotting (Fig.3) the dependencies of the collection efficiency Q_i on the thickness W of solid solution layers of different compositions at fixed value of the exciting radiation wavelength: $\lambda_{exc} = 0.4$ μm. It is clear from curves 2 and 3 in Fig.3 that reduction in thickness W from 30 nm to 5–7 nm increased Q at λ=0.4 μm and the rate of increase was greater for the $Al_xGa_{1-x}As$ layer with $x = 0.5$ (curve 3 in Fig.3), which was due to a higher rate of reduction of the absorption losses on reduction in W in layers with a smaller width of the band gap. When the solid-solution layer was characterized by $x = 0.9$ (curve 1 in Fig.3) a reduction in the thickness W in the range 30–20 nm increases slightly Q_i. Then, between 20 and 5–7 nm the value Q_i remained at the maximum level due to the practically complete transparency of the $Al_{0.9}Ga_{0.1}As$

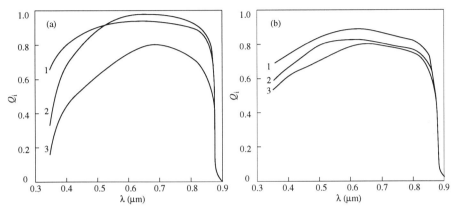

Fig. 2 Spectral dependences of the collection efficiency Q_i of SC's based on n-GaAs-pGaAs (2 mm)-pAl$_x$Ga$_{1-x}$As structures (Fig.1,b) with the wide-gap layer thickness: $W = 20$ nm(a) and 5 nm (b) AlAs concentration (x): 1–0.9; 2–0.7; 3–0.5. thick wide-gap layers.

layer of thickness $W < 20$ nm to the $\lambda = 0.4$ µm radiation. In the case of samples with $x = 0.8$–0.85 of the wide-gap layer Al$_x$Ga$_{1-x}$As we investigated a more detailed dependence $Q = f(W)$ in the range of thickness $W = 30$–2 nm at exciting radiation wavelength $\lambda_{exc} = 0.63$ µm (He-Ne laser). A constant high value of $Q_i >0.95$ was exhibited by these samples right down to thickness $W = 6$–7 nm (Fig.4). The fall of Q_i in the range $W < 5$–6 nm was due to an increase in the probability of tunneling of photocarriers from p-type GaAs to the surface of the structure and an increase in the fraction of electrons trapped in potential "well" formed by the surface band bending.

These results made it possible to establish experimentally the dependence of the photocarrier collection efficiency on the thickness ($W = 2$–50 nm) and composition ($x = 0.5$–0.9) of Al$_x$Ga$_{1-x}$As solid solution layers and to find the optimal parameters of these layers ensuring widening of the photosensitivity toward shorter wavelengths while maintaining the maximum value of Q_i in the transparency band of the solid solution.

The first type of heterostructures (Fig.1,a) with thickness of window layer about 30–50 nm was used for fabrication of 0.074 cm^2 total illuminatted area (3.07 mm diameter) concentrator solar cells. Highly-doped p$^+$-GaAs front layer in this structure provides low resistance contact preparation. This

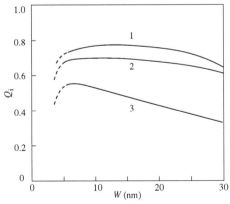

Fig. 3 Dependences of the collection efficiency Q_i of the thickness of wide-gap layers obtaind using exciting radiation of wavelength $\lambda_{exc} = 0.4$ µm. AlAs concentration (x):1–0.9; 2–0.7; 3–0.5.

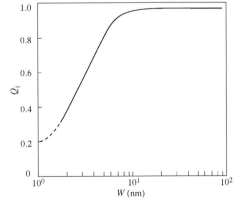

Fig. 4 Dependence of the internal collection efficiency Q_i on the thickness of the wide-gap Al$_x$Ga$_{1-x}$As (x=0.8–0.85) layers at the excitation wavelength $\lambda_{exc} = 0.63$ µm.

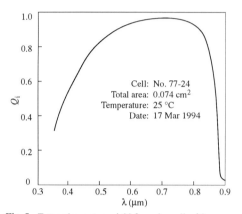

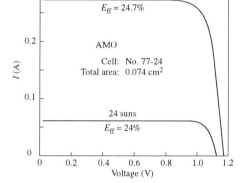

Fig. 5 External quantum yield for solar cell with parameters shown on Fig. 6 and Fig. 7.

Fig. 6 Current vs voltage for an experimental GaAs-concentrator space cell No. 77–24 with prismatic cover under AMO, 25 °C conditions (measurements at NASA LeRC).

p+-GaAs layer is selectively removed between the grid lines prior the preparation of antireflection coating. Ta_2O_5 layers was deposited for preparation of antireflection coating. Silicone prismatic cover on the solar cell surface was used to reduce the contact shadowing. External quantum efficiency curve of one of these cells is shown on Fig.5.

Fig.6, Table 1 and Fig.7 show the parameters of the solar cell with silicone prismatic cover. I_{sc} calculated on total designated illuminated area I_{sc}=34.7 mA×cm^{-2} was measured under AMO simulator. The efficiency value is 24% at 24 suns and 24.7% at 103 suns (AMO, 25°C). We believe that achieved efficiency is among of the best published for single junction concentrator solar cells.

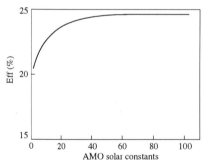

Fig. 7 Solar energy conversion efficiency versus concentration ratio (AMO, 25 °C) for solar cell No. 77–24 with prismatic cover.

Table 1

Parameter	24 Suns	103 Suns
I_{SC} [mA]	64	272.6
V_{OC} [mV]	112	1161
I_{max} [mA]	60.7	257
V_{max} [mV]	967	1004
P_{max} [mW]	58.7	258
FF	0.82	0.815
Eff. [%]	24	24.7

Summary

The investigation made it possible to find experimentally the dependencies of quantum yield on the thickness (W = 2–50 nm) and composition (x = 0.5–0.9) of $Al_xGa_{1-x}As$ solid solution layers and to find the optimal parameters of AlGaAs/GaAs heterostructures ensuring widening of the photosensitivity toward shorter wavelengths. Solar cells with efficiency 24–24.7% at 20–100 suns (AMO, 25 °C) have been manufactured on the bases of developed heterostructures.

Acknowledgement

This work has been partly supported by the Department of Defense, USA. Critical solar cell measurements were performed at NASA LeRC (Dr.David J.Brinker and Mr.David A.Scheiman).

References

[1] V. M. Andreev, V. R. Larionov, V. D. Rumyantsev, O. M. Fedorova, Sh. Sh. Shamukhamedov. *Sov. Tech. Phys. Lett.* 9, 537(1983).

[2] V. M. Andreev, V. S. Kalinovskii, O. V. Sulima. *Proc. of 10th European Photovoltaic Solar Energy conference,* 8–12 April 1991, Lisbon, p.52–54.

[3] V. M. Andreev, A. A. Vodnev, V. R. Larionov, V. D. Rumyantsev, K. Ya. Rasulov, V. P. Khvostikov. *Sov. Tech. Phys. Lett.* 14, 623(1988).

[4] V. M. Andreev, V. S. Kalinovskii, M. M. Milanova, A. M. Mintairov, V. D. Rumyantsev, K. E. Smekalin, E. O. Strugova. *Semiconductors* 27, 82(1993).

[5] Y. T. Chering, D. H. Jaw, G. B. Stringfellow. *J. Appl. Phys.* 65, 3285(1989).

[6] R. Fukasawa, M. Wakaki, K. Ohta, H. Okumura. *Jpn. J. Appl. Phys.* 25, 652(1986).

[7] V. M. Andreev, V. R. Larionov, A. M. Mintairov, T. A. Prutskikh, V. D. Rumyantsev, K. E. Smekalin, V. P. Khvostikov. *Sov. Techn. Phys. Lett.* 16, 325(1990).

Materials Science Forum Vols. 173-174 (1995) pp. 171-176

USE OF PHOTOINDUCED MICROWAVE REFLECTION FOR THE NON-DESTRUCTIVE CHARACTERIZATION OF SOLAR CELL MATERIALS AND DEVICE STRUCTURES

J.M. Borrego

Electrical, Computer and Systems Engineering Department, Rensselaer Polytechnic Institute
Troy, New York 12180-3590, USA

Keywords: Microwaves, Non-Destructive, Lifetime, Surface Recombination Velocity

Abstract. Photoinduced microwave reflection using lasers as the source of light excitation has proven to be a powerful non-destructive technique for the characterization of important material and device parameters of solar cells or similar devices. In this paper we describe some applications in which this technique can be used to extract either material or device parameters of solar cells type structures.

Introduction

The important material and device parameters which affect the efficiency of a solar cell are the doping concentration, the mobility, the excess carrier lifetime and the recombination velocity at any of the interfaces of the device. Traditionally doping concentration and mobility have been determined using Hall effect techniques. Measurement of excess carrier lifetime is usually measured by using photoconductivity decay or by measuring the recovery time of a forward bias pn junction. Recently microwave transmission or reflection has been used for the evaluation of doping concentration and mobility [1,2,3]. The addition of pulsed laser light to a microwave reflectance set-up for increasing the conductivity of the sample allows the measurement of the excess carrier lifetime and of the interface recombination velocity [4,5]. The attractiveness of using microwave reflection for determining the above material parameters is that the sample can be outside the waveguide and it is not necessary to provide any ohmic contacts to the sample so the technique can be used to monitor those parameters as the device goes thru several processing steps. It is the purpose of this paper to present results we have obtained in our laboratory by using photoinduced microwave reflection for measuring the lifetime and interface recombination velocity in materials like GaAs and InP which are being used for the fabrication of high efficiency solar cells. The paper starts first with a short review of the microwave reflectance technique and of the measurement set up used. This is followed by the results obtained in our laboratory of several solar device structures of GaAs, AlGaAs and InP and their implications for obtaining high efficiency solar cells.

Microwave Reflection Measurement Set Up

In the microwave reflection technique continuous wave (CW) microwave power is incident on the sample and the reflected power is measured. The reflected power depends on the conductivity, the dielectric constant, the sample thickness and the microwave frequency. The conductivity of the sample can be determined from the measurement of the reflection coefficient. The excess carrier lifetime can be measured by monitoring the conductivity of the sample thru the reflected microwave signal after generation of hole-electron pairs by a pulsed monochromatic light source.

Figure 1 is a schematic diagram of the microwave reflection measurement system used in our laboratory. It is essentially a microwave bridge using a hybrid T capable of measuring the reflection coefficient (magnitude and phase) or the total reflected power at the sample according to the setting of the attenuator. The source of microwave power is a Gunn diode operating at 35 GHz at a power level of around 100 mW. A digital voltmeter is used to measure the output voltage of the microwave detector for steady state measurement. A Tektronix digital oscilloscope is used for acquiring transient signals during photoconductivity decay measurement of excess carrier lifetime. An IBM PS/2 Mod 55 computer is used for storing and analyzing the data acquired by the oscilloscope. To facilitate measurements it is desirable to have the sample outside the waveguide. This allows for ease sample placement and removal and avoids any sample shaping. We have used two different types of antenna for making measurements. One of them is the opened end of WR-28 waveguide which gives a spatial resolution of 3mm x 3mm. The other one is a tapered parallel plate antenna whose resolution is determined by the near field of the antenna and gives a spatial resolution of around 1mm x 1mm. The advantage of the WR-28 waveguide antenna is that it is easier to analyze the wave propagating in the semiconductor sample placed near the open end of the waveguide.

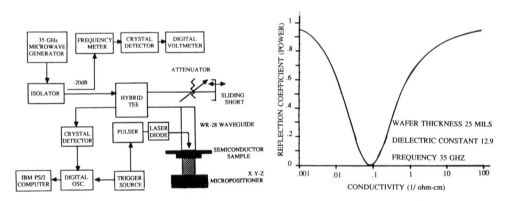

Fig. 1 Schematic Diagram of
Measurement Set-Up

Fig. 2 Power Reflection Coefficient as
a Function of Wafer Resistivity

The light sources we have used in our set up are GaAs and AlGaAs pulsed lasers emitting light at 904 nm and 880 nm respectively. These lasers have an optical fiber coupled to them so the laser light can be easily directed to the sample under test. We have used other lasers like a dye laser with pulse durations of hundred picoseconds and a CW YAG laser emitting light at 1.06 μm. The GaAs and AlGaAs are mainly used for measuring excess carrier lifetime, and the shortest lifetime which can be measured with them is determined by the decay constant of the light pulse which is of the order of 1 to 2 nsec which sets a limit of around 3 nsec the shortest lifetime which can be measured. The advantage of using a monochromatic light source is that the photoconductivity induced in the test structure is much easier to calculate than in the case of having a light source with a broadband spectrum. In that situation it is necessary to know the photon flux distribution as a function of wavelength for the source which makes the calculation more difficult.

Microwave Reflection Measurement Theory

For the experimental set-up of Fig. 1 with a high attenuation in one of the ports of the hybrid T, the relationship between the wave incident in the detector b_D and the forward wave b_G produced by the microwave signal source is given by:

$$b_D = K \, \Gamma_S \, b_G \tag{1}$$

where Γ_S is the reflection coefficient at the sample terminals and K is a constant of the hybrid T (3dB in our case). The RF power P_D incident in the detector is $|b_D|^2$ and it is given by:

$$P_D = |K|^2 \, P_G \, |\Gamma_S|^2 \tag{2}$$

where P_G is the power of the microwave source. If there is a change $\Delta\Gamma_S$ in the reflection coefficient of the sample produced by light excitation, then the change in the detector power is given by:

$$\Delta P_D = 2|K|^2 \, P_G \, |\Gamma_S| \, \Delta|\Gamma_S| \tag{3}$$

In general the relationship between the reflection coefficient of the sample and its conductivity is given by a curve similar to the one shown in Fig. 2. This relationship can be calculated for a sample of given geometry if the form of wave propagation in the sample is known. The simplest case is to assume plane wave propagation and this is approximately the case when the open end of a waveguide is used as an antenna. Notice that the curve of Fig. 2 shows that the reflection coefficient becomes 1 for the case of zero conductivity (insulator) as well as for the case of a perfect conductor. The relationship shown in Fig. 2 indicates that for small changes in conductivity there is a linear relationship between the change in the reflection coefficient and the change in the sample conductivity $\Delta\sigma$. Therefore the change in the RF power at the detector port is linearly related to the change in sample conductivity:

$$\Delta P_D = K_T \, \Delta\sigma \tag{4}$$

where K_T is a constant which takes into account the constant in Eq. (3) as well as the relationship expressed in Fig. 2. Therefore if the conductivity induced in the sample decays exponentially with time after the light excitation, the time dependence of this decay can be obtained by monitoring the detector voltage in the set up of Fig. 1. For a more detailed treatment of the experimental requirements for the measurement of excess carrier lifetime the reader is referred to a previous paper dealing with this topic [6].

Recombination Velocity in GaAs and AlGaAs Interfaces

The bulk lifetime and recombination velocity of NN$^+$ GaAs and NGaAs-N$^+$ AlGaAs layers was measured in the test structures shown in Fig. 3. The layers were grown in a conventional OMVPE atmospheric pressure reactor and the details of the growth parameters have been given elsewhere [7]. Two dopants were used in the N$^+$ GaAs layers sulphur and silicon. The N-layers were undoped and had a carrier concentration of 3×10^{15} cm^{-3}. The N$^+$ AlGaAs layer had a composition of Al$_{0.15}$ Ga$_{0.85}$ As and the dopant was sulphur.

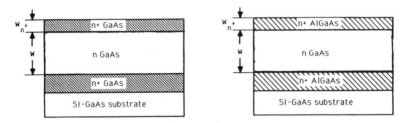

Fig. 3 Schematic Diagram of Tests Structured Measured

In order to obtain the bulk lifetime and the interface recombination velocity from the effective lifetime measured by the photoinduced microwave reflection, the thickness W of the epilayer and the doping ratio N/N$^+$ were varied as the test structures were grown. The bulk lifetime τ_B and the interface recombination velocity S$_i$ were obtained from the relationship:

$$\frac{1}{\tau_{eff}} = \frac{1}{\tau_B} + \frac{2 S_i}{W} \tag{5}$$

where τ_{eff} is the effective lifetime measured. The relationship between the interface recombination velocity S$_i$ and the recombination velocity S of either the top surface of the cap layer or of the cap-layer substrate interface is given by:

$$S_i = S \frac{N}{N^+} \tag{6}$$

where N is the doping concentration of the epi-layer and N$^+$ is the doping concentration of the cap layer. Analysis of the data obtained gave the following expressions for the interface recombination velocity of there different structures:
NGaAs - N$^+$GaAs (Silicon Doped)

$$S_i = 415 + 6.4 \times 10^5 \frac{N}{N^+} \; cm/sec \tag{7}$$

NGaAs - N⁺GaAs (Sulphur Doped)

$$S_i = 3500 + 6.5 \ x \ 10^5 \ \frac{N}{N^+} \quad cm/sec \qquad (8)$$

NGaAs - N⁺AlGaAs (Sulphur Doped)

$$S_i = 1900 + 5.6 \ x \ 10^5 \ \frac{N}{N^+} \quad cm/sec \qquad (9)$$

Notice that the value of 6.4×10^5 cm/sec of Eq. (7) is consistent with the value of 6.5×10^5 of Eq. (8). This value is also consistent with the value of the surface recombination velocity of bare GaAs which is estimated to be of the order of 10^6 cm/sec. The results obtained from our measurements and analysis of the data show that the best interface of an N-GaAs layer is obtained when the cap layer is a N⁺ GaAs layer doped with silicon. In fact the data shows that its interface recombination velocity is lower than the one which can be obtained with an AlGaAs cap layer. This result has implications for high efficiency solar cell design. As we have pointed out in a recent paper an all GaAs cell having an N⁺N emitter (shallow N⁺) and a PP⁺ base such that the light absorption takes place mainly in the N emitter and P base regions can have an efficiency as large as the one of a GaAs cell using an AlGaAs cap layer [8]. The doping in the N and P regions should be less than 10^{16} cm⁻³ in order to take advantage of the large lifetime in GaAs material as to reduce the dark saturation current for increasing the open circuit voltage.

Lifetime and Surface Recombination in InP

Solar cells made of InP are attractive for outer space applications because they are more radiation damage resistant than GaAs or Si cells. In addition InP has been claimed to have a lower value of surface recombination velocity than GaAs. Because of these two characteristics research activity on InP solar cells has increased in the last few years. In our laboratory we have fabricated N⁺PP⁺ InP solar cells using both bulk P material as well as growing the N⁺ and the P layers by OMVPE. In both cases, as we have reported in the literature, the quantum efficiency measured as a function of wavelength indicates that the recombination velocity at the bare surface of InP is of the order of the 10^6 cm/sec which is not consistent with the initial reports of low surface recombination velocity in InP [9]. For this purpose we decided to investigate the surface recombination velocity and the bulk lifetime of N-InP epitaxial layers grown on SI-InP substrates. Effective lifetime was measured by transient photoinduced microwave reflectance using epitaxial layers of different thickness in order to separate bulk lifetime from surface recombination velocity. The results for a bare N-InP layer doped to 3×10^{15} cm⁻³ grown in SI-InP gave a bulk lifetime of 30 nsec and a surface recombination velocity of around 10^4 cm/sec.

The above value of surface recombination velocity, when compared to the value of 10^6 cm/sec needed to explain the quantum efficiency of InP solar cells at short wavelengths, may be due to surface band bending. If the surface Fermi level is pinned close to the conduction band then lightly doped N material should show a low value of surface recombination velocity but highly doped N⁺ material should have a larger value. The low value of lifetime of 30 nsec for a doping of 3×10^{15} cm⁻³ indicates that the recombination is thru recombination centers since for that doping level the expected radiative lifetime is greater than 500 nsec.

Conclusions

In this paper we have shown that photoinduced microwave reflection using lasers as light excitation is a very powerful technique for determining important material parameters of solar cells. Interface recombination and bulk lifetime can be determined using this technique by growing special structures. In addition it is possible to use the microwave reflection technique for many more applications like determination of trapping levels using similar techniques as DLTS, but without the need of making contacts, as well as mobility measurements. Many of these applications have been documented in the literature by us as well as by many other authors.

Acknowledgement

The author wishes to acknowledge some of his past students Drs. H. Bhimnathwala, S. Bothra, K.K. Parat, S. Tyagi, R. Venkatasubramanian and M.S. Wang who were key participants in obtaining the results presented here. The diligence of Mrs. Sandi Laviolette in manuscript preparation is gratefully appreciated.

References

[1]   N. Braslau, Inst. Phys. Conf. Ser. **74**, 269, (1984).

[2]   W. Jantz, T.L. Frey and K.H. Bachem, App. Phys. A **45**, 225, (1988).

[3]   S. Bothra and J.M. Borrego, *Proc. of the 6th Conf. on Semi-Insulating III-V Materials*, ed. by C. Miener (IOP Publishing LTD 1990) 263.

[4]   M. Kunst and G. Beck, Journal App. Phys. **60**, 3558, (1986).

[5]   J.M. Borrego, R.J. Gutmann, N. Jensen and O. Paz, Solid State Elec. **30**, 195, (1987).

[6]   M.S. Wang and J.M. Borrego, IEEE Trans on Inst. and Meas. **39**, 1054, (1990).

[7]   J.M. Borrego and S.K. Ghandhi, Solid State Elec. **33**, 733, (1990).

[8]   S. Bothra and J.M. Borrego, Solar Cells **28**, 95, (1990).

[9]   S. Bothra, H.G. Bhimnathwala, K.K. Parat, S.K. Ghandhi and J.M. Borrego, *Proc. of the 19th IEEE Photovolt. Spec. Conf.*, ed. D. Flood (IEEE Press. 1987), 261.

Materials Science Forum Vols. 173-174 (1995) pp. 177-182
© 1995 Trans Tech Publications, Switzerland

PHOTOTHERMAL DEFLECTION SPECTROSCOPY ON AMORPHOUS SEMICONDUCTOR HETEROJUNCTIONS AND DETERMINATION OF THE INTERFACE DEFECT DENSITIES

F. Becker[1], R. Carius[1], J.-Th. Zettler[2], J. Klomfass[1], C. Walker[1] and H. Wagner

[1] ISI-PV, Forschungszentrum Jülich GmbH, D-52425 Jülich, Germany

[2] Institut für Festkörperphysik, TU Berlin, Hardenbergstr. 36, D-10623 Berlin, Germany

Keywords: Photothermal Deflection Spectroscopy, Amorphous Semiconductor Heterojunctions, Defect Density

Abstract. Photothermal deflection spectra showing interference fringes are compared with calculated spectra. In certain cases discussed in this paper the best fit reveals information on the position and the density of a defect layer embedded in an amorphous semiconductor layer stack.

Introduction

To investigate thin amorphous semiconductor films photothermal deflection spectroscopy (PDS) is usually used to measure the defect density by determing the absorption coefficient in the energy range where the defect absorption predominates.

PDS-spectra of a-Si:H can be divided into three photon energy intervals (see Fig. 2). At high photon energies the spectra show a saturation value at $\alpha \cdot d \geq 2$ (product of absorption coefficient α and thickness of the film d). The low energy region (E<1.5eV), where absorption due to defects dominates, gives information on the defect density of the sample. In the energy interval between these two regions an exponential absorption due to band tail transitions determines the spectrum.

Spectra of films deposited on flat substrates contain interference fringes. To obtain the absorption coefficient the fringes are averaged out by various methods. However, the interference fringes contain important information. By comparison with optical multilayer calculations information on the defect density of the surface can be obtained (Amato et al. [1], Asano et al. [2], G. Grillo and L. De Angelis [3]).

We investigated the defect density of thin defect layers in bulk films and the interface defect density in amorphous semiconductor heterojunctions. The knowledge of the interface defect density is important for device physics, because defect layers degrade the electronic properties of the device.

Additionally phase shift measurements of the laser-deflection with respect to the pump-light-source are carried out. Measurements of the phase shift have already been used to obtain information on undesirable substrate absorption. In this study the influence of thin defect layers on the phase lag is shown.

PDS-spectra of amorphous semiconductor samples containing intentionally introduced defect layers are compared with spectra derived from optical multilayer calculations.

Experimental and calculated spectra of amorphous silicon films containing thin (10 nm) defect rich layers and of an amorphous silicon / amorphous chalcogenide heterojunction are compared and the requirements and limitations of this technique (pumplight aperture, optical flatness of the sample, ratio of defect-layer and bulk defects etc.) are discussed.

For the cases discussed in this paper, the calculations are carried out to accurately model the position and density of these defect layers. The model is also used to evaluate interface defects at an amorphous semiconductor heterojunction. We show that interference fringes in PDS-spectra can be used to extract useful information about defect layers in homo- and heterojunction structures.

In the following section measurements of samples containing different defect layers are shown. These spectra are compared with calculated ones to obtain information on the sensitivity of the defect density determination.

Measurements

Experimental. The PDS-spectra were taken by a conventional PDS experimental set up [4]. The sample is irradiated by a pump-light source. Depending on the absorption this results in different temperature gradients in the deflection medium (usually CCl_4) which is in contact with the film surface. The temperature gradient induces a refractive index gradient in which a laser beam is deflected. The deflection angle is measured by a position sensitive detector. An intensity-stabilized HeNe or a temperature-stabilized semiconductor laser was used to improve the point stability of the laser beam. The aperture ratio of the pump-light source was minimized (1:2) to prevent suppression of the interference fringes. In addition the spectral resolution of the monochromator was adjusted to 6nm, corresponding to 5meV at a photon energy of 1eV. The interaction length of the laser beam with the irradiated sample was reduced to less than 4mm to avoid influences of inhomogenities of the film thickness.

Thin (10nm) p-doped a-SiC:H or n-doped a-Si:H layers, which are used as the inner p- or n-contact of stacked solar cells, are imbedded at various positions in an approximately 1.8µm thick a-Si:H film. Fig. 1 is an illustration of the position of the defect layers and the experimental arrangement of the pump-light and the laser beam.

Thin defect layers at various positions and defect densities in an a-Si:H film.

Variation of the position of the defect layer. Fig. 2 shows three PDS-spectra of films with a defect layer at the surface (0), interface (1) between film and substrate and in the middle (1/2) of the a-Si:H film. In the case of a surface defect layer (0) the peak to valley ratio of the interference fringes becomes enhanced compared to the bulk ratio. If the defect layer lies at the interface, then the interference modulation is the same as for the bulk film. Every second interference maximum is supressed when the film contains a defect layer in the middle.

In the case of spectrum 1 in Fig. 2 a large phase shift is observed, which is due to the lower dc-level of the temperature compared to curve 1/2 in the low absorption region. The modulation of the phase below about 1.5eV follows the PDS-signal, a result of the pump-light intensity profile and therefore the total absorption.

If the film is illuminated by the pump-light through the substrate, the assignments 0 and 1 have to be exchanged.

From spectra 0 and 1 in Fig. 2 it is obvious that an estimate of the average absorption coefficient by averaging out the PDS-spectra on a linear scale would lead to a wrong result for spectrum 0. This is also confirmed by calculations. The difference of the position of the interference fringes and the total defect density of the sample labeled 1/2 are caused by different growth conditions.

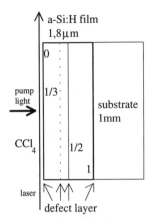

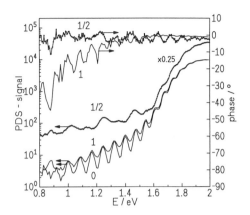

Fig. 1: Relative position referring to the surface of the imbedded defect layer in an a-Si:H film.

Fig. 2: PDS-spectra and phase of a-Si:H films containing defect layers at position 0, 1/2, 1 mentioned in Fig. 1 with pump-light incident from the film side.

Variation of the defect density. The sensitivity of the PDS-spectra on the defect density of the layer is investigated by varying the defect densities. Fig. 3 shows three samples containing thin layers of different defect density placed in the middle of the film.

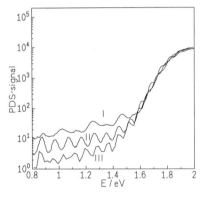

Fig. 3: PDS-spectra of a-Si:H films containing defect layers in the middle of the a-Si:H film of different defect densities N_D:
I: n-doped a-Si:H, $N_D=5\cdot10^{18}cm^{-3}$
II: p-doped a-SiC:H, $N_D=2\cdot10^{18}cm^{-3}$
III: a-Si:H, $N_D=10^{16}cm^{-3}$

The defect densities of these defect layers are obtained by PDS-measurements of thicker films prepared under the same conditions.

Obviously every second interference maximum is supressed in curve I and II, i. e. when $\alpha_D\cdot d_D$ of the defect layer is about equal (curve II) or three times as large as $\alpha_F\cdot d_F$ of the a-Si:H film (curve I). The sensitivity of the modulation depth on the defect density of the layer has been calculated. The results are shown in Fig. 6.

Heterojunction. The effect of interface defects between two different materials having different refractive indices on the PDS-spectrum is demonstrated using a glass-substrate/a-Si:H/a-As$_2$Se$_3$ heterojunction. a-As$_2$Se$_3$ films are characterized by a lower defect absorption, a larger bandgap by about 50 meV and a lower refractive index compared to a-Si:H.

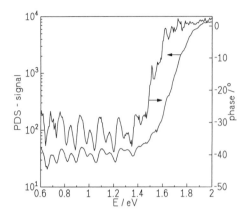

Fig. 4: PDS-spectrum and phase of the substrate/a-Si:H/a-As$_2$Se$_3$ heterojunction.

Fig. 4 shows the PDS-signal and the phase of the laser deflection.

The spectrum reveals interference fringes due to the combined layer thickness composed of a 1070nm thick a-Si:H and a 1270nm thick a-As$_2$Se$_3$ film. The slight change in the modulation depth is due to the fact that the defect layer is optically not perfectly located in the middle of the heterojunction. A large phase shift is observed which is partly due to the higher dc-level of the temperature in the high absorption region. The phase shift effect is enhanced by the higher defect absorption of the a-Si:H film compared to the high quality a-As$_2$Se$_3$ film on top of the layer stack. The modulation of the phase below about 1.5eV shows the same interference structure as the PDS-signal, i. e. the same argument holds as in case of Fig. 2.

Optical multilayer calculations

The PDS-spectra are compared with calculated spectra to gain more insight into the effect of the defect layers on the spectra and to test how more quantitative information can be obtained.

For simplification a constant defect absorption coefficient in the energy range 0.6-1.5eV and only one dominating defect layer is assumed.

The calculations are based on an optical multilayer program using a complex 2×2-matrix formalism [5]. Each layer of the layer stack is described by such a complex matrix. These matrices account for the absorption in each layer and the reflection and transmission at the corresponding interface. A layer system is described by a product of these matrices [2,6,7]. The dielectric functions needed as the input parameters for the calculation are obtained from PDS and transmission measurements of individual films. Fig. 5 is an illustration of the required real n_i and imaginary k_i parts of the corresponding refractive indices.

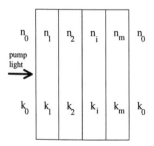

Fig. 5: Refractive index n_i and attenuation k_i of the layer stack.

Thin defect layers at various positions and defect densities in an a-Si:H film. In Fig. 6a) the dependence of the shape of the interference fringes on the position of the defect layer is shown. If the relative position of the embedded defect layer in an a-Si:H film referring to the surface is 1/n (n is an integer), then every n'th interference fringe is supressed except for n=0 (at the surface) and 1 (at the interface). However, the latter positions can be distinguished by the modulation depth as shown.

Fig. 6b) illustrates the dependence of the defect density in the middle of the a-Si:H film. In figure 7 the modulation depth of the interference fringes is shown as a function of the defect density of a 10nm thick defect layer in the middle of the film. As can be seen from this figure the defect layer is still observable when the absorption of the defect layer $\alpha_D d_D$ is less than the bulk absorption $\alpha_F d_F$. At a defect density of $10^{18} cm^{-3}$ αd for bulk and defect layer are equal.

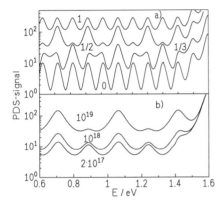

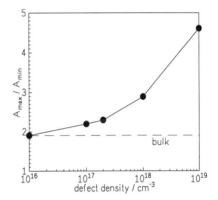

Fig. 6: Calculated spectra for a) A 10nm thick defect layer with defect density $10^{19} cm^{-3}$ at positions indicated in Fig. 1 in a 2μm thick a-Si:H film with defect density $10^{16} cm^{-3}$. The spectra are shifted vertically for illustration.
b) Spectra of layers with indicated defect densities (in cm^{-3}) in the middle of a 1μm thick film.

Fig. 7: Relation A_{max}/A_{min} (interference maximum/mimimum) versus defect density of a 10nm thick defect layer at 1/2 in a 1μm thick film with defect density $10^{16} cm^{-3}$, partially from Fig. 6b).

Dependent on the signal to noise ratio the detection limit for the defect layer can be as low as $\alpha_F d_F/3$. By taking the ratio of the supressed and the undisturbed fringe the sensitivity can be even enhanced. The modulation depth of the calculated spectra are more pronounced compared to the experimental ones, because of inhomogeneities of the film thickness and additional defect layers at the interface or the surface.

Heterojunction. The modelling of the a-Si:H/a-As$_2$Se$_3$ heterojunction is shown in Fig. 8 taking into account a defect absorption only in the amorphous films (lower curve) and an additional defect layer (here 10nm thick, defect density $6 \cdot 10^{18} cm^{-3}$) at the interface between the two films. A reasonably good agreement with the experimental spectrum is only obtained by introducing the defect layer as can be seen by comparing Fig. 8 and Fig. 4. A problem is the thickness inhomogenity of the chalcogenide layer, which is responsible for the smaller modulation observed in the experiment.

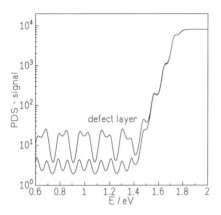

Fig. 8: Calculated spectra of the heterojunction without and with a defect layer.

Conclusions

PDS-spectra exhibiting interference fringes reveal more information than usually extracted. By comparison with calculated spectra the positions of defect layers in films can be determined and under certain conditions also quantitative information on the defect density can be obtained. The method is sensitive enough to detect defect absorption caused by thin layers which is even less than the bulk absorption.

The validity of the results have been demonstrated using defect layers at various positions and of different defect density.

A remaining problem seems to be inhomogenities of the film thicknesses which supress the modulation of the interference fringes.

The phase shift depends on the integrated absorption itself. To quantify the data the heat transport equation has to be solved, which is part of our current work.

Acknowledgement

We gratfully acknowledge helpful discussions and provision of samples by F. Finger, Jülich and R. P. Barclay, Swansea.

References

[1]   G. Amato, G. Benedetto, I. Boarino and R. Spagnolo, Appl. Phys. A **50**, 503 (1990).

[2]   A. Asano, M. Stutzmann, J. Appl. Phys. **70**, 5025 (1991).

[3]   G. Grillo, L. De Angelis, J. of Non-Cryst. Solids **114**, 750 (1989).

[4]   W. B. Jackson, N. M. Amer, A. C. Boccara, D. Fournier, Appl. Optics **20**, 1333 (1981).

[5]   L. I. Epstein, J. of opt. Soc. Amer. **42**, 806 (1952).

[6]   P. Grosse, R. Wynands, Appl. Phys. B **48**, 59 (1989).

[7]   L. Schrottke, thesis, Humboldt Uni Berlin (1987).

Materials Science Forum Vols. 173-174 (1995) pp. 183-184
© 1995 Trans Tech Publications, Switzerland

SUMMARY ABSTRACT: LASER SURFACE PHOTOVOLTAGE SPECTROSCOPY - A NEW TOOL FOR DETERMINATION OF SURFACE STATE DISTRIBUTIONS AND PROPERTIES

L. Kronik, L. Burstein, E. Fefer, M. Leibovitch, Y. Shapira

Department of Electrical Engineering, Physical Electronics, Faculty of Engineering, Tel-Aviv University, Ramat-Aviv 69978, Israel

Keywords: Surface States, Laser Spectroscopy, Surface Voltage

Abstract. A new experimental technique, which utilizes a tunable laser as the illumination source for surface photovoltage spectroscopy measurements is presented. The technique determines the distribution function of gap states observed at semiconductor interfaces, makes it possible to distinguish between surface and bulk states, and to find the thermal and optical cross-sections for electrons and holes. This method may be used for both in-situ monitoring and measurements in any ambient. Moreover, it is contactless, nondestructive, and simple to apply and interpret.

Introduction

Surface photovoltage spectroscopy (SPS) is a well-established surface-sensitive technique, in which sub-bandgap illumination is used to determine the energy positions of semiconductor surface states [1]. However, SPS cannot evaluate the energy distribution of the latter. Our new laser SPS technique (LSPS) uses sufficiently high illumination intensities such that all thermal transitions may be neglected with respect to the optical ones. Hence, the population of a local level, n_t is determined by its optical parameters only.

Resolution of the energy distribution of surface states

If the photon energy is increased by dE over an initial value E corresponding to a certain surface state, the observed SPV signal increase may be attributed to either an increase in the optical cross-section of the level, or to a transition of dn_t electrons from levels situated between E and $E + dE$ to E_c. LSPS provides the necessary intensities to rule out the first option. Thus, a derivative of (the square root) of the spectrum results in the energy distribution of the observed surface state (see Fig. 1).

Distinction between surface and bulk states

We have carried out a theoretical analysis which shows that SPS is inherently more sensitive to surface states then to bulk states. Moreover, we have shown that examination of the relation between the change in the surface potential, ΔV_s, and the illumination intensity shows a way to distinguish between bulk and surface states by means of SPS experiments. The use of laser radiation combined with appropriate optical attenuators provides a convenient means of obtaining an illumination intensity which is variable over many orders of magnitude.

Under the depletion approximation, analytical expressions for ΔV_s were obtained in two limiting cases, where either the surface or bulk states are optically inactive. When the surface states are optically active, the results of a numerical simulation are in an excellent agreement with the predictions of the simple analytical formula. In the opposite case, the increase of ΔV_s with illumination intensity is significantly slower than that predicted by the simple theory. Thus, the distinction of surface and bulk states may be carried out using intensity resolved SPS. A 'soft' saturation regime is characteristic of bulk states, and a super-linear $\Delta V_s(I)$ is characteristic of surface states. These conclusions have been supported by experiments on two samples of InP: One with a high density of bulk states, and the other with a high density of surface states.

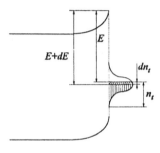

Fig. 1 Schematic band diagram of a semiconductor surface.

Additional Features

Experimentally, laser SPS is well-suited for the evaluation of the optical and thermal cross-sections of the surface states (and hence trap lifetimes), by means of intensity- and time-resolved SPS. In addition, it may also provide similar information about buried interfaces and thin films, which cannot be obtained using the conventional SPS technique.

We are currently conducting experiments in order to characterize the lateral distribution of the surface states and the surface potential on a microscopic scale by combining SPS with scanning tunneling microscopy (STM). This is made possible because the STM tip is sensitive to changes in the surface potential [2].

Conclusions

Energy-, time- and intensity-resolved LSPS provide a powerful means for quantitative characterization of local electronic states at semiconductor interfaces. A complete discussion of all aspects of the approach presented here may be found in refs. [3-5].

References

1. H. C. Gatos and J. Lagowski, J. Vac. Sci. Technol. **10**, 130 (1973).
2. Y. Kuk, R. S. Becker, P. J. Silverman and G. P. Kochanski, Phys. Rev. Lett. **65**, 456 (1990).
3. L. Kronik, L. Burstein, Y. Shapira and M. Oron, Appl. Phys. Lett. **63**, 60 (1993).
4. L. Kronik and Y. Shapira, J. Vac. Sci. Technol. **A11**, 3081 (1993).
5. M. Leibovitch, L. Kronik, E. Fefer and Y. Shapira, Phys. Rev. B., in press.

Materials Science Forum Vols. 173-174 (1995) pp. 185-190
© 1995 Trans Tech Publications, Switzerland

PHOTOREFRACTIVE TWO WAVE MIXING AND STEADY STATE PHOTOCARRIER GRATINGS FOR SEMICONDUCTOR CHARACTERIZATION

B. Smandek, Y. Ding, V. Hagemann, S. Gohlke, A. Wappelt and R. Wendt

Technical University Berlin, Optical Institute, Strasse des 17. Juni 135,
D-10623 Berlin, Germany

Keywords: Photorefractive Effect, Photocarrier Grating, Crystal Orientation, Deep Levels, Wafer Mapping, Diffusion Length

Abstract: Photorefractive Two-Wave Mixing (TWM) and the Steady State Photocarrier Grating method (SSPG) use the interference of two coherent cw laser beams at mW/cm^2 intensities to modulate the charge density of the semiconductor. In TWM a refractive index grating is created, leading to efficient diffraction of the beams. In III-V and II-VI single crystal semiconductors the orientation, the effective trap density and the homogeneity of wafers can be deduced. In the SSPG case the change of photocurrent with the grating period results in the determination of the diffusion length of semiconductors. Possible applications are the characterization of new photovoltaic materials and quality testing in the semiconductor industry.

Photocarrier Gratings for Semiconductor Characterization

Photocarrier gratings created by ps to ns pulses have been used to determine diffusion coefficients D and lifetimes τ of photocarriers for more than two decades by a contactless method [1]. In these cases the modulation of the free carrier density in the semiconductor led to a modulation of the refractive index and an efficient beam diffraction. However, due to the high excitation densities in the MW/cm^2-range the deduced quantities are not necessarily the physical constants close to thermal equilibrium or, in the case of photovoltaic materials, similar to AM 1 illumination conditions. As an example, the diffusion constant can be different by more than one order of magnitude [2].

These difficulties can be removed, as well as considerable experimental simplifications achieved, if cw laser beams in the

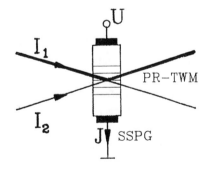

Fig.1: Schematic of experimental setup for SSPG and photorefractive TWM

mW/cm^2 intensity range are used (Fig.1). For contactless characterization the photorefractive effect is suitable, creating a refractive index grating via strong internal electerical fields. In the case of SSPG, determining mainly the diffusion length of the material, electrical contacts have to be applied, but they may not necessarily be Ohmic in character. Whereas TWM is restricted to our current knowledge to compound, single crystal semiconductors, there is no restriction for the applicability of SSPG.

The Photorefractive Effect and Two Wave Mixing

Use in Nonlinear Optics. The photorefractive effect [3] was first detected in GaAs and InP in 1984[4,5], with representatives of the II-VI compounds following in 1987[6]. Within this last decade the optical nonlinearity due to the photorefractive effect has been used for information processing [7], beam coupling[8], or phase conjugation [9], reaching the high efficiencies of photorefractive oxides.

    Origin of the Effect. To initiate the photorefractive effect, the photon energy has to be smaller than the band gap of the material, leading to the excitation of deep trap levels(Fig.2). The excited carriers will diffuse in their respective bands, leaving localized ionized traps in the material. As the semiconductor is illuminated by the interference pattern of two laser beams, this induced anisotropy leads to a loss of local charge neutrality. The resulting electrical field produces refractive index changes via the Pockels effect. As this requires off-diagonal elements of the electro-optic tensor, single crystals of compound semiconductors are necessary.

    In the case of the $\overline{4}$3m-space group, comprising all III-V and most of the II-VI compounds, the refractive index changes Δn are given by

$$\left[\Delta \left(\frac{1}{n^2} \right)_{ij} \right] = r_{41} \begin{pmatrix} O & E_z & E_y \\ E_z & O & E_x \\ E_y & E_x & O \end{pmatrix} \tag{1}$$

where r_{41} is the electro-optic coefficient, $E_{x,y,z}$ are the components of the space charge field in the kV/cm-range along the crystallographic axes, and n is the refractive index of the material.

Fig.2: The photorefractive effect occurs, if deep traps are ionized inhomogeneously by interfering laser beams, breaking local charge neutrality via diffusion.

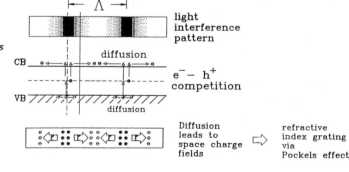

Two Wave Mixing (TWM). This refractive index grating can be probed by the diffraction of one of the laser beams into the direction of the other. In the simplest form the solution of the coupled wave equations for the intensities of the beams I_1 and I_2 takes the form of an exponential function for the weaker beam $I_1 \ll I_2$:

$$\frac{I_1(z = 1)}{I_1(z = 0)} = e^{(\Gamma - \alpha) 1} \qquad , \qquad (2)$$

where we defined the probe length 1 in the z-direction along the intersect of the two beams, the photorefractive gain coefficient Γ and the absorption coefficient α. The gain coefficient Γ is given by

$$\Gamma = \frac{1}{2} \left(\frac{2\pi}{\lambda} \right) n^3 r_{eff} E_{sc} \qquad , \qquad (3)$$

where the orientation dependent terms of Eq.1 are contained in r_{eff} and E_{sc} is the magnitude of the space charge field.

One of the main features of the photorefractive effect is its high sensitivity, as the photorefractive material integrates the excitation over time. Fig.3 shows the onset of photorefractive beam coupling at $\mu W/cm^2$ intensity levels in the case of the – new – photorefractive ZnSe:Cr [10]. The smaller saturation value for the 488 nm excitation can be explained by a strong competition between hole and electron excitation of the trap level.

The effective ionized trap density at a particular excitation wavelength is obtained by determining the dependence of the gain coefficient vs. the grating period Λ, Fig.4.

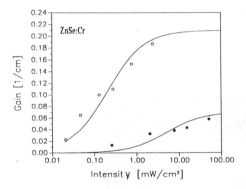

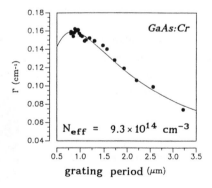

Fig.3: Photorefractive gain vs. laser intensity for photorefractive ZnSe at 670 nm (top curve) and 488 nm (bottom curve).

Fig.4: Photorefractive gain coefficient vs. grating period.

Applications of Photorefractive Gratings

Crystallographic Surface Characterization. Due to the tensorial nature of the photorefractive effect, the gain coefficient is strongly anisotropic; the induced birefringence leads to an additional polarization dependence. By rotating the sample, cut at a particular orientation and illuminated by a specifically polarized light, the orientation of the surface can be determined all optically. Fig.5 gives the example of a GaAs(111)-surface. Fig.5a shows the gain coefficient Γ for p-polarization in polar coordinates with respect to the axes of the crystal lattice, fig.5b the experimental results and a theoreticel fit and fig.5c the Laue diagram of the same surface. The photorefractive data, as well as the X-ray data depict the three-fold symmetry of the (111)-surface.

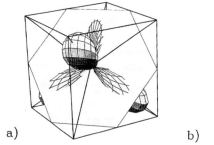
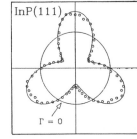

a)   b)   c)

Fig.5: Orientational dependence of the coefficient Γ in polar coordinates for p-polarization for the (111)-surface of the $\overline{4}$3m space group. a) theory b) experimental data and theoretical fit c) Laue diagram

Homogeneity of Wafers. GaAs and InP semiinsulating substrates are of increasing interest for optoelectronic device production. To reduce the processing failure losses of such microelectronic devices, the homogeneity of wafers is of paramount importance. ontactless determining trap densities via the photorefractive effect is a possible alternative to photoluminescence characterization. Contrary to the latter, it is a pure bulk effect, offering the possibility of quality assesment of ingots before cut. Fig.6 shows a representative scan of a GaAs(111) 2" wafer, showing high trap densities in the vicinity of the center.

Fig.6: Photorefractive gain vs. lateral position of a GaAs(111) wafer showing strong inhomogeneities of the trap density.

Steady State Photocarrier Grating Method

With the introduction of the Steady State Photocarrier Grating Technique (SSPG) by
Weiser et al. in 1987 [11] a direct determination of the diffusion length

$$L = \sqrt{D\,\tau} \qquad\qquad (5)$$

was possible under mW/cm^2 excitation. The SSPG-method determines the difference in
photocurrent in the absence and presence of a photocarrier grating, created by two
interfering cw-laser beams. The absence of the grating under the same integral illumination intensity is realized by a polarization rotator, e.g. a $\lambda/2$ in the classical case.
In this simple form, the method was used to determine diffusion length in α-Si and
semiinsolating GaAs. Since the initial introduction of the method, the validity of the
method was checked under a variety of operating conditions [11-14], including a separate treatment of electrons and holes and the incorporation of space charge effects
[14].

   Improved SSPG. As already pointed out in the original work, in samples with substantial dark conductivity, such as photovoltaic CuInS$_2$, determining small current differences in separate experiments rapidly approaches the error margin and impedes an
accurate measurement. By using a dynamic polarization rotator, such as a Kerr-cell we
accessed directly the decisive quantity

$$\Delta J = J(I_1 \| I_2) - J(I_1 \perp I_2) \qquad , \qquad\qquad (6)$$

where the first term represents the photocurrent J, produced by the beams I_1 and I_2,
in the case of paralell polarization and the second term for polarization crossed. ΔJ is
related to the diffusion length by

$$\Delta J / J(I_2) \;=\; \frac{2\,m\,\gamma_0}{\left[\,1 + [2\,\pi\,L/\Lambda]^2\,\right]^2} \qquad , \qquad\qquad (7)$$

where m is the exponent of the intensity vs. photocurrent charactristic in the range
between 0.5 and 1, γ_0 is a coherence parameter and Λ is the grating period, which can
be varied by the angle of the incoming laser beams. In this way we achieved an increase of the operating range of the SSPG method by several orders of magnitude, ma-
king it suitable for samples with dark conductivity $\sigma < 10\ \Omega^{-1}\mathrm{cm}^{-1}$. This is the relevant range for photovoltaic but also electronic grade material.

   Diffusion Length in CuInS$_2$. As an example, the determined diffusion length for the
recently developed photovoltaic material CuInS$_2$ are shown in Fig.7. The Cu-doped
samples, grown at the Hahn Meitner Institute Berlin proven to be solar grade material
with 10.9 % efficiency [15], show three times larger diffusion lengths than the In-do-
ped samples[16]. Under the assumption of infinite surface recombination velocity the
Cu-doped sample has a diffusion length of 0.7 µm.

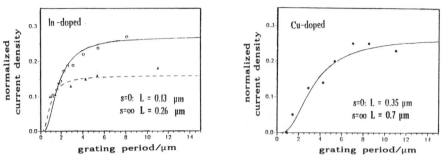

Fig.7: *Diffusion length in In- (left) and Cu doped (right) CuInS$_2$.*

Summary

By creating inhomogenous excitations in a semiconductor via interfering cw laser beams crystallographic-, deep level- and electronic properties of the material can be determined. Wafer homogeneity in two and possibly three dimensions can be analyzed. Whereas photorefractive gratings find potential industrial applications in III-V characterization, diffusion lengths determination by the SSPG method is particular useful for analysis of newly created semiconductors, such as photovoltaic compounds.

References

[1] H.J.Eichler, P.Günther, D.W.Pohl, *Laser Induced Dynamic Gratings* (Springer Verlag, Berlin, Heidelberg, New York, 1986)

[2] H.Weinert, M.Petrauskas, J.Kolenda, A.H.Galeckas, F. Wang and R.Schwarz , Proc. at the MRS Spring Meeting, San Francisco, April 12-16, 1993

[3] P.Günther, J.P.Huignard, *Photorefractive Materials and Their Applications, vol.I,II* (Springer Verlag, Berlin, Heidelberg, New York, 1988)

[4] M.B.Klein, Opt.Lett. **9**, 350 (1984)

[5] A.M.Glass, A.M.Johnson, D.H.Olson, W.Simpson, A.A.Ballman, Appl.Phys.Lett. **44**,948 (1984)

[6] R.B.Bylsma, P.M.Bridenbaugh, D.H.Olson, A.M.Glass, Appl.Phys.Lett. 51, 889 (1987)

[7] P.Yeh, J.Opt.Soc.Am. B5, 1811 (1988)

[8] H.J.Eichler, Y.Ding, B.Smandek, Opt.Comm. **94**, 127 (1992)

[9] Y.Ding, R.Röpnack, H.J.Eichler, Proc.Int.Conf. on LASERS 93, 5.-10.Dec.1993.

[10] I.Rückmann, B.Hanke, B.Smandek, Y.Ding, S.Juodkazis, M.Petrauskas, 8th Int. Symp. on Ultrafast phenomena, 22.-24.Sept. 1992, Vilnius, Lituania

[11] D.Ritter, E.Zeldov, K.Weiser, J.Appl.Phys. **62**, 4563-4570 (1987)

[12] I.Balberg, A.E.Delahoy, H.A.Weaklim, Appl.Phys.Lett. **53**, 992 (1988)

[13] I.Balberg, A.E.Delahoy, H.A.Weakliem, Appl.Phys.Lett 70, 2204 (1991)

[14] Y.-M. Li, Phys.Rev.**B 4**, 9025 (1990)

[15] R.Scheer, T.Walter, H.W.Schock, M.L.Fearheiley, H.J.Lewerenz, Appl.Phys.Lett. 63, 3294 (1993)

[16] R.Wendt, B.Smandek, A.Wappelt, H.J.Eichler, R.Scheer, H.J.Lewerenz, Proc. XII European Photovoltaics Conference, Amsterdam, 11.4.-15.4.1994.

Materials Science Forum Vols. 173-174 (1995) pp. 191-196
© *1995 Trans Tech Publications, Switzerland*

A NOVEL CONTACTLESS APPROACH FOR ACCURATE MEASUREMENTS OF ELECTRON-HOLE RECOMBINATION LIFETIME

G. Breglio, A. Cutolo, P. Spirito and L. Zeni

Dip. Ing. Elettronica, Via Claudio 21, I-80125 Napoli, Italy

Keywords: Contactless Diagnostics, Semiconductor, Optical Characterization

Abstract. We report on a novel interferometric technique for the measurement of the recombination lifetime of electron-hole (eh) pairs as a function of their concentration.

Introduction.

The recombination lifetime (τ) is independent of the way by which the non equilibrium condition has been generated thus permitting the development of many different techniques either electrical or optical for its measurement [1-4]. In particular, optical methods are very important because they are contactless so that they can be used at any process step required for the fabrication of electronic devices [4-8]. All of them share the common feature of monitoring the variations of the absorption coefficient induced by the free carriers and nobody has tried to measure the recombination time through a measurement of the variation of the refractive index, that can extend the measurement range to very low injection levels.

As τ depends on the excess carrier concentration, a well designed technique must measure the injection level at which the measurement is performed. Nevertheless all the methods, based on the use of optical or microwave radiation, do not measure the injection level but they typically provide only a rough estimate of the generated carrier concentration with a relative error always greater than 70-80%.

On this line of argument, we have devised and tested a novel interferometric method [9] which measures τ as a function of the carrier injection level which is measured without making any "ab initio" hypothesis on the pump laser beam parameters and on the quantum efficiency. Our experimental results demonstrate that our set up has a sensitivity much greater than that of the other methods described in the literature.

Theoretical background

As well known, the recombination process in a semiconductor slab depends on both bulk and surface effects and the observed temporal decay of the excess carriers ($\Delta N(t)$) generated by a light pulse, can be described in terms of exponential terms sum:

$$\Delta N(t) = \frac{1}{d} \int_{-d/2}^{+d/2} \sum_{n=1}^{\infty} A_n \cos(\alpha_n x) \exp\left[-\left(\frac{1}{\tau_b} + \alpha_n^2 D\right)t\right] dx$$

each term of which is referred to as recombination modes [10]. It is characterized by a temporal decay described by the factor:

$$\exp\left[-\left(\frac{1}{\tau_b} + \alpha_n^2 D\right)t\right] \qquad (1)$$

$$\alpha_n = \frac{2\delta_n}{d} \tag{2}$$

$$ctg\delta_n = \frac{2D}{Sd}\delta_n \tag{3}$$

where τ_b is the bulk lifetime, S is the surface recombination velocity, D is the diffusion constant, d is thickness of the sample. The validity of this model rests on the hypothesis of small electric fields induced by the carrier gradients and of low injection levels. The comparison between the results obtained by this analytical approach with those provided by using the numerical simulator PISCES [11] has shown an excellent agreement. The decay of the excess carrier is characterizied by several time constants and an inspection of Eqs.(1-3) reveals us that the n-th recombination mode is characterized by its decay time τ_n which depends on the bulk decay time (τ_b) and on the surface recombination effects. This fact generates several problems in the correct use of contactless techniques in order to distinguish between the surface and the bulk effects. In fact, any contact less allows a fairly easy measurement only of the effective recombination lifetime. In order to clarify this point, we report, in Fig. 1, the main effective recombination constant as a function of the bulk lifetime for different values of the surface recombination velocity. It is clear that we can neglect the surface effects only in a restricted range of bulk lifetimes and surface recombination velocities.

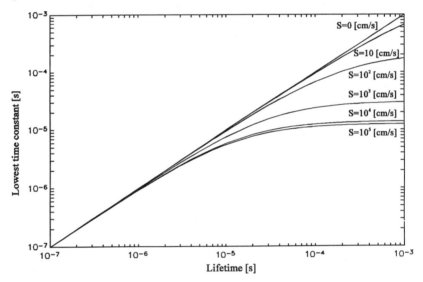

Fig.1 The main time constant of the recombination process versus the bulk lifetime for various surface recombination velocities. In this numerical example, the sample has a thickness equal to 400 μm and a doping level $N_A = 6*10^{14} cm^{-3}$

To get rid of this problem we have two choices:
a) To reduce the surface recombination velocity to a negligible value by an appropriate surface treatment, so that the decay process is described by a single time constant dominated by the bulk lifetime.
b) To measure at least two main recombination constants and, through the knowledge of the diffusion constant and the sample thickness, to calculate the bulk recombination lifetime and the surface recombination velocity.

Description of the technique

When in a semiconductor there is an excess of free carriers ΔN then its optical properties change [6] according to

$$\Delta n = -\frac{n\Delta N q^2}{2\varepsilon\omega^2 m^*} \tag{4}$$

$$\Delta\alpha = \frac{\gamma n \Delta N q^2}{2\varepsilon c \omega^2 m^*} = (-\Delta n)\frac{\gamma}{c} \tag{5}$$

where $\Delta\alpha$ and Δn are the variations of the absorption coefficient α and of the refractive index n at the optical angular frequency ω, m^* is the "parallel" equivalent between the effective masses of the holes and of the electrons, e is the dielectric constant, q is the electron charge, c is the speed of light and γ is the scattering rate of the excess free carriers.

With reference to Fig.2, our set-up allows the measurement of either Δn or $\Delta\alpha$ in order to compare the sensitivity of our apparatus, based on the measurement of Δn, with that of the systems described in literature, based on the measurement of $\Delta\alpha$. A train of high power laser pulses, with a duration time shorter than 100ns, delivered by a Q-switched Nd:YAG laser (λ_p=1.06μm) (pump) is focused on the under test device (a silicon wafer in our case) in order to generate e-h pairs. Then, their concentration is measured by monitoring the variations of the refractive index by a Michelson interferometer powered by an infrared linearly polarized c.w. He-Ne laser (λ=1.5μm) (probe) incident on the wafer at Brewster's angle. We have chosen this configuration because it does not require to know in advance the quality of the interfaces. In addition, our set-up can be easily arranged to work also in a configuration for the measurement of the variation of the absorption coefficient by simply darkening the mirror (b) (see Fig.2). The moving mirror of the interferometer has been mounted on a piezoelectric translation stage in such a way to easily control, by an appropriate feedback loop, the position of the mirror by keeping the interferometer in its maximum sensitivity configuration.

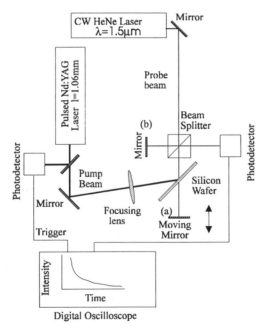

Fig.2 Schematic of the measurement set-up. A pulsed Nd:YAG laser (pump) is used to excite eh pairs inside a semiconductor which, in our case, is a silicon wafer. Then an infrared HeNe laser (probe) is exploited to monitor the decay of the excess free carrier concentration through a measurement of the variation of either the absorption coefficient or of the refractive index by an interferometer.

The accurate measurement of the injection level is based on the use of the Michelson interferometer. Referring to ref. [9] for more details, here we limit to notice that the excess carrier concentration ΔN can be measured from the phase delay $\Delta\phi$ between the two arms of the interferometer through the relation:

$$\Delta N = \frac{\varepsilon \omega^2 m^* \lambda}{q^2 n \pi d} \Delta \phi \qquad (6)$$

which requires only the knowledge of both the refractive index n semiconductor and the equivalent effective mass m* which can be considered well known from the literature.

Experimental results

Taking advantage of our technique, we have characterized a P-Silicon wafer of thickness d=370μm and doping level N_A=6*10^14 cm^-3. The surfaces of the sample have been treated in order to reduce the surface recombination velocity. In this way we could observed a decay process described by a single time constant directly related to the bulk lifetime according to the discussion of previous section. The results of this measurement (Fig.3) show that, at an injection level equal to 10 (corresponding, in our case, to an injection of excess eh pairs equal to 6*10^15) our apparatus is much more sensitive than those reported in literature based on the measurement of the free carrier absorption coefficient. In fact, by our interferometric technique, it is possible to measure, at this injection level, a recombination lifetime equal to about 15μs where the signal provided by the technique based on the measurement of the free carrier absorption coefficient can not provide any usefull result as the signal is completely darkened by the noise.

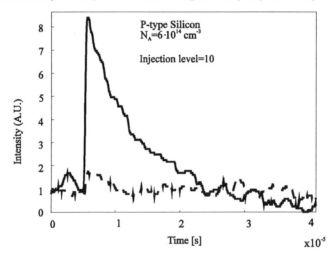

Fig.3. Comparison between the decay curves measured with the interferometric technique (full line) and the free carriers absorption technique (dotted line) for a silicon wafer with thickness equal to 370 μm and doping level N_A=6*10^14cm^-3

Then, we have measured the lifetime of the same sample as a function of the injection level (see Fig.4). The results of this measurement are in agreement with the SRH theory (1) for injection level up to 10^16 where the lifetime begins to decrease because of the Auger recombination mechanism.

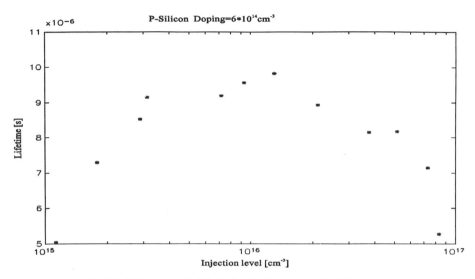

Fig.4 The lifetime versus the injection level for the same sample of Fig.3.

Summary

We have described an accurate contactless technique for the measurement of the recombination lifetime of excess eh pairs as a function of their concentration. The main advantages of this technique, when compared with those reported in literature, are an higher sensitivity and the possibility of performing an accurate measurement of the injection level without making any hypotheses or measurement on the input laser pulse. Work is now in progress to improve our set-up in order to perform more accurate measurements of the bulk lifetime in more general conditions for any value of the surface recombination velocity.

Acknowledgements

The kind assistance of Mr. Antonio Zecchino, Dr. Salvatore Schettino, Mr. Aniello Matrone is sincerely appreciated.

References

[1] D. K. Schroder, *Semiconductor materials and device characterization*, John Wiley and Sons (1990).

[2] P.Spirito, G. Cocorullo, IEEE Trans. Electron. Dev., **ED-32**, 1708 (1985)

[3] P.Spirito, S.Bellone, C.M.Ramson, G.Busatto, G.Cocorullo, IEEE Electron. Dev. Lett., **EDL-10**, 23 (1989)

[4] S.Perkowitz, *Optical characterization of semiconductors*, Academic Press (1993)

[5] F. Sanh, F.P. Gills, J.L. Gray, Solid State Electron., **35**, 311 (1992)

[6] C.M. Horwitz, R.M. Swanson, Solid State Electron., **23**, 1191 (1980)

[7] A.M. Afromowitz, M. Di Domenico, J. Appl. Phys., **42**, 3205 (1971)

[8] J. Waldmeyer, J. Appl. Phys., **63**, 1977 (1988)

[9] G.Breglio, A.Cutolo, P.Spirito, L.Zeni, IEEE Electron Devices Letters, **EDL-14**, 487 (1993)

[10] K.L. Luke, L. Cheng, J. Appl. Phys. **61**, 2282 (1987).

[11] Pisces Simulator, User Manual, 2B Version 9033, by Tech. Model. Assoc., 300 Hamilton Ave., 94301 Palo Alto (California)

Materials Science Forum Vols. 173-174 (1995) pp. 197-202
© 1995 Trans Tech Publications, Switzerland

RAMAN STUDIES OF DOPED POLY-Si THIN FILMS PREPARED BY PULSED EXCIMER LASER ANNEALING

A. Compaan[1], M.E. Savage[1], U. Jayamaha[1], T. Azfar[2] and A. Aydinli[2]

[1] Department of Physics and Astronomy, The University of Toledo, Toledo, OH 43606, USA

[2] Department of Physics, Bilkent University, Ankara 06533, Turkey

Keywords: Polycrystalline, Silicon, Excimer, Laser, Annealing, Raman Scattering, Fano Interference

Abstract. Hydrogenated amorphous silicon films (a-Si:H) with dopants of phosphorous and boron were grown on glass by plasma-enhanced CVD. In a-Si:H:P films the Raman signal shows that recrystallization begins with the first laser pulse but that multiple pulses are needed to generate the highest electron concentrations of $\sim 6 \times 10^{20}$ cm^{-3}. In a-Si:H:B the hole concentration reaches $\sim 1 \times 10^{21}$ cm^{-3} after laser anneal which produces a dip rather than a peak near the phonon line as a consequence of a negative Fano-interference parameter, q. The results show that Raman scattering can be used to obtain carrier concentrations in poly-silicon provided that wavelength-dependent Fano interference effects are properly included.

Introduction

Amorphous silicon (a-Si) photovoltaics as well as thin film transistors usually depend on heavily doped layers of a-Si to form the required p-i-n structures. In solar cells the p-type a-Si:H is part of the window layer which must transmit the sunlight into the intrinsic a-si:H layer. Ideally it must have high optical transparency and high mobility. For this reason, the a-Si:H:B is sometimes replaced by a layer of heavily doped (p-type) polycrystalline silicon (poly-Si). Such a layer can be prepared by modifying the glow discharge parameters [1] and also by pulsed laser annealing [2]. The properties of such poly-Si layers are important to measure and Raman scattering has been shown to be an important tool for studies of both a-Si:H and of poly-Si. For example Raman scattering has been used to measure the polycrystalline/amorphous fraction [1]. In this study we report on the Raman properties of the very heavily doped poly-Si which we have prepared by excimer laser annealing of glow-discharge a-Si:H films which we prepared with both doping types [3].

The linewidth and asymmetry of phonon Raman lines have been used to obtain grain sizes in polycrystalline semiconductors. The asymmetry arises from the relaxation of wavevector selection rules due to finite wavevector effects in very small crystallites (≤ 100 Å). This technique was first used in undoped polycrystalline silicon thin films [4] and has subsequently been extended to several other semiconductor systems and also applied to the case of alloy broadening and ion-implantation-produced damage effects [5]. However, doped semiconductors can present problems for such lineshape analyses due to line shape changes and broadening arising from effects of single particle scattering from free carriers. In doped Si, free particle scattering interferes with the discrete phonon Raman scattering to produce asymmetric and broadened lines producing a Fano-type lineshape [6]. Understanding the various contributions to the Raman line shift and shape can improve the utility of Raman scattering as a tool to obtain carrier concentrations. This is particularly useful in polycrystalline materials where transport measurements are complicated by grain boundary effects. Transport properties of doped polycrystalline Si are of considerable interest due the material's use in photovoltaics and thin-film transistors [7].

Previous studies of pulsed-laser-annealed films of undoped a-Si:H have shown that a layered structure exists after annealing. Annealing with a picosecond YAG laser [8] and with 10-20 nsec excimer lasers [2] have been shown to form a thin (30 - 100 nm) surface layer which is polycrystalline (grain sizes ≥ 50 nm). Underneath this lies a microcrystalline or fine-grained layer (usually observed as a region with grains of 20 nm diameter or less surrounded by amorphous material [8,9]) of about the same thickness. In addition the laser annealing may produce a layer of a-Si with greatly reduced hydrogen content immediately under the microcrystalline Si. Finally, if the original films are thick enough, there may remain a residual layer of unannealed a-Si:H. We shall use this four-layer model [8] to interpret our Raman data.

Experimental

The Raman scattering was performed with two argon laser lines (λ = 457.9 nm and 514.5 nm) and the krypton laser line at λ = 647.1 nm using a cylindrical focus to avoid beam heating effects. The power density on the sample was below 10 W/cm^2. The Raman spectra obtained at Toledo utilized an ISA S3000 triple spectrometer operated in the scanning mode, whereas the data obtained at Bilkent used a JY U1000 double spectrometer. All data were obtained at room temperature with the samples held in air. The Raman probe depths for backscattering [$(2\alpha)^{-1}$, where α is the absorption coefficient] in crystalline or polycrystalline Si are 0.15 μm for λ=458 nm, 0.30 μm for 514 nm, and 1.5 μm for 647 nm.

Results and Discussion

n-type films--The progress of sub-melt-threshold laser annealing with increasing numbers of laser pulses in P-doped a-Si:H is shown in Figure 1. With the pulse energy density of 150 mJ/cm^2, one annealing pulse produces a distinct peak near 520 cm^{-1} corresponding to the Raman-active (Brillouin-zone-center) mode $\Gamma_{25'}$ in crystalline Si. However, with 10, 100, and 1000 pulses, a second peak develops which is shifted downward by ~4.5 ± 0.5 cm^{-1}. It is also broader and asymmetric.

The shift, broadening, and asymmetry of this second peak in annealed a-Si:H:P films is a result of improved carrier activation with additional pulses. These are characteristic signatures in Raman scattering of n-type electrical activation of the phosphorous dopants. The amount of asymmetry and phonon softening (downward frequency renormalization) increases with dopant concentration.

The effects of n-type doping on the Raman spectra from crystalline Si have been reviewed by Abstreiter, Cardona, and Pinczuk [10] and the frequency shift in extremely heavily doped crystalline Si has been measured in Raman scattering by Compaan, Contreras, and Cardona [11]. For the case of heavily doped n-Si, the continuum of single-particle excitations which interferes with the discrete phonon line arises from inelastic scattering of electrons near the X-point of the conduction band. In general the peak shift and line shape can be used to estimate the doping density. From the shift of 4.5 ± 0.5 cm^{-1} observed in Fig. 1 and by comparison with the data of

Ref. 11, we infer that the electron concentration is $6 \pm 2 \times 10^{20}$ cm^{-3}.

Our interpretation is that in the initial stages of laser annealing (first few pulses), recrystallization occurs with the associated evolution of hydrogen from the film. However, the initial crystallites must possess sufficiently high numbers of defects that carrier activation is suppressed. Incomplete H evolution could be a factor leading to suppression of dopant activity. Further annealing with additional pulses removes carrier traps and activates the dopants. The Raman data show that even after 1000 pulses a recrystallized but electrically inactive layer persists deep in the film, presumably near the boundary with the microcrystalline and/or amorphous material. This boundary layer is simply driven deeper and deeper into the film with increasing numbers of pulses.

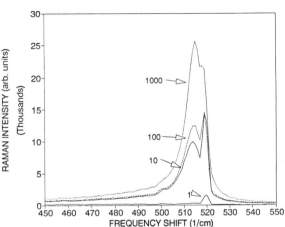

Figure 1: Raman scattering at λ=514.5 nm from a-Si:H:P films after 1, 10, 100, and 1000 XeCl excimer laser pulses at 150 mJ/cm^2.

p-type films--Pulsed laser annealing of boron-doped a-Si:H yields extremely heavily doped p-type poly-Si. Figure 2 gives the Raman spectra obtained from these films for three different wavelengths. As observed in p-type crystalline Si, the phonon Raman lineshape near 500 cm^{-1} is a strong function of the exciting wavelength [12]. Two additional peaks can be identified just above 600 cm^{-1} which are known to arise from the local vibrational modes of ^{11}B and ^{10}B in the Si lattice [13]. The existence of these local modes is clear evidence that the boron is reaching substitutional lattice locations, and the interference lineshape is an indication of the electrical activity of the boron. Both results confirm that a major fraction of the boron in the poly-Si material is active even though in the amorphous state only a small fraction of the B is electrically active [14].

The asymmetries in the intrinsic phonon mode and in the boron local modes arise from discrete-continuum interactions in which the discrete phonon or local modes interact with the continuum of excitations from inelastic scattering of the holes between valence bands (light-hole, heavy-hole, and spin-orbit split). The Raman lineshapes have been extensively analyzed for doped crystalline Si by Cerdeira, *et al.* [12], and by Chandrasekhar, *et al.* [13]. The solid curve in Figure 2 is a fit of the data to the resulting Fano-type [6] lineshape following Ref. 13:

$$I(\omega) = A(q_p + \varepsilon_p)^2/(1+\varepsilon_p^2) + B[0.8(q_i+\varepsilon_i)^2/(1+\varepsilon_i^2) + 0.2(q_j+\varepsilon_j)^2/(1+\varepsilon_j^2)] + C;$$

$$\text{with } \varepsilon_{p,i,j} = (\omega - \omega_{p,i,j} - \Delta\omega_{p,i,j})/\Gamma_{p,i,j}. \tag{1}$$

Here the subscripts p, i, and j refer to the intrinsic phonon and the two local modes of ^{11}B and ^{10}B respectively. $q_{p,i,j}$ are the asymmetry parameters which relate to the discrete-continuum mixing, ω is the scattering frequency, $\omega_{p,i,j}$ are the original phonon and local mode frequencies, $\Delta\omega_{p,i,j}$ are the phonon or local mode frequency shifts (renormalizations), and $\Gamma_{p,i,j}$ are the imaginary parts of the phonon or local mode self energies (not including the usual phonon broadening). From Ref. 13 we take the unshifted frequencies as $\omega_p = 520$ cm^{-1}, $\omega_i = 620$ cm^{-1}, and $\omega_j = 644$ cm^{-1}. Note that the amplitude factors of 0.8 and 0.2 simply reflect the relative abundances of the two boron isotopes. The parameters used for the fits shown in Fig. 2 are given in Table 1.

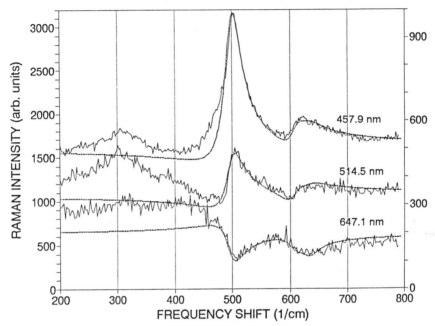

Figure 2:   Raman scattering from laser-annealed a-Si:H:B.   Dashed curves are the described fits to the Fano lineshape. Baseline +1000 for 457.9 nm; right scale for 647.1 nm data.

For the fits shown in Fig. 2, we have forced the parameters ($\Delta\omega_{i,j}$, $q_{i,j}$, and $\Gamma_{i,j}$) of the [11]B and [10]B local modes to be the same. Thus $\Delta\omega_i = \Delta\omega_j$, $q_i = q_j$, and $\Gamma_i = \Gamma_j$. Furthermore we have used the same values of $\Delta\omega_p$, $\Delta\omega_i$, $\Delta\omega_j$, and of Γ_p, Γ_i, Γ_j, for all three excitation wavelengths. By comparing the values of the best fit parameters given in Table 1 with those of the highest concentration sample of Ref. 12 (4×10^{20} cm^{-3}), it is easily seen that the free hole concentration is larger than this value and near 1×10^{21} cm^{-3} as discussed below.

Table 1: Interference Equation Parameters

λ(nm)	$\Delta\omega_p$	q_p	Γ_p	$\Delta\omega_{i,j}$	$q_{i,j}$	$\Gamma_{i,j}$	A:B:C
457.9	-23±1	4.0±0.5	18±1	-16	1.3±0.5	18	1:1.3:3.8
514.5	-23±2	2.2±0.3	18±2	-16	0.7±0.3	18	1:1.5:7.7
647.1	-23±3	-0.5±0.2	18±5	-16	-0.3±0.2	18	1:0.5:0.15

Perhaps the most striking feature of Fig. 2 is the shape of the spectrum obtained with λ = 647.1 nm. The Raman peak shows up as a dip in the spectrum and the corresponding values of the parameters q_p, q_i, and q_j are <u>negative</u>. The fit to Equation 1 for the 647.1 nm data is not as good as for the shorter wavelengths; and this may indicate that the film is not uniformly activated deeper than about 0.3 µm, as found above for the phosphorous-doped films. This may account for the larger value of Γ for 647 nm but a more uniform film is unlikely to change the q-parameters substantially. To our knowledge a negative value of the q-parameter in p-type Si has not previously been observed in Raman scattering [15] although its physical meaning is readily apparent, as discussed below.

Following the notation of Ref. 13 the q_p parameter is given by [16]:

$$q_p = (V_p T_p / T_e + \Delta\omega_p) / \Gamma_p, \tag{2}$$

where V_p is the matrix element of the hole-phonon interaction. Because the phonon scattering matrix element T_p (a third order process) resonates more strongly at the E_1 critical point (3.4 eV) than does the hole scattering matrix element T_e (a second order process), the first term in Eq. 2 dominates with 457.9 nm excitation (hv = 2.70 eV) and must therefore be positive since q_p is positive as seen in Table 1. On the other hand the frequency renormalization term $\Delta\omega_p$ is seen to be negative from Table 1. As the photon frequencies move away from the 3.4 eV E_1 resonance the first term decreases and hence q_p decreases. In fact Table 1 shows that q_p is much smaller for 514.5 nm than for 457.9 nm excitation and eventually becomes negative for 647.1 nm. Previous studies of the Fano lineshape in p-type silicon utilized doping levels and wavelengths for which q_p was positive [10-13,17]. The present study, using films doped above the solid solubility limits [18], have provided an interesting opportunity to extend the range of application of this interaction.

Conclusions

One of our original objectives in performing the Raman scattering on these films was to determine the suitability of Raman scattering as a method for obtaining the free carrier concentration. The interference effects described above complicate the analysis somewhat; however, the use of near-resonant conditions (hv=2.7 eV or higher) selectively enhances the phonon amplitude over the hole scattering amplitude and permits "cleaner" analysis of the frequency shift as well as the amplitude of the boron local modes. Shorter wavelengths than 457.9 nm are even more advantageous as shown in Ref. 11. Unfortunately the parameters q_p, $q_{i,j}$, and the linewidths $\Gamma_p, \Gamma_{i,j}$ do not scale simply with carrier concentration for p-type Si. However, for short wavelengths the amplitude of the boron local modes do scale approximately with the density of boron incorporated in the lattice. The ratio of the local mode intensity to that of the intrinsic phonon is approximately twice as large in these films as it is in the 4×10^{20} cm^{-3} sample studied by Cerdeira, et al [12]. Considering this and the values of the Fano parameters we obtain the density of substitutional boron and of the free carriers to be $\sim 1 \pm 0.2 \times 10^{21}$ cm^{-3}.

We conclude that careful analysis of the Raman line shift and shape can yield good estimates of free carrier concentrations in polycrystalline silicon thin films. However, the wavelength-dependent interference effects must be appropriately included.

Acknowledgments

This work was supported in part by the Thomas Edison program of the State of Ohio and Glasstech, Inc. The work at Bilkent University was supported in part by TUBITAK. ADC wishes to thank TUBITAK for a Distinguished Senior Visitor award under which this work was completed.

References

[1]   S.C. Saha, A.K. Baruan, and Swati Ray, J. Appl. Phys. **74**, 5561 (1993).

[2]   K. Winer, G.B. Anderson, S.E. Ready, B.Z. Bachrach, R.I. Johnson, R.A. Ponce, & J.B. Boyce, Appl. Phys. Lett. **57** 2222 (1990).

[3]   A. Compaan, M.E. Savage, A. Aydinli, and T. Azfar, Solid State Commun. **90**, 77 (1994); M.E. Savage, U. Jayamaha, A. Compaan, A. Aydinli, and D. Shen, Mat. Res. Soc. Symp. Proc. **283**, 321 (1993).

[4]   H. Richter, Z.P. Wang, and L. Ley, Sol. State. Commun. **39**, 652 (1981).

[5]   K.K. Tiong, P.M. Amirtharaj, F.H. Pollak, and D.E. Aspnes, Appl. Phys. Lett. **44**, 122 (1983).

[6]   U. Fano, Phys. Rev. **124**, 1866 (1961).

[7]   K. Sera, F. Okumura, H. Uchida, S. Itoh, S. Kaneko, and K. Hotta, IEEE Trans. El. Devices ED-36, 2868 (1989); H. Kuriyama, S. Kiyama, S. Noguchi, T. Kuwahara, S. Ishida, T. Nohda, K. Sano, H. Iwata, H. Kawata, M. Osumi, S. Tsuda, S. Nakano, & Y. Kuwano, Jpn. J. Appl. Phys. **30**, 3700 (1991).

[8]   P.H. Liang, C.J. Fang, D.S. Jiang, P. Wagner, and L. Ley, Appl. Phys. **A 26**, 39 (1981).

[9]   B. Goldstein, C.R. Dickson, I.H. Campbell, & P. M. Fauchet, Appl. Phys. Lett. **53**, 2672 (1988).

[10]   G. Abstreiter, M. Cardona, and A. Pinczuk in <u>Light Scattering in Solids IV</u> (Springer-Verlag, Berlin, 1984), pp. 5-150.

[11]   A. Compaan, G. Contreras, and M. Cardona, Mat. Res. Soc. Symp. Proc. **23**, 117 (1984); G. Contreras, A.K. Sood, M. Cardona, and A. Compaan, Sol. State Commun. **49**, 303 (1984).

[12]   F. Cerdiera, T.A. Fjeldly, and M. Cardona, Phys. Rev. B **8**, 4734 (1973); F. Cerdiera, T.A. Fjeldly, and M. Cardona, Phys. Rev. B **9**, 4344 (1974).

[13]   M. Chandrasekhar, H.R. Chandrasekhar, M. Grimsditch, and M. Cardona, Phys. Rev. B **22**, 4825 (1980).

[14]   P.G. LeComber, D.I. Jones, and W.E. Spears, Phil. Mag. **35**, 1173 (1977).

[15]   In n-Si and in p-Ge the sign of the q-parameter is negative and large for a wide range of wavelengths and dopant densities. This results from a different sign in the electronic scattering matrix T_e. See Ref. [10].

[16]   The q-parameter is often written in the form $q_p = [R_p/(\pi V_p R_e)]$ which does not readily display the possibility of sign reversal. However the phonon scattering matrix element R_p involves the phonon state *dressed with the hole scattering interactions*. See Refs. [6] and [10].

[17]   R. Beserman and T. Bernstein, J. Appl. Phys. **48**, 1548 (1977).

[18]   F. A. Trumbore, Bell Sys. Tech. Journal, **39**, 205, (1960); J. Chikawa and F. Sato, Jpn. J. Appl. Phys. **19**, L577 (1980).

Materials Science Forum Vols. 173-174 (1995) pp. 203-208
© *1995 Trans Tech Publications, Switzerland*

NEAR INFRARED QUASI-ELASTIC LIGHT SCATTERING SPECTROSCOPY OF ELECTRONIC EXCITATIONS IN III-V SEMICONDUCTORS

B.H. Bairamov[1,2], I.P. Ipatova[2], G. Irmer[3], J. Monecke[3], V.K. Negoduyko[2], V.A. Voitenko[2] and V.V. Toporov[2]

[1] Department of Physics, Cavendish Laboratory, Cambridge, CB3 OHE, U.K.

[2] A.F. Ioffe Physico-Technical Institute, St. Petersburg 194021, Russia

[3] Fachbereich Physik der Bergakademie Freiberg, D-09599 Freiberg, Germany

Keywords: Near Infrared Quasi-Elastic Light Scattering Spectra, Raman Scattering, Electronic Excitation Fluctuations, III-V Semiconductors

ABSTRACT. We present results of the first observation of the concentration dependencies of integrated intensities and line shapes of quasi-elastic scattering from the free electron gas in III-V semiconductors in the concentration range from $\sim 10^8$ up to $\sim 10^{19}$ cm^{-3} under a condition of non-resonant excitation by using the 1064 nm line of a cw Nd^{3+}:YAG laser, when the observation of light scattering from energy- and momentum-density fluctuations is unambiguously demonstrated.

Introduction

One of the most important subjects of semiconductor physics remains the electron gas and inelastic light scattering by electronic excitations, which is of strong current interest and is a fast developing branch of optical spectroscopy. In particular, the papers concerning the electronic light scattering cover one of the most exciting chapters in the optics of microstructures and superlattices [1].

A common feature of all types of single particle electronic Raman scattering (RS) processes comes from conservation laws, which govern them and make the energy $\hbar\omega$, and the momentum $\hbar q$, transferred during the scattering process, small. Within the framework of the effective-mass approximation, the light scattering mechanisms in semiconductor quantum wells and superlattices are similar to those of the parent 3D systems [2].

There are all conditions for the observation of the majority of known single-particle excitations by the RS technique in III-V semiconductors. These excitations form the basic mechanisms of scattering which include charge-density fluctuations, energy-and momentum-density fluctuations as well as spin-density fluctuations .

Most of the recent inelastic electronic light scattering experiments (Pinczuk et al 1989) have been carried out in GaAs with excitation photon energies $\hbar\omega_i$ in the range 1.5-1.9 eV close to the resonance with the fundamental E_o or spin-orbit split-off $E_o + \Delta_o$ optical gaps, or with quantum well excitons providing large scattering enhancement required for sufficient sensitivity. On the other hand the simultaneous appearance of strong hot luminescence in such cases allows one to carry out measurements only in a limited electron concentration range and prevents temperature dependent measurements which, as we will show below, are essential for the observation of new features in the light scattering spectra from electronic excitations in semiconductors.

In this paper we examine the different scattering mechanisms from electronic excitations by using non-resonant near-infrared excitation of n-InP: $E_o + \Delta_o = 1.43 + 0.11 = 1.54$ eV (at T = 10K) and $\hbar\omega_i = 1.17$ eV. The absence of intense background luminescence in this case has provided us with an opportunity to carry out not only the conventional concentration dependent polarisation studies but also temperature dependent measurements of the integrated intensities and spectral linewidths of the quasi-elastic electronic scattering spectra. We find rather different effects of electronic excitations on polarised and depolarised scattering and undertake both theoretical and experimental comparison between intensities of scattering from energy- and spin-density fluctuations. We find these to be approximately equal at room temperature which removes any possible doubts concerning scattering from energy-density fluctuations, since scattering from spin-density fluctuations is well established.

Experimental set-up

Our polarised quasi-elastic light scattering measurements were performed by using the 1064.2 nm line of a *cw* Nd^{3+}:*YAG* laser and doped n-InP samples with carrier concentration from 10^8 up to 10^{19} cm^{-3} and the temperature range 27-300K. All measurements were conducted in right angle scattering geometry with the incident light always along $(11\bar{2})$ and scattered light along $(\bar{1}11)$ directions: for polarised spectra, hereafter labelled by $\bar{e}_i \| \bar{e}_s$ and $(11\bar{2})\{(\bar{1}10),(11\bar{2})\}(\bar{1}11)$ for depolarised spectra, labelled as $\bar{e}_i \perp \bar{e}_s$. The scattered light was analysed by double monochromator with a spectral resolution of 2.1 cm^{-1} and detected with a cooled photomultiplier tube with a photon-counting electronic system.

Results and discussion

The important aspect in the discussion of RS from charged excitations is the electrical screening of fluctuations. If the concentration of free carriers is low (n ≤ 10^{16} cm^{-3}) and the screening radius r_s is large enough to satisfy the condition $qr_s \gg 1$, then every excitation created by the scattered light with the wave vector q is not screened. On the other hand, the quasi elastic light scattering spectra are recorded at room temperature when there are enough carriers to provide screening of the charged impurities. The condition of linear screening holds if $nr_s^3 \gg 1$, which means that isolated ionised impurities are screened. Under this condition there still remain the long-range fluctuations in impurity potential with the mean square value $\gamma(r_s) = u(r_s)\sqrt{Nr_s^3}$, where $u(r) = \frac{e^2}{\varepsilon r}$ is the potential of an isolated impurity atom, and N is the total concentration of impurities. So single particle electronic excitations, created by light, are scattered by the long-range fluctuation potential and one can get the following expression for the light scattering cross section

$$\frac{d^2\sigma}{d\omega d\Omega} = n\left(\frac{e}{m^*c^2}\right)^2 (e_i \cdot e_s)^2 \int \frac{dt}{2\pi} \cos(\omega t) \ \exp\left[-\frac{(qv_T t)^2}{4} - \frac{\rho_0 q^2 t^4}{24m^{*2}}\right], \quad (1)$$

where $v_T^2 = 2T / m^*$ and ρ_0 is the root mean square fluctuation of the random force acting on the electron. The line shape of the spectrum differs slightly from a Gaussian and gives an approximately linear dependence of linewidth Γ on the concentration, in good agreement with the experimental observations (Figure 1).

Fig.1 Raman spectra of n-type InP samples in the low-concentration range. The solid curves obtained after subtraction of the difference frequency combination two-phonon contribution, represent spectra for single-particle scattering from conduction electrons. The theoretical fit with a Gaussian approximation is shown by open circles with half-width $\Gamma=35.0 \text{cm}^{-1}$ for the sample with n = 5 x 10^{15} cm^{-3} and Γ = 24.9cm^{-1} for the sample with n = 7 x 10^{14} cm^{-3}. $e_i \| e_s$, $\lambda_i = 1.06$ μm, T = 300K.

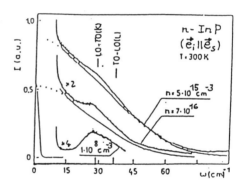

With increasing concentration up to n = 3.0 x 10^{16} cm^{-3} when the condition of frequent collisions holds, (i.e. $q\ell \ll 1$, where ℓ is the electron mean free path) and due to the realisation of conditions for scattering from spin-density fluctuations $(e_i \perp e_s)$ the Gaussian type line shape transforms into a Lorentzian [3,4].   In the concentration range from 7.2 x 10^{16} up to 3.6 x 10^{17} cm^{-3}, collision controlled narrowing of the Lorentzian single particle spectra from 52 to 32 cm^{-1} is found experimentally. The narrowing of the linewidth with the concentration increase is a direct indication on the diffusion nature of the scattering mechanism.

Figures 2 (a and b) show typical Raman spectra in the polarised scattering configuration $(e_i \| e_s)$ obtained on n-InP sample with n = 1.1 x 10^{18} cm^{-3} in the frequency range from 150 up to 450 cm^{-1} at different temperatures at 300K and 27. Figures 3 (a and b) show the same for the depolarised scattering configuration $(e_i \perp e_s)$. The scan is linear in wavelength and the positions of the lines are indicated in wave numbers.

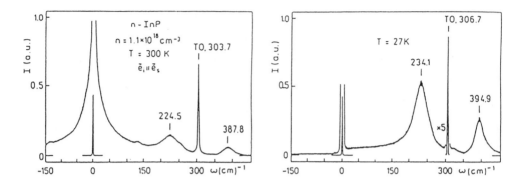

Fig. 2(a and b). Experimental Raman spectra of n-InP with n = 1.1×10^{18} cm^{-3} for polarised scattering configuration $\tilde{e}_i \parallel e_s$ at different temperatures which are indicated in the figures. The quasi-elastic Lorentzian contours discussed in the text are near the laser line.

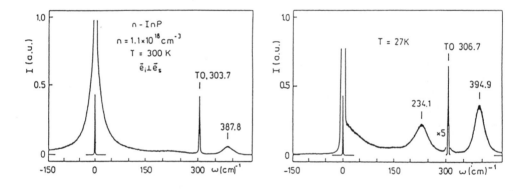

Fig. 3(a and b). Same as in Fig.2 but for the depolarised scattering configuration for crossed polarisations $e_i \perp e_s$. The quasi-elastic Lorentzian contours correspond to spin density fluctuations.

We unambiguously identify in all the spectra, sharp quasi-elastic parts of Raman scattering from the free electron gas, located within ± 150 cm^{-1} with the required characteristic finite intensity at frequencies close to zero. The Stokes and anti-Stokes components at near room temperatures are equal in intensity due to the response of the detector and spectrometer used.

In order to determine the integrated values of the intensities of the quasi-elastic electronic scattering we used uncoupled TO(Γ) phonon spectra for calibration, so that the spectra at different temperatures may be directly compared. Figure 4 compares the corresponding experimental integrated intensities. It is particularly striking, that a pronounced difference in the temperature dependencies of the intensity, as well as the linewidth, of quasi-elastic electronic scattering between the polarised and depolarised scattering configurations is readily observable.

To understand the data obtained it is necessary to briefly consider results of the developed hydrodynamic model of the electron gas (Bairamov et al 1993). In order to describe the temperature dependence of the cross-section in the case of parallel polarisations $e_i \parallel e_s$ one should take into account both contributions from energy and momentum density fluctuations

$$I_{\varepsilon,p} = \left(\frac{e^2}{mc^2}\right)^2 \left[\varsigma B_p(\omega_1)\right]^2 V \left(\frac{\delta n}{\delta \varsigma}\right)_T T\left[1+\left(\frac{10T}{\varsigma}\right)^2\right], \qquad (2)$$

where V is the crystal volume and ς is the chemical potential of electrons. One more contribution to the quasi-elastic scattering in the case of the depolarised scattering configuration is determined by spin-density fluctuations

$$I_\sigma = \frac{3}{2}V\left[\frac{e^2}{mc^2}B_\sigma(\omega_1)\right]^2 |e_1 \times e_s^*|^2 n\frac{T}{\varsigma}\left(1+\frac{\alpha\,T}{2\,\varsigma}\right) \qquad (3)$$

There is a small parabolic correction to usual the linear temperature dependence in this expression. Here $B_{p,\sigma}$ are band structure dependent coefficients. The ratio of the cross-sections in parallel and crossed polarisations of incident and scattered light is given by

$$\frac{I_\varepsilon}{I_\sigma} = \left(\frac{T}{\hbar\omega_1}\right)^2 D^2 \text{ where } D = \frac{\displaystyle\sum_j \frac{E_{gj}^2+(\hbar\omega_1)^2}{\left[E_{gj}^2-(\hbar\omega_1)^2\right]^2}\left(1+\frac{m_e^\cdot}{m_j^\cdot}\right)}{\dfrac{1}{E_g^2-(\hbar\omega_1)^2}-\dfrac{1}{(E_g+\Delta)^2-(\hbar\omega_1)^2}} \qquad (4)$$

The estimation for n-InP and with $\hbar\omega_1 = 1,17eV, E_g = 1,43eV, E_g + \Delta = 1.54eV$ gives

$$D = \frac{25.77}{0.48} = 53.47; \left(\frac{T}{\hbar\omega_1}\right)^2 = 3.88\cdot10^{-4} \text{ and thus } \frac{I_\varepsilon}{I_\sigma} = 53.47^2 \times 3.88\cdot10^{-4} = 1.11 \qquad (5)$$

So equations (4) and (5) show that mechanisms of spin and energy density fluctuations give comparable intensities for the conditions of our experiments. Theoretical temperature dependencies of the integrated intensities $I_{\varepsilon,p}$ and I_σ from equations (2) and (3) are plotted in Figure 4 by solid and dot-dashed lines respectively, while filled and open circles represent the corresponding experimental points. Adjustable parameters α which is the temperature derivative $d\varsigma / dT /_{T=0}$ and $\alpha = 4, \varsigma = 99 meV$. The dashed line represents theoretical curve for $I_{\varepsilon,p}$

$$I_{\varepsilon,p} = \left(\frac{e^2}{mc^2}\right)^2 \left[\xi B_p(\omega_i)\right]^2 V \left(\frac{\delta n}{\delta \varsigma}\right)_T T \left(1 + \frac{100T}{\varsigma}\right), \tag{6}$$

which is obtained using temperature independent classical value of the electron heat capacity $C_v = \frac{3}{2} K_B n$, K_B being the Boltzmann constant. It gives a poor fit to the experimental points as compared with the solid line. The temperature dependence of C_v is therefore significant in accordance with the rather large value of $\varsigma \gg T$.

Fig.4. Temperature dependence of the integrated cross-section from the same sample as in Figures 1 and 2. Solid and dashed line-theory for energy-momentum density fluctuations corresponding to equations (2) and (6),filled circles are corresponding experimental points available from polarised scattering. Dash-dotted line-theory for spin density fluctuations (equation 3), open circles are corresponding experimental points available from depolarised scattering.

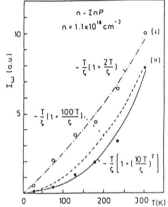

Summary

Finally the new approach to the problem and the observed results provide clear evidence for the existence of strongly temperature dependent free electron gas fluctuations in semiconductors with nonparabolic dispersion of energy bands. This proves unambiguously the observation of light scattering from energy- and momentum-density fluctuations and gives a sensitive measure of the magnitude of the different relaxation times of the electron gas.

References

[1] E.Burstein, M.Cardona, D.J.Lockwood, A.Pinczuk and J.F.Young, In: *Light Scattering in Semiconductor Structures and Superlattices*, by D.J.Lockwood and J.F.Young, Proc. of the NATO Symposium on Light Scattering in Superlattices and Microstructures, Quebec, Plenum Press, New York, 1991, pp 1-17.

[2] A.Pinczuk and G.Abstreiter, In: *Light Scattering in Solids* ed. by M.Cardona and G.Guntherodt, Topics Appl. Phys. V.66, Springer-Verlag, Berlin, 1989, pp 153-211.

[3] B.H.Bairamov, V.A.Voitenko, I.P.Ipatova and V.V.Toporov, Proc.Int.Conf.held in connection with of the birth centenary of C.V.Raman and Diamond Jubilee of the Discovery of Raman Effect, Recent Trends in Raman Spectroscopy ed S.S.Banerjee and S.S.Sha (Singapore: World Scientific 1989) pp 386-398.

[4] B.H.Bairamov, V.A.Voitenko, I.P.Ipatova, Phys.Reports, 229,222(1993).

Materials Science Forum Vols. 173-174 (1995) pp. 209-214
© *1995 Trans Tech Publications, Switzerland*

LASER PULSE INDUCED MICROWAVE CONDUCTIVITY AND SPECTROSCOPIC ELLIPSOMETRY CHARACTERIZATION OF HELIUM AND HYDROGEN PLASMA DAMAGE OF THE CRYSTALLINE SILICON SURFACE

H.C. Neitzert[1], N. Layadi[2], P. Roca i Cabarrocas[2], R. Vanderhaghen[2] and M. Kunst[3]

[1] Centro Studi e Laboratori Telecomunicazioni, Via G. Reiss Romoli 274, I-10148 Torino, Italy
[2] LPICM, Ecole Polytechnique, F-91128 Palaiseau Cedex, France
[3] Abt. Solare Energetik, Hahn-Meitner Institut, Glienicker Str. 100, D-14109 Berlin, Germany

Keywords: Plasma Etching, Crystalline Silicon, Electronic Defects, *In-Situ* Characterization, Microwave Reflection, Spectroscopic Ellipsometry

Abstract. The decay of excess charge carriers, optically generated by an excimer laser pulse at the surface of monocrystalline silicon, has been determined by in-situ microwave reflection measurements. The influence of hydrogen and helium plasmas on the excess charge carrier kinetics has been determined. Both plasma treatments increase the decay rate and decrease the amplitude of the microwave reflection transients dramatically. Spectroscopic ellipsometry measurements performed during the plasma exposure have been fitted assuming the creation of a damaged surface layer. The thickness of this surface layer is increasing with increasing plasma exposure time. While for the helium plasma the creation of electronic defects is mainly an instant effect after plasma ignition, in the case of the hydrogen plasma exposure a continuously increasing number of recombination centers can be observed.

Introduction

Plasma treatments of semiconductor surfaces serve a variety of purposes. Besides plasma etching with reactive gases, passivation of defects [1] and surface cleaning [2] before further wafer processing are pursued. An inherent problem of plasma surface treatments of semiconductors is the introduction of electronic defects (see for example [3,4]). Two classes of plasmas have been intensively studied: firstly the noble gas plasmas, and secondly reactive gases, like CF_4, SF_6, CCl_2 and hydrogen. Hydrogen is used for passivation of electronic defects like silicon dangling bonds [5], but may also influence other properties of semiconductors like the dopant efficiency [6]. Hydrogen has also been reported to etch selectively the disordered phase in the silicon surface [7].
Spectroscopic ellipsometry [8] is a commonly used technique for the characterization of plasma damage at the surface of crystalline semiconductors. This technique gives, via the determination of the optical constants of the investigated layers, detailed information about the surface structure of these materials [9]. For application in electronic and optoelectronic devices it is important to have a characterization technique of the electronic defects at the surface of the material. Angle resolved photoemission spectroscopy has been used to directly determine the surface states introduced by a hydrogen plasma on he crystalline silicon surface [10]. Another possibility to get information about surface defects is the characterization of the kinetics of excess charge carriers in the semiconductors. Besides transient photoluminescence studies [11], the change in the reflection of electromagnetic waves at the semiconductor after the creation of excess charge carriers has also been used for the characterization of plasma-processing. A wide range of the electromagnetic spectrum has been used for the probing, from radio frequencies [12] to microwave frequencies [13] and to the infrared [14].
Some techniques have been demonstrated to enable an in-situ characterization already during the processing (for an overview see for example [15]). This is advantageous because a direct process control is achievable and also contamination of the samples before the measurements can be minimized.
Here we present in-situ microwave detected transient photoconductivity (TRMC) measurements [16] and spectroscopic ellipsometry measurements performed during hydrogen and helium plasma processing of monocrystalline silicon wafers. The combination of both techniques

can give more insight into the surface degradation processes by determining the change in recombination and correlating it to the structure of the processed crystalline silicon surface.

Experimental

Boron doped monocrystalline silicon wafers, <111> oriented, 510μm thick with a nominal resistivity of 8Ωcm, have been stripped of their natural oxide by a short dip in 40% HF, then rinsed with deionized water, dried, and loaded into a plasma reactor within 5min. The plasma reactor has been pumped down to a pressure of 10^{-6} mTorr before plasma start. The substrate temperature was 250°C. For the hydrogen plasma the following parameters have been used: a hydrogen flow rate of 30sccm, a pressure of 120mTorr, and an RF-power of 80W. The self bias voltage of the capacitively coupled RF (13.56MHz) electrode was -140V. For the helium plasma the following parameters have been used: a helium flow rate of 40sccm, a pressure of 300mTorr and an RF-power of 80W. More details about the reactor configuration are given in [17].

Spectroscopic ellipsometry measurements and microwave detected transient photoconductivity measurements have been performed before and after different times of plasma exposure. Both types of in-situ measurements can be performed without interrupting the plasma, but in the present case we stopped the plasma for the measurements in order to compare both for the same time of plasma exposure. For the restart of the plasma a shutter has been used to enable a stabilization of the plasma conditions before the next exposure of the crystalline silicon.

Ellipsometry measurements have been performed in the spectral range 2-4.8eV, using a UV-visible phase modulated ellipsometer described elsewhere [18]. From the fitting of the spectral dependence of the pseudo-dielectric function (as derived from the ellipsometry data) the surface structure of the investigated samples has been determined.

We recall briefly that ellipsometry measures the complex quantity ρ defined by:

$$\rho = r_p/r_s = tg\ \Psi\ e^{i\Delta} \tag{1}$$

where Ψ and Δ are the conventional ellipsometric angles. r_p and r_s are the complex Fresnel reflectivities for light polarized parallel and perpendicular to the plane of incidence, respectively. In the case of reflection at an angle of incidence φ from a sample treated as a semi-infinite medium, ρ is directly related to the pseudo-dielectric function $< \varepsilon >$.

$$< \varepsilon > = < \varepsilon_1 > + < \varepsilon_2 > = \sin^2\varphi(1+((1-\rho)^2/(1+\rho)^2)tg^2\varphi) \tag{2}$$

The morphology (density and surface roughness) and microstructure (amorphous and crystalline phase, microvoids) of the films were computed through the use of the Bruggeman effective medium approximation [9].

TRMC measurements have been done in a configuration where the silicon substrate is terminating the open end of a waveguide, separated only by a small quartz vacuum window, which has been built into the stainless steel substrate electrode. The sample is illuminated homogeneously by 40ns long pulses of an excimer laser at a wavelength of 308nm and an energy of 5μJ from the top side of the substrate, which is exposed to the hydrogen plasma. Because the UV light is strongly absorbed in a thin region of the silicon surface (about 70Å), in this experiment we are mainly probing the change in the surface recombination rate. The microwave setup operates in the frequency range 26.5-40GHz (Ka-band). A detailed description of the setup and the principle of operation are given in [16].

Results and Discussion

In Fig.1 TRMC-transients of monocrystalline silicon substrates have been compared before and after different times of exposure to a helium or to a hydrogen plasma. It can clearly be seen that hydrogen as well as helium plasmas decrease the lifetime of optically generated excess charge carriers drastically.

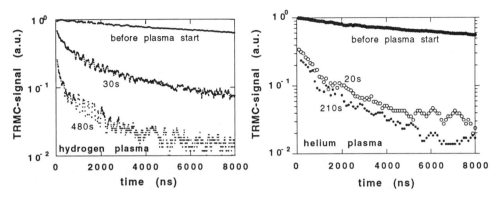

Fig.1   TRMC-transients measured on p-type monocrystalline silicon at 250°C before and at different times of hydrogen and helium plasma exposures

After 30s of hydrogen plasma exposure of the crystalline silicon a much faster decay of the TRMC signal is already observed, with a fast initial decay within less than 1000ns followed by a slower decay for times longer than 1000ns. The transient taken after 480s of hydrogen plasma exposure shows a further decrease of the decay times for times shorter than 1000ns, while the decay on a longer time range does not significantly change compared to the signal taken after 30s. After a short time of helium plasma exposure (in this case 20s), an even more pronounced change in the TRMC-decay can be seen. Also in this case we find two regimes with different decay rates, a very fast decay below 1000ns and a slower one for longer times. The transient measured after 210s of helium plasma exposure, however, shows a similar decay behaviour on both the short and the longer time range like the one taken after 20s.

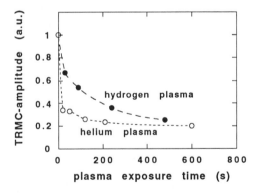

Fig.2   Normalized amplitude of the TRMC-signal measured on p-type monocrystalline silicon at 250°C before and at different times of hydrogen (full circles) and helium plasma (open circles) exposures

The transient microwave conductivity signal under the present measurement conditions has been shown [16] to be directly proportional to the product of the number of excess charge carriers multiplied by their respective mobilities. For both types of plasma treatments the maximum value of the TRMC transients decreases with increasing plasma exposure time. Due to the small absorption length of the excimer laser light in the crystalline silicon, this cannot be explained by a change in the number of absorbed photons [19]. Assuming that the excess charge carrier mobility changes after plasma treatment can be neglected, the lower TRMC amplitudes must be due to fast recombination of a fraction of the optically generated electron-hole pairs on a time scale shorter than the present time resolution. Therefore the value of the TRMC amplitude can be used as a simple measurement for these fast recombination processes.

In Fig.2 the amplitude of the TRMC signals is plotted versus the time of hydrogen and helium plasma exposure. A clearly different temporal evolution of the TRMC-amplitudes during the both plasma treatments is seen. While in the case of the helium plasma there is an immediate drop of the amplitude to a value of about 36% of the original value, in the case of the hydrogen plasma the amplitude at a comparable time drops only to 73% of its original value. During further plasma exposure the amplitude measured during helium plasma exposure remains fairly constant and reaches a saturation value of about 24% of its original value. In the case of the hydrogen plasma however, a more gradual continuous decrease of the TRMC amplitude is seen. The amplitude value, after 500s of plasma treatment has about the same value as in the case of a helium plasma exposure of the same time. In conclusion it can be stated, that the creation of defects responsible for the fast interface recombination has a very different temporal evolution for the two types of plasma , but the final result for long plasma exposure times is about the same.

At the same times at which the microwave reflection transients have been taken, we performed spectroscopic ellipsometry measurements as well. From the results of these measurements we calculated the spectral dependence of the pseudo-dielectric function of the investigated silicon samples. The results for the imaginary part of this pseudo-dielectric function (ε_2) are displayed in Fig.3 for different times of plasma processing.

Before plasma start the spectrum shows two characteristic peaks at 3.4eV and 4.2eV. During the hydrogen plasma exposure the peaks at 3.4eV and 4.2eV decrease and shift slightly to lower energies. A continuous increase of the values of ε_2 is seen for energies below 3eV.

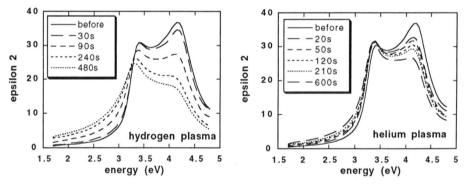

Fig.3   Imaginary part of the pseudo-dielectric function versus photon energy determined by spectroscopic ellipsometry measurements on p-type crystalline silicon measured at 250°C before and after different times of hydrogen and helium plasma exposures.

The spectrum derived from the measurements during helium plasma exposure changes in a different manner. The peak in the ε_2 spectrum at 3.4eV does almost not change with plasma exposure, while the peak at 4.2eV decreases and shifts slightly to lower energies. But the final change of the height of the latter peak is smaller than in the case of the hydrogen plasma exposure. The obtained spectra were fitted with a model assuming the creation of an overlayer with different phases during plasma processing. In Fig.4 the resulting thickness of the overlayer is shown for both plasma treatments. As can be seen, the evolution of the overlayer thickness is quite different for hydrogen and helium plasma treatments. The helium plasma instantaneously creates a 10Å thick damaged surface layer, which in the course of the further plasma exposure increases slightly to a thickness of 25Å. The hydrogen plasma, however, creates a surface layer, which directly after plasma start is thinner than the one created by the helium plasma. During longer hydrogen plasma exposure times the overlayer thickness increases continuously and reaches a saturation value of about 60Å after 500s of exposure time.

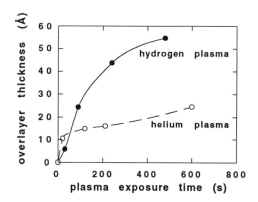

Fig.4   Thickness of the overlayer, created during hydrogen (full circles) and helium (open circles) plasma exposure of p-type crystalline silicon at 250°C as a function of the plasma exposure time.

Conclusion

At long times of exposure, helium and hydrogen plasma exposures of crystalline silicon result in the same decrease of the TRMC amplitude which indicates the creation of similiar electronic defects. However the kinetics of defect creation is faster in the case of a helium plasma.
    This is supported by analysis of the spectroscopic ellipsometry measurements performed at different exposure times. Indeed, the helium plasma immediately creates a defective surface layer while in the case of a hydrogen plasma the thickness of the overlayer increases smoothly. At longer times of exposure the hydrogen plasma results in a thicker damaged layer.

References

[1]   I. Perichaud, H. El Ghitani and S. Martinuzzi, Physica B, **170**, 553 (1991)

[2]   P. Raynaud, J.P. Booth and C. Pomot, Appl. Surf. Sci., **46** , 435 (1990)

[3]   K. L. Seaward and N. J. Moll, J. Vac. Sci. Technol. B, **10**, 46 (1992)

[4]   R. G. Frieser, F. J. Montillo, N. B. Zingerman, W. K. Chu and S R. Mader, J. Elechtrochem. Soc.**130**, 2237 (1983)

[5]   N. H. Nickel, N. M. Johnson and W. B. Jackson, Appl. Phys. Lett., **62**, 3285 (1993)

[6]   G. G. DeLeo, Physica B, **170**, 295 (1991)

[7]   I. Solomon, B. Drevillon, H. Shirai and N. Layadi, J. Non-Cryst. Solids, **164-166**, 791 (1991)

[8]   Y. Z. Hu, M. Li, K. Conrad, J. W. Andrews, E. A. Irene, M. Denker, M. Ray and G. McGuire, J. Vac. Sci. Technol. B, **10**, 1111 (1992)

[9]   D.E. Aspnes, Thin Solid Films, **89**, 249 (1982)

[10]   J. Cho, T. P. Schneider, J. Vanderweide, H; Jeon and R. J. Nemanich, Appl. Phys. Lett., **59**, 1995 (1991)

[11]   C. H. Wang and A. Neugroschel, J. Appl. Phys., **74**, 3257 (1993)

[12]   T. Tiedje, J. I. Haberman, R. W.Francis and A. K. Ghosh, J. Appl. Phys., **54**, 2499 (1983)

[13]   H. C. Neitzert, W. Hirsch and M. Kunst, J. Appl. Phys., **73**, 7446 (1993)

[14]   A. Esser, G. Maidorn and H. Kurz, Appl. Surf. Sci., **54** , 482 (1992)

[15]   G. M. W. Kroesen, F. J. de Hoog, Appl Phys. A, 56, 479 (1993),

[16]   M. Kunst and G. Beck, J. Appl. Phys., **60**, 3558 (1986)

[17]   N. Layadi, P. Poca i Cabarrocas, J. Huc, J.-Y. Parey, and B. Drevillon,
        in: Polycrystalline Semiconductors III - Physics and Technology, Solid State Phenomena,
        Ed.: H. P. Strunk, J. H. Werner, B. Fortin and O. Bonnaud, (Trans Tech, Zürich, 1994),
        to be published

[18]   B. Drevillon, J.-Y. Parey, M. Stchakovsky, R. Benferhat, Y. Josserand and B. Schlayen,
        SPIE Symp. Proc. , Vol. **1188**, (1989), p. 174

[19]   H. C. Neitzert, N. Layadi, R. Vanderhaghen, P. Roca i Cabarrocas and M. Kunst,
        unpublished results

Materials Science Forum Vols. 173-174 (1995) pp. 215-220
© 1995 Trans Tech Publications, Switzerland

LASER DIAGNOSTICS OF InSb SURFACE

G.V. Beketov and A. I. Liptuga

Institute of Semiconductor Physics, Academy of Sciences of the Ukraine, Prospect Nauki 45, Kiev, 252028, Ukraine

Keywords: Electron-Hole Plasma, IR-Laser Diagnostics, Passivating Layer, Crossed Electric and Magnetic Fields, Surface Recombination Rate, InSb

Abstract. The low surface recombination rate in the InSb crystals has been obtained using a new technique of surface treatment. The contactless optical method has been developed to measure the surface recombination rate in the narrow-gap semiconductor at T ~ 300 K. It is shown that lowering the surface recombination rate enables to change within a wide range the surface carrier concentration in the sample placed in the crossed electric and magnetic fields. The modulation of the carrier concentration in a crystal may serve as a basis for development of optical IR-devices.

Introduction

Investigation of the physical phenomena related with generation of non-equilibrium charge carriers in semiconductors have led to development of a number of efficient devices with a wide range destination. Among them an important place is possessed by the devices developed on the basis of the magnetoconcentration effect, the operation element of which is a semiconductor wafer with a thickness of about a charge carrier diffusion length. The parameters of such devices (modulators, light sources, magnetosensors, etc.) are mainly determined by the surface recombination rate. This follows from the fact that the recombination in a thin wafer occurs at the surface where defects and crystal structure damages are concentrated. Therefore, at low values of the surface recombination rate (S) one can achieve the maximum change of the surface and total concentrations of non-equilibrium charge carriers. The operation element surface of the optoelectronic device should be optically perfect, preferably provided with a passivating layer which is transparent in a wide spectral region.

The methods of surface treatment known at present for InSb, which is a material widely used for IR-optoelectronics, do not meet all the demands mentioned above. Good results are obtained using the electrolytic etching method first widely used in manufacturing the InSb diode structures. A comparison between the high-quality chemical etching process [1] and the electrolytic one shows that the latter provides higher performances of the p-n-junction and the InSb surface state close to stoichiometric. Significant investigations of the anodic dissolution process of this material and of the optimum electrolyte composition were obtained in [2]. It was shown that there is a possibility to obtain an undamaged and optically smooth surface with the surface recombination rate of 10^3 cm/s at temperatures ≤ 200 K. In the room temperature region, which is of importance from practical point of view, the surface recombination rate is by more than an order of magnitude higher.

In this paper we discuss the results of the fabrication of high quality surface of InSb with low S in the room temperature range of T≈300 K. The technique of studying S in the room temperature range based on the analysis of laser radiation parameters in the case of radiation interaction with semiconductors is described. We have carried out the measurements of S and the non-equilibrium charge carrier concentration near the sample surface which has been induced by placing the sample in the crossed electric E and magnetic H fields .

Experimental procedure

To obtain the perfect surface of InSb single crystals we used the method of electrolytic polishing. The conditions of etching and the electrolyte composition were similar to those used in [2]. After electropolishing the passivating layer was formed on the sample surface by the method of anodizing in the polysulphide electrolyte. An investigation of the single crystal surface microprofile, using the electron microscopy, gives the microhill height on the surface to be no more than a few tens of $\overset{0}{A}$. Earlier [3] it was found that a similar passivating layer, which is transparent in the 7 to 15 μm spectral range, leads to a reduced surface recombination rate in InSb down to 2 to $5 \cdot 10^2$ cm/s at T=77 K. The data known from published papers for the surface recombination rate of narrow-gap semiconductors are mainly presented for low temperatures. An insufficient information on studies of the surface generation-recombination at higher temperatures is probably related to the problems of measuring the non-equilibrium processes in low-resistance crystals. Indeed, in order to determine the surface generation rate in wide-gap semiconductors a charge depletion region is created near to surface whith the use of the field effect or the reverse bias of a p-n-junction. This method is widely used for the determination of the surface recombination rate, for instance in Ge, Si, and it can not be used for an investigation of narrow-gap semiconductors at higher temperatures.

Let us consider the method we have developed for measuring the surface recombination rate in narrow-gap semiconductors.

The quasineutral distribution of carriers (n=p) in a semiconductor wafer with the intrinsic conductivity in the crossed E and H fields is described by the equation [4,5]:

$$f'' + \gamma f' = \beta^2 (f)(f - 1) , \tag{1}$$

$$[f' + \gamma f \pm S_\pm (f - 1)]_{\xi = \pm 1} = 0 , \tag{2}$$

where

$$f(\xi) = \frac{n}{n_i}, \quad \xi = \frac{y}{d}, \quad \gamma = \frac{ed(\mu_n + \mu_p)EH}{2ckT}, \quad S_\pm = \frac{s_\pm d}{D},$$

$$\beta^2(f) = \frac{d^2}{D} \left[\frac{1}{\tau_1} + \frac{f+1}{2\tau_2} + \frac{f(f+1)}{2\tau_3} \right],$$

$$D = \frac{2kT}{e} \frac{\mu_n \mu_p}{\mu_n + \mu_p} \left[1 + \frac{\mu_n \mu_p}{c^2} H^2 \right]^{-1},$$

$S_\pm$ is the surface recombination rate at the surfaces $y = \pm d$ ($S_+ = S_{max} \gg S_{min} = S_-$), μ_n and μ_p are the carrier mobilities at $H=0$, D is the bipolar diffusion coefficient, τ_1, τ_2, τ_3 are the times of linear, quadratic and cubic recombination, n_i is the intrinsic concentration, k is the Boltzmann constant, n is the electron concentration.

In the crossed E×H fields, the concentration of non-equilibrium carriers near the surface with low S may be either more than the equilibrium one (enrichment regime, i.e., $np/n_i^2 > 1$) or less than it (depletion $np/n_i^2 < 1$). Consequently, the band-to-band recombination radiation intensity is modulated and, with regard to an area unit of the surface $y=-d$ of the semiconductor wafer under consideration ($-d \le y \le d$), it can be given as:

$$P = \frac{hq}{\pi^2 c^2 (1+q)^2} \int_{E_g/h}^{\infty} d\omega \int_{-d}^{d} dy\, \alpha\omega^3 \frac{np}{n_i^2} \exp\left[-(y+d)\alpha - \frac{h\omega}{kT}\right]. \tag{3}$$

Under weak fields and under the conditions of depletion, i.e., when the pair concentration differs slightly from its equilibrium value, we obtain from (1)-(3) in the linear approximation with regard to γ:

$$\frac{P - P_0}{P_0} = \frac{2\gamma}{S_- + \beta}, \tag{4}$$

where P_0 is the intensity of the equilibrium ($np/n_i^2 = 1$) radiation from a wafer in the spectral region of $\omega \ge E_g/h$. We obtain from (4) the expression which enables, using experimental data, to determine the surface recombination rate:

$$S_- = \frac{2\gamma P_0}{P - P_0} - \beta. \tag{5}$$

It should be noticed that the results of measurements in this case will correspond to the values of the surface recombination rate averaged over the crystal surface. However, frequently, for a number of cases, including practical ones, it is of importance to get information on the local value of S in a selected surface area. One can solve this task using the method which is similar to that described above and which is based on laser probing the semiconductor surface. It consist of the following.

Let us consider light incident on the semiconductor surface at $y=-d$ which is a monochromatic wave with a frequency ω ($\omega \ll E_g/h$). Interaction of radiation with semiconductor is characterized by the fact that due to effects of the crossed fields the non-equilibrium carriers are distributed non-uniformly over a sample. Under such circumstances the refraction index can be expressed as $q(y)=q_0 + \Delta q(y)$, where q_0 is the equilibrium value of the refraction index, $\Delta q(y)$ is the value added to it and caused by an external perturbation. Let us assume that a characteristic non-uniformity scale $\Delta q(y)$ exceeds the wavelength λ in a medium, i.e., $(\lambda/2\pi)|d\ln q/dy| \ll 1$. Then the expression for relative changes of the reflection coefficient R is:

$$\frac{R - R_0}{R_0} = \frac{4\Delta q(0)}{q_0^2 - 1}. \tag{6}$$

where $q(0)$ is the refraction index at the semiconductor boundary, R_0 - is the equilibrium value of the reflection coefficient.

. In the intrinsic semiconductor with the electron-hole pair concentration n_i the refraction index is given (for $\omega\tau \gg 1$) as:

$$q_0 = \left[\kappa_0 \left(1 - \frac{\omega_p^2}{\omega^2} \right) \right]^{1/2} \tag{7}$$

where

$$\omega_p^2 = \frac{4\pi e^2 n_i}{\kappa_0 m*}, \qquad \frac{1}{m*} = \frac{1}{m_n} + \frac{1}{m_p}, \tag{8}$$

κ_0 is the lattice permittivity, $m_{n,p}$ are the effective masses of an electron and a hole. Now we can easily obtain the change of $q(0)$ in consequence of perturbation:

$$\Delta q(0) = \frac{\kappa_0}{2 q_0} \frac{\Delta \omega_p^2}{\omega^2}, \tag{9}$$

$$\Delta \omega_p^2 = \frac{4\pi e^2 \Delta n(0)}{\kappa_0 m*}. \tag{10}$$

It is worth mentioning that (7) remains valid if $\omega_c \ll \omega$, where $\omega_c = eH/m*c$ is the cyclotron frequency.

Substituting (9) into (6), we obtain the expression which connects the relative change of the reflection coefficient with the surface concentration of non-equilibrium carriers and enables to determine its value:

$$\frac{R - R_0}{R_0} = -\frac{8\pi e^2 \Delta n(0)}{q_0 (q_0^2 - 1) \omega^2 m*}. \tag{11}$$

Since the reflection coefficient modulation and the recombination radiation intensity are controlled by the non-equilibrium carrier concentration, then one can obtain Eq. (5), where the values of R, R_0 appear instead of P, P_0.

Experimental Results

The measurements were carried out at T=300 K for the n-type InSb samples with the impurity concentration $N_d - N_a = 1.2 \cdot 10^{14}$ cm^{-3}. The samples had dimensions of 2.5×7×0.04 mm^3. The CO_2 laser with the wavelength of λ=10.6 μm was used as a probing radiation source.

Presented in Fig.1 are the results of measurements of the surface recombination rate, obtained using the technique described above, and the data of other authors. It can be seen that in our case the surface recombination rate has the value of about $7 \cdot 10^3$ cm/s. This value is comparable to the results of other authors who applied other treatment procedure and only for considerably lower temperatures. It should be noticed that after short (about 30 min) low-temperature heating (T=360 K) we observed small increase of the surface recombination rate. We have not investigated this phenomen in this paper. The most probable reason for the decrease of the surface recombination rate is a decrease of the surface electron state density, a part of which is of a recombination nature [11].

Let us further consider the results of studying the non-equilibrium charge carrier concentration at the crystal surface under the conditions of crossed fields (cf. [6]). Shown in Fig. 2 are the field dependences of the surface carrier concentration in the chemically etched InSb sample (curve 1) and electrolytically etched one with subsequent passivation of its surface (curve 2). It can be seen that, in spite of the highly developed Auger-recombination process (T=300 K),it is possible to achieve, because of the low surface recombination rate, the maximum value of the crystal surface carrier concentration, which is $1.8 \cdot 10^{17}$ cm^{-3} , i.e., one observes a tenfold excess of the surface concentration above its equilibrium magnitude. Creation of such a high non-equilibrium carrier concentration in InSb is not a simple task. One way to solve it is the use of the laser excitation. In this case one may use, for instance, the ruby or neodimium laser with the intensity of about 10^{23} photons/cm·s, which works in the Q-switching regime. However the duration of holding the non-equilibrium state in the case of laser excitation (20-50 ns) is only as long as a laser pulse duration. In the case of crossed fields the exciting pulse duration is considerably longer (5-10 µs) and it is limited only by the Joule heating. Application of the described technique for treatment of the InSb crystal surface can serve as a basis for the development of new devices with improvement of parameters. In particular, a possibility to vary in a wide range the surface concentration of non-equilibrium carriers should enable to increase the modulation amplitude of the interference modulator, to widen the spectral operation range of the modulator based on the principle of semiconductor reflectance control in the plasma resonance region [12], to increase the IR-source radiation intensity, etc.

References

[1] H.L. Heneke, J.Appl.Phys. **36**, 2967 (1965).

[2] E.P. Matsas, V.I. Chaikin, V.K. Malyutenko, O.V. Snitko, Elektrokhimiya **2**, 1873 (1975).

[3] E.A. Salkov, G.V. Beketov, O.A. Yakubtsov I.A. Molchanovskii, A.I.Noskov,: *Composition and Photoelectric Properties of the InSb Surface Passivated in the Polysulphide Electrolyte* (Trans All-Union Conf. Photoelectric Phenomena, USSR, Ashkhabad,1991).

[4] S.S. Bolgov, V.K. Malyutenko, V.I. Pipa, Ukr. Fiz. Zhur. **31**, 247 (1986).

[5] S.S. Bolgov, V.K. Malyutenko, V.I. Pipa, Pis'ma Zh. Tekhn. Fiziki, **5**, 1444 (1979).

[6] L.A. Almazov, A.I. Liptuga, V.K. Malyutenko, Fizika i Tekhnika Poluprovodnikov, **13**, 52 (1979).

[7] A.P. Medvid, V.I. Chaikin, Izv. AN Lat. SSR, N2, 76 (1977).

[8] H. Fuyisada, J. Appl. Phys. **45**, 3530 (1974).

[9] A.F. Kravchenko, B.V. Morozov, E. I. Skok, Fizika i Tekhnika Poluprovodnikov, **8**, 2035 (1974).

[10] E.L. Davis, Bul. Amer. Phys. Soc. **6**, 18 (1961).

[11] V.I. Chaikin, V.K. Malyutenko: *Properties of MOS Structures Based on InSb* (Trans III Intern. Conf. Phys. Narrow-Gap Semicond., Warsaw, Poland, 1977).

[12] L.I. Berezhinsky, A.I. Liptuga, V.K. Malyutenko, Infr.Phys. **23**, 33 (1983).

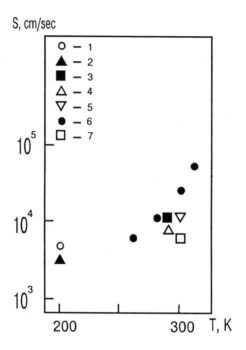

Fig.1. Temperature dependence of the surface recombination rate in InSb :
1 - [6]; 2 - [7]; 3 - [8]; 4 - [9]; 5 - [10];
6, 7 - our data, chemical and electrolytic etching, respectively

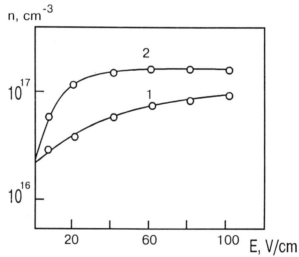

Fig.2. Field dependence of the surface carrier concentration at T=300 K and H=7 kOe:
1 - etching in CP - 4; 2 - electrolytic etching

Materials Science Forum Vols. 173-174 (1995) pp. 221-226
© 1995 Trans Tech Publications, Switzerland

CONTACTLESS CHARACTERIZATION OF SEMICONDUCTORS USING LASER-INDUCED SURFACE PHOTO-CHARGE VOLTAGE MEASUREMENTS

A. Abbate[1], P. Rencibia[2], O. Ivanov[3], G. Masini[4], F. Palma[4] and P. Das[2]

[1] Benet Labs, Watervliet, NY 12189-4050, USA

[2] Electrical, Computer and System Engineering Department, Rensselaer Polytechnic Institute, Troy, New York, USA

[3] Bulgarian Academy of Sciences, Georgi Nadjakov Institute of Solid State Physics, Sofia, Bulgaria

[4] Department of Electronics, University „La Sapienza", I-00100 Roma, Italy

Keywords: Semiconductor Characterization, Surface Charge, Surface States, ZnSe, Spectroscopy

Abstract. A new technique for evaluating the electrical properties of semiconductor wafers and devices using Surface Photo-Charge Voltage (SPCV) measurements is here presented. SPCV measures the change in the surface electrical charge induced by a chopped laser light whose photon energy exceeds the bandgap energy of the semiconductor sample. This charge is measured capacitatively, thus SPCV measurements do not require the fabrication of metal contacts. In Photo-Charge Voltage Spectroscopy measurements the SPCV is measured as a function of the energy of a sub-bandgap monochromatic steady-state illumination, and its derivative spectrum is associated with the density of surface states. A qualitative analysis of the proposed measurement is here presented along with experimental results performed on GaAs samples passivated with a thin ZnSe film of variable thickness. The proposed technique is completely contactless and it can be used as an in-line nondestructive characterization of semiconductor wafers during the various stages of integrated circuits fabrication.

Introduction

Characterization of semiconductor materials and devices plays an important role in the area of solid state device processing. As the level of integration of electronic circuits increases, the performance of such devices is heavily affected by the presence of impurities or defects. Many techniques have been developed for characterizing defects states at semiconductor surfaces and interfaces, i.e. Capacitance-Voltage[1] and Deep Level Transient Spectroscopy[2] (DLTS). Various optical techniques have also been used for many years for investigating semiconductors. In particular, Optical-DLTS[3], Surface Photovoltage[4] and Photocapacitance[5] measurements are the most interesting. These methods are commonly based on the detection of a voltage or of the measurement of a capacitance from a p-n junction or Schottky barrier with the measured signal normally depending on three parameters: diffusion length of minority carriers, surface recombination rate, and the resistivity of the sample.

A new method for the investigation of the surface properties of semiconductor samples is here introduced. The photo-induced surface charge is capacitatively measured as a voltage, here referred to as the Surface Photo-Charge Voltage (SPCV), while the sample is illuminated with a steady monochromatic light of variable wavelength. This monochromatic light is used to create a variation in the steady state population of trap levels in the space charge region. This variation does

*This work was partially supported by NATO joint program with Grant No. 0750/87.

result in a change in the measured voltage and from the derivative of the measured SPCV as a function of the energy of the monochromatic light, the energy position and the relative density of trap levels can be estimated.

This paper is devoted to a description of the Photo-Charge effect and to an experimental demonstration of the Photo-Charge Voltage Spectroscopy technique for the investigation of surface states in semiconductors. In particular results are shown for GaAs samples on which ZnSe films were grown by MOCVD process. For small thicknesses of the film, it is found that ZnSe is a good passivating insulator for GaAs.

Experimental Setup

A block diagram of the experimental set-up used in the photo-charge voltage measurement is shown in Fig. 1. The He-Ne laser, used to optically excite the sample, is modulated as a sequence of bursts, using either a mechanical chopper or an acousto-optic modulator, with variable pulse length and period. The typical repetition rate for the chopper is 15 Hz with light pulses of 4-ms in duration, and the amplitude of the detected signal is of the order of mVolts. The light is incident on the sample placed in a metal box to shield external electrical disturbance. The design of the metal box holding the sample is very important for obtaining large signal-to-noise ratio[7]. The output electrical signal is obtained from a metal plate which is pressed against the back surface of the sample. The front contact is made by a transparent metal plate with a dielectric spacer.

The instrumentation to monitor the SPCV is relatively simple. The output of the sample is connected to a very high input impedance amplifier with adjustable band pass filter. The output of the amplifier is connected to a digital oscilloscope (Hewlett Packard 54100D) and to a lock-in amplifier (EG&G Princeton Applied Research 186A). The synchronization to the oscilloscope and the lock-in amplifier is supplied by the chopper. The wavelength of the monochromatic light is varied and the relative amplitude of the SPCV is monitored with the Lock-In amplifier. All data is memorized in the computer. The He-Ne laser has a measured output power of 7 mWatt, with a calculated photon flux Φ_L of 2×10^{18} cm^{-2}sec^{-1}. A High-Intensity Baush & Lomb monochromator, with output power of 0.18 mWatt, was used in the presented experiments. The calculated photon flux Φ_M is 5×10^{14} cm^{-2}sec^{-1}, which is high enough to generate a variation in the surface charge, but still is order of magnitudes less than the laser probe.

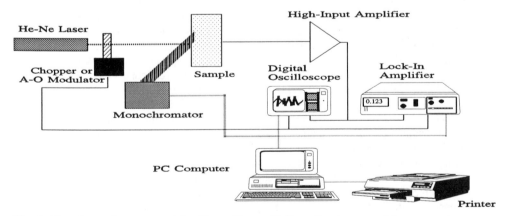

Fig. 1 Experimental set-up used for Photo-Charge Voltage Spectroscopy Measurements.

Qualitative Description of the Photo-Charge Effect

Light incident on a semiconductor can cause the generation of carriers, provided the energy of incident photons is larger than a critical value[6]. The photo-charge effect[7], here described, is observed in semiconductors with lights of all photon energy and also in conducting materials[8].   In metals, the effect can be explained by considering the attenuation of an electromagnetic wave incident to a conductive surfaced[9].   For semiconductors, the photo-charge effect is also dependent on the presence of surface charges due to trap levels.

SPCV is generated by the modulated laser beam, which induces a variation in the surface charge. Since the SPCV is measured capacitatively, no net current is present across the sample, thus the photo-induced charge is constrained within the space charge region of the semiconductor sample, resulting in a redistribution of the total charge compensated by a decrease in the potential inside the semiconductor space charge region. In steady state, the optical generation of free charge carriers is thus balanced by the recombination at the surface and by the diffusion in the bulk. The amplitude of the detected voltage is calculated as:

$$V_L = \frac{q \cdot w \cdot \Phi_L}{\left[S_p + \frac{L_p}{\tau_p} \exp\left(\frac{q\psi_s}{KT}\right) \right]} \cdot \frac{1}{C_{SC} + q \cdot N_{SS}} \tag{1}$$

where q represents the electron charge. From eq. 1, it is clear that V_L is function of: (a) the surface space region, through the surface band bending ψ_s, the width of the space charge region w and its capacitance C_{sc}; (b) the surface states, through the surface recombination velocity and the density

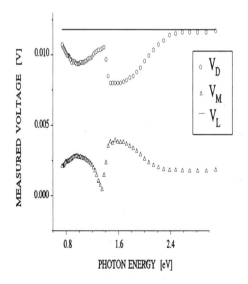

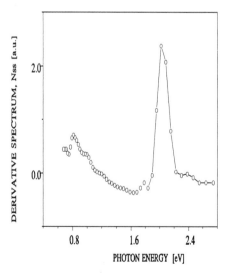

Fig. 2 Plot of the experimental Photo-Charge Voltage measured on a GaAs sample. A detailed explanation is found in the text.

Fig. 3 Plot of the relative density of the surface states N_{SS}, calculated using the derivative spectrum A_P.

of surface states (S_p, N_{ss});(c) the bulk trap levels, through the lifetime and average depth (τ_p, L_p).

The response of the semiconductor sample to the laser illumination is thus influenced by the equilibrium conditions existing in the material prior to the laser pulse. Under illumination due to a monochromatic light of energy below bandgap, a voltage V_M is also measured using our set-up. In this case, free carriers are generated by optical transitions assisted by trap levels, and the term $q\Phi_L$ in eq. 1 is replaced by the product $q\Phi_M\sigma^\circ N_t$ where σ° and N_t represent the optical cross section and the density of the trap level involved in the photo-generation process, respectively.

Photo-Charge Voltage Spectroscopy

In Photo-Charge Voltage Spectroscopy measurements, the sample is illuminated by a steady state monochromatic light, and the laser is pulsed. When the sample is illuminated by a constant monochromatic light of energy $E_M = h\nu_M$, and with an intensity $\Phi_M < \Phi_L$, the conditions of the semiconductor sample, prior to the laser pulse, are varied. In this case, the measured voltage is referred to as V_D. In Fig. 2 the Photo-Charge voltage V_M is plotted as a function of the energy E_M of the monochromator, and for comparison also the amplitude of the SPCV due to the laser (V_L) is plotted as a straight line. If the sample is illuminated by a steady monochromatic light, the amplitude of SPCV is related to the conditions in which the semiconductor is found prior to the laser pulse. If the intensity of the monochromator is high enough to create an excess of free carriers in the space charge, but is still less than the intensity of the laser, the amplitude of the measured voltage V_D can be related to the difference between V_L and V_M. A plot of the experimental V_D as a function of E_M is also shown in Fig. 3. The relationship between the three SPCV measured in different conditions is clearly perceived, and it is used in the following to evaluate the density of trap levels. To extract the density of the trap levels using Photo-Charge Voltage Spectroscopy measurements, a new quantity A_P is introduced:

$$A_P(E_M) \doteq \frac{V_L - V_D}{V_L} = \frac{\Phi_M}{\Phi_L} \int \pm \sigma^\circ(E,E_M)\frac{dN_t}{dE}\, dE \qquad (2)$$

where the integral is extended to all energies. The $\pm$ sign is due to the fact that a transition can either increase or decrease the total concentration of minority carriers in the space charge region. It has also to be pointed out that the quantity A_P is dimensionless, whose first derivative as a function of the energy E_M can be represented as:

$$\frac{dA_P}{dE_M} \propto \int \pm \frac{dN_t}{dE}\frac{d\sigma^\circ(E,E_M)}{dE_M}\, dE. \qquad (3)$$

As a first approximation, the optical cross section $\sigma^\circ(E,E_M)$ can be represented by a step function with its derivative given by a delta function at $E=E_M$. Hence the observed derivative spectrum of A_P plotted as a function of E_M is a representation of the trap density of states. A plot of the derivative spectrum of A_P is given in Fig. 3 for the V_D data of Fig. 2. The transition due to the GaAs bandgap is easily discerned along with a distribution of states in the 0.8 eV range.

Characterization of the ZnSe/GaAs interface using Photo-Charge Voltage Spectroscopy

An increasing interest has been given lately to compound semiconductors such as GaAs for the possibility of obtaining faster electronic devices. Unfortunately the surface properties of GaAs are

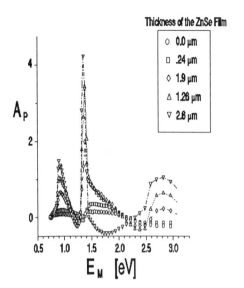

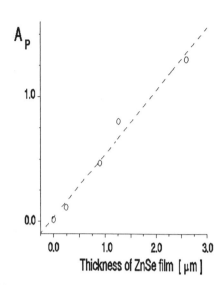

Fig. 4 Experimental SPCV spectra measured for GaAs samples with different ZnSe thickness.

Fig. 5 Amplitude of the Ap term as a function of the thickness of the ZnSe film.

relatively poor compared to Si, with a detrimental effect on the performance of surface oriented devices. Proper passivation of the unsatisfied bonds at the GaAs surface is increasingly important as the device sizes shrink. Many attempts have been made to improve the surface quality of GaAs using anodic oxides, photochemical treatments and other insulating materials. However these efforts have not yet yielded completely satisfactory results[10]. An alternate method of passivating the GaAs surface is to grow an epilayer of a high quality semiconductor which ties up the dangling bonds at the interface.

ZnSe, with its large bandgap, and compatible thermal expansion coefficient and lattice constant (0.25% mismatch), is an attractive choice for use as a passivation layer. It has been shown that thin pseudomorphic ZnSe films, grown by epitaxy on GaAs, reduce the surface recombination velocity (S.R.V.) at the interface. The effect of a ZnSe epitaxial layer on the surface properties of GaAs was studied by comparing the SPCV derivative spectra for samples of different ZnSe epilayer thicknesses. The relative density of trap levels was thus determined as a function of the thickness of the ZnSe film. The experimental SPCV curves are shown in Fig. 4. The bandgap of the ZnSe is estimated to be 2.667 eV [11,12], in that range of energies, the V_D is a linear function of the thickness of the ZnSe film. A plot of the peak value of A_P for energies higher than 2.7 eV as a function of the thickness of the ZnSe film is given in Fig. 5, this result can be explained by replacing the quantity w with the film thickness d in eq. 1. The maximum value of the SPCV derivative spectrum for energies below the bandgap of the GaAs is plotted in Fig. 6 as a function of the thickness of the ZnSe film. For comparison, the density of the interface defect density[11] and the surface recombination velocity[12], measured as a function of ZnSe thickness, are also plotted in figure. It is interesting to note that the density of surface states, and thus the surface recombination velocity, initially decreases for the pseudomorphic ZnSe layer and increases for thicker samples. This is an expected result, since the small lattice mismatch (0.25%) between ZnSe and GaAs generates a uniform elastic strain between the layer and the substrate. The amount of strain is proportional to the thickness of the ZnSe film until a "critical thickness" is reached and the strain is relieved by the formation of defects such as misfit dislocations [13].

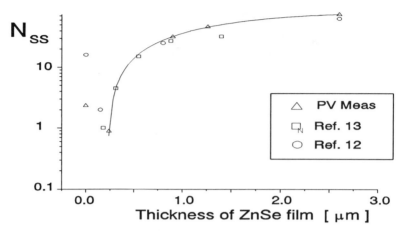

Fig. 6 Plot of the relative values of the density of surface states as a function of ZnSe thickness, obtained by SPCV measurements, and compared with other results.

Conclusions

A new technique for the characterization of surface properties of semiconductor samples using Surface Photo-Charge Voltage measurements, was introduced. The change in the surface electrical charge induced by illumination, is capacitatively measured as a function of the wavelength of a monochromatic steady-state illumination. GaAs samples passivated with thin ZnSe films were analyzed using SPCV technique and an increase in the density of surface states was observed.

References

[1]   C.N. Berglund, IEEE Trans. Electron Dev. 13, 701 (1966).
[2]   D.V. Lang, J. Appl. Phys. 45, 3023 (1974).
[3]   A. Chancre, G. Vincent and D. Bois, Phys. Rev. B 23, 3335 (1981).
[4]   R.H. Bube, *Photo-Conductivity of Solids*, (ed. by Krieger, Huntington, 1978), p. 135
[5]   A.M. White, P.J. Dean and P. Porteous, J. Appl. Phys. 47, 3230 (1976).
[6]   H.C. Grimmeis and C. Ovren, J. Phys. E: Sci. Instrum. 14, 1032 (1981).
[7]   P. Das, V. Mihailov, O. Ivanov, V. Gueorgiev, S. Andreev and V.I. Pustovoit, Electron Dev. Lett. 13, 291 (1992).
[8]   V.I. Pustovoit, M. Borrisov and O. Ivanov, Solid State Commun. 72, 613 (1989).
[9]   V.I. Pustovoit, M. Borrisov and O. Ivanov, Phys. Lett. 135, 59 (1989).
[10]   C.W. Wilmsen, *Physics and Chemistry of III-V Compound Semiconductor Interfaces* (Plenum New York 1985), p. 165.
[11]   D.J. Olego, J. Vac. Sci. Technol B 6, 1193 (1988).
[12]   K.J. Han, A. Abbate, I.B. Bhat, S. Akram and P. Das, J. Appl. Phys. 74, 364 (1993).
[13]   D.J. Olego and D. Cammack, J. Cryst. Growth 101, 546 (1990).

Materials Science Forum Vols. 173-174 (1995) pp. 227-230
© 1995 Trans Tech Publications, Switzerland

NARROWING OF PHOTOLUMINESCENCE LINE FROM SINGLE QUANTUM WELL UNDER HIGH EXCITATION LEVELS

D.G. Revin and V.Ya. Aleshkin

Institute for Physics of Microstructures of the Russian Academy of Sciences,
603600 Ulyanov str. 46, N. Novgorod, Russia

Keywords: Superluminescence, Quantum Well

Abstract. Photoluminescence from heterostructures made of single quantum well as GaAs/In$_x$Ga$_{1-x}$As and In$_x$Ga$_{1-x}$P/GaAs/In$_x$Ga$_{1-x}$As systems was studied at excitation powers from 10^3 to 10^6 W/cm^2 normally to the quantum well plane. The narrowing of the photoluminescence line was observed with the heterostructure made of In$_x$Ga$_{1-x}$P layers. The anisotropy of the optical losses in In$_x$Ga$_{1-x}$P/GaAs/In$_x$Ga$_{1-x}$P waveguide leads to a high polarization of this line.

Introduction

Photoluminescence (PL) is one of the most useful techniques for investigation the electron spectrum in a quantum well. Usually the conditions of PL observation are such that only a small part of the electron states in the quantum well is filled with photoexcited carriers. This occurs either because the excitation powers used do not exceed 10^3 W/cm^2, or when the quantum well collects photoexcited carriers from a small area around it [1-3]. At strong excitation intensities superior to 10^3 W/cm^2, a considerable part of the electron states in the quantum well was occupied by photocarriers. In this case the many-particle interaction in a dense of two-dimensional electron-hole system in the quantum well leads to renormalization of energy gap, as well as a change of the particle energy-momentum dispersion and transition probabilities [4].

In this work we have investigated PL emitted by the weakly doped In$_x$Ga$_{1-x}$As single quantum well at strong excitation powers (from 10^3 to 10^6 W/cm^2) in two types of the heterostructures: with and without dielectric waveguide.

Experiments and materials

The heterostructures were grown by MOCVD epitaxy under atmospheric pressure on semi-insulating (001) GaAs substrates. The samples of one type contained 1 μm In$_{0.48}$Ga$_{0.52}$P layer, 0.05÷0.1 μm GaAs layer, ~ 10 nm In$_{0.2}$Ga$_{0.8}$As quantum well, 0.05÷0.1 μm GaAs layer and 1 μm In$_{0.48}$Ga$_{0.52}$P layer. InGaP layers in these structures served as a dielectric waveguide. The other samples (without waveguide) contained ~ 10 nm In$_{0.15}$Ga$_{0.85}$As quantum well and covered with a 0.4 μm thickness of a GaAs layer. The structures were doped to 10^{16} cm^{-3} donor concentration. The 0.53 μm laser of 100 nsec pulse duration was used as a PL excitation source. The PL signal was passed through a monochromator and was detected by cooled photomultiplier with a boxcar integrator for increasing the signal-to-noise ratio. The measurements were carried out at 4.2 and 77 K. PL was investigated normally to the quantum well plane. The heterostructure chips of

5×5 mm² size were used. The laser beam was focused to a spot of 0.5 mm in diameter in the center of the chip.

Results and discussion

The typical 77 K PL spectra are represented in Fig.1 and Fig.2 for the two types of heterostructure at five different excitation powers (P1 - 10^3, P2 - 5×10^3, P2 - 2×10^4, P3 - 10^5, P4 - 10^6 W/cm²). The arrows in the figures indicate the position of calculated optical transitions between different electron and heavy hole quantum-size subbands in these quantum wells. The mismatch between the lattice constants of GaAs and quantum well material gives rise to an elastic deformation of the quantum well lattice. Therefore, the light hole levels were pushed out from the quantum well.

The PL spectra of both types of heterostructure at excitation power of less than 10^4 W/cm² were similar. At the excitation power of 10^3 W/cm², when the filling of the electron states in the quantum well with photoexcited carriers is small, only one maximum in the PL spectra was observed. This maximum corresponded to electron transitions between the nearest quantum-size subbands in the quantum well (e1-hh1 transition). The increasing of the excitation power (more than 10^4 W/cm²) gives rise to new PL maxima appeared in the region of the transition between the second electron and hole quantum-size subbands (e2-hh2).

At the excitation power superior to 10^5 W/cm² when almost all states in the quantum well were occupied by the photoexcited carriers, the PL spectrum from the structures without InGaP layers (Fig.1) was a broad band which included the radiation from the GaAs layers and the transitions between different quantum-size subbands in the quantum well [4,5].

It is clearly seen from Fig.2 that in the case of the structures made of InGaP layers the narrowing of the PL line was observed at strong excitation power. This narrow peak lay at the frequency near the transition between the first electron and heavy hole quantum-size subbands but was not equal to it.

The density of the excitation power under which this narrow peak appeared decreased with increasing diameter of the laser spot on the sample, what leads us to propose if as a superluminescence. In spite of the superluminescence energy propagating along the optical waveguide some scattering in the structure allowed us to observe superluminescence normally to the quantum well plane. At 4.2 K the superluminescence already appeared at excitation levels less than 10^4 W/cm².

The PL polarization measurements carried out under excitation power of 10^5 W/cm² showed a strong (up to 80%) polarization of the superluminescence (Fig.3). On the other hand, a spontaneous luminescence was not polarized. The measurements of waveguide losses showed that polarization of the superluminescence corresponded to the direction of the wave with the largest optical losses. If InGaP lattice constant was more than that of GaAs (the positive mismatch - $\Delta a/a > 0$), the superluminescence intensity was the largest if the vector of the electric field of the wave was directed along [110] (curves 2), in the case of the negative mismatch - along [$\bar{1}$10] (curves 1). Note that for curves 1 only weak long-wavelength peaks correspond to the superluminescence, but short-wavelength peaks correspond to the spontaneous PL. We did not observe the dependence between the polarization of the superluminescence and the exciting laser.

According to the selection rules the electron transitions from the conduction band to the valence band in the quantum well may proceed only under excitation by the electric field component being in the quantum well plane. Thus in this dielectric waveguide mainly TE-wave was excited.

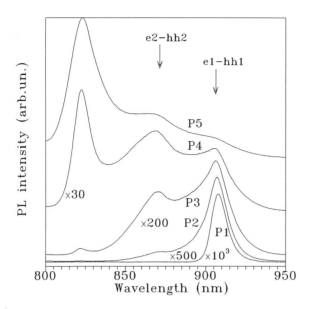

Fig.1 Photoluminescence spectra of the heterostructure without dielectric waveguide at different excitation levels.

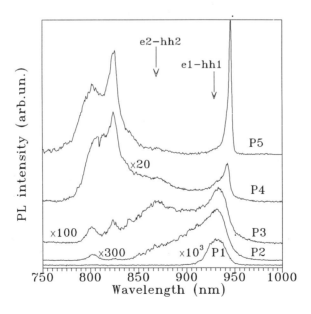

Fig.2 Photoluminescence spectra of the heterostructure with dielectric waveguide at different excitation levels.

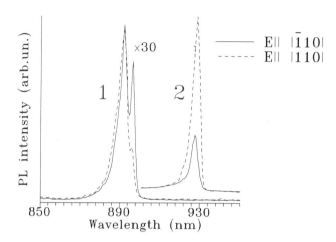

Fig.3 Superluminescence spectra of the heterostructures with negative $\Delta a/a = -2.5 \times 10^{-3}$ (1) and positive $\Delta a/a = 6 \times 10^{-3}$ (2) mismatche between the lattice constants of GaAs and InGaP layers. Solid curves correspond to the direction of the electric field along $[\bar{1}10]$, dashed curves - along $[110]$. Excitation level - 10^5 W/cm².

Conclusion

The superluminescence obtained from $In_xGa_{1-x}P/GaAs/In_xGa_{1-x}As/GaAs/In_xGa_{1-x}P$ heterostructure at 77 K under excitation powers of more than 10^4 W/cm² was observed. The polarization of the superluminescence corresponded to the direction of the wave with the largest optical losses. If InGaP lattice constant was more than GaAs, the direction of the smallest optical losses was $[110]$, if less - $[\bar{1}10]$.

References

[1]　G. Trancle, E.Lach, A.Forchel and F.Scholz, Phys. Rev. B **36**, 6712 (1987).
[2]　G.Bongiovanni and J.L.Staehly, Phys. Rev. B **39**, 8359 (1989).
[3]　E.Lach, G.Lehr, A.Forchell, K.Ploog and G.Weimann, Surf.Sci. **228**, 168 (1990).
[4]　L.V.Butov, V.D.Kulakovskii and T.G.Andersson, Phys. Rev. B **44**, 1692 (1991).
[5]　V.Ya.Aleshkin, Z.F.Krasil'nik and D.G.Revin, Semiconductors **27**(7), 655 (1993).

Materials Science Forum Vols. 173-174 (1995) pp. 231-236
© 1995 Trans Tech Publications, Switzerland

RESONANCE RAMAN SPECTROSCOPY OF LANGMUIR-BLODGETT FILMS FROM METALLO-PHTHALOCYANINES

L.L. Larina[1], N.N. Melnik[2], V.P. Poponin[1], O.I. Shevaleevskii[1] and A.A. Kalachev[3]

[1] N.N. Semenov Institute of Chemical Physics, Russian Academy of Sciences, 4 Kosygin st., 117977 Moscow, Russia

[2] P.N. Lebedev Physics Institute, Russian Academy of Sciences, 53 Leninsky pr., 117924 Moscow, Russia

[3] Institut für Mikrotechnik, Ackermannweg 10, D-55128 Mainz, Germany

Keywords: Phthalocyanines, Langmuir-Blodgett Films, Resonance Raman Spectroscopy

Abstract. Low frequency resonance Raman spectra (RRS) (from 50 cm^{-1} to 800 cm^{-1}) of Langmuir-Blodgett (LB) films from cupper- iron- and cobalt- phthalocyanines (CuPc; FePC and CoPc) have been measured for the first time. Several new lines in the low frequency part of phonon spectra have been identified. The problem of the low frequency Raman lines assignment to metal-nitrogen stretching and bending vibrational modes is discussed. The results are analyzed in the frame of the D$_{4h}$ symmetry model.

Introduction

In recent years there has been much interest in study of metallo-phthalocyanines as an important organic semiconductor material [1,2]. This particular interest is owing to its obvious relationships with the primary quantum conversion process in photosynthetic systems as well as due to its potential applicability in solar energy conversion systems [2,3]. Recently, the photovoltaic properties of phthalocyanines and its metal complexes have been the object of intesive research [3]. Metallo-phthalocyanine derivatives have unique property to exist in the different polymorphic forms in liquid crystalline as well as in solid crystalline phases, depending on the type of peripherial substituents on the aromatic rings and crystallization conditions [2]. One of the most powerful tools to determine MePc structure is vibrational (Infrared - IR and resonance Raman - RR) spectroscopy [4]. A lot of papers have been devoted to the vibrational (IR and Raman) spectroscopy of Metal-Phthalocyanines in the high frequency (from 500 cm^{-1} to 2000 cm^{-1}) part of vibrational spectra. The problems of spectral lines (IR and Raman) assignment for this region of vibrational spectra as well as normal mode analyzes for several types of MePc were addressed in the literature (See [5,6] and references sited therein). At the same time, very little is known about the low frequency part (below 500 cm^{-1}) of MePc vibrational spectra. The assignment of spectral lines to metal - nitrogen stretching and bending vibrational modes in MePc is of particular importance. These lines in MePc as well as in other coordination compounds can provide valuable information about relative bond strength, symmetry of metal surrounding (binding site), substitution on the ligand and changes in coordination number. Because of the complexity of the low frequency vibrational spectra of these compounds, their assignment and interpretation still remains a challenging problem. Since these lines are usually relatively weak the problem of experimental registration and positive identification of these lines in Raman spectra is also of primary importance. So far only few lines in the low frequency part of Raman spectra, which could be assigned to metal-nitrogen stretching and bending vibrations, have been observed and identified for thick evaporated films [7]. Monolayer and multilayer L-B films from MePc have been studied by Surface Enhanced Raman Scattering (SERS) technique [8] as well as by Resonance Raman Scattering (RRS) technique [9], but low frequency bands assignable to metal - nitrogen vibrations have not been positively identified.

In the present contribution we report about the measurements of the low frequency (50 cm^{-1} up to 800 cm^{-1}) resonance Raman spectra (RRS) of Langmuir-Blodgett (LB) films from metal-substituted tetra-*tert*-butil phthalocyanine compounds. Special attention is payed to the study of the low frequency region (below the 500 cm^{-1}) of RRS. Several new Raman lines in the low frequency part of spectra which coud be positively identified and assigned to normal modes related with metal - nitrogen stretching and bending vibrations for CuPc, FePc and CoPC LB films are presented. In conclusion, the problem of assignment and interpretation of the low frequency lines in the Resonance Raman Spectra is discussed in the frame of the D$_{4h}$ symmetry model.

Experimental Procedures

Multilayer LB-films were prepared from CuPc, CoPc and FePc derivatives (Me-Pc[C(CH$_3$)$_3$]$_4$) - tetra-*tert*-butil phthalocyanines) by standard methods described in the literature [10]. The LB-films were deposited on the support from silica plates. Argon - krypton ion laser, which can generate several lines, have been used for excitation of Raman spectra. Two lines of argon-krypton laser (514.5 nm and 647.1 nm) have been used in our experiments. Raman spectra were registrated and analysed with U1000 JOBIN IVON Raman spectrometer. Typical experimental conditions included slit width with spectral resolution of 2 cm^{-1} and s-polarized light 45^0-incidence angle and 90^0- collection of scattered light. The reflection geometry of experimental set up have been used. Laser power level was set in the range from 50 mW to 100 mW.

Experimental results

Me-Pc[C(CH$_3$)$_3$]$_4$ is a derivative of the tetraazabenzoporphyrin in which one hydrogen of each isoindole moiety has been replaced by a *tert*-butyl group. It is well known, that visible absorption spectra of the MePc has two bands near 615 nm and near 700 nm [7]. The 700 nm bands is attributed to monomers, while 615 nm band is assigned to aggregates. One of the two frequencies used for excitation in our experiments (647.1 nm) could be very close to one of the red resonance absorption bands of MePc, so resonance conditions for excitation of Raman spectra have been fulfilled for this excitation frequency. While second line (514.5 nm) used for excitation was far from resonance frequency for all the MePc under the investigation. It should be emphasized, that for more complete study of Raman spectra of MePc it is useful to compare several spectra obtained with different excitation frequences. In particular, in the study reported in the paper [7] three lines 514.5 nm, 593.2 nm and 603.2 nm of argon ion laser have been used for excitation. Unfortunately, all these lines are far from the resonance absorption bands for MePc. Main advantage of the line (647.1 nm) used in our experiments for excitation is due to the fact, that this line is close to resonance absorbtion bands. Due to realization of the resonance excitation conditions it was possible to detect some new lines and to observe some new features of the low frequency part of Resonance Raman spectra. Since most intense Raman lines have been obtained for the CoPc LB films in our study, one should assume that in this case excitation conditions were more close to resonance, then in the case of the CuPc and FePc LB films.

Raman spectra for cupper phthalocyanine L-B film

In order to demonstrate the effect of excitation frequency on the Raman spectra in this section the samples of Raman spectra for the CuPc L-B films obtained with two different excitation wavelengths are presented. In the first case the excitation wavelength (514.5 nm) was far from resonance absorption band for CuPc, while in the second one (647.1 nm) the excitation wavelength was rather close to resonance conditions.

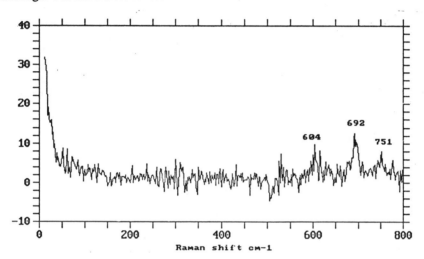

Fig. 1. Raman Spectra for the 30 layer CuPc L-B film on the silica slide. The wavelength of the excitation radiation is 514.53 nm (nonresonant excitation). Silica slide spectra have been subtracted.

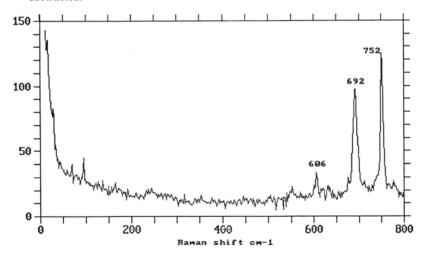

Fig.2. Raman spectra for the 54 layer CuPc L-B film on the silica slide. The wavelength of the excitation radiation is 647.1 nm (resonant excitation). Silica slide support spectra have been subtracted.

In the case of Fig. 1 the intensities of the lines in Raman spectra are rather weak. There are only three lines in that region of spectra: 602cm-1, 694cm-1, 750cm-1 - which could be positively identified. In the low frequency part of the spectra there are no lines wich could be positively identified, so, from these data one could only suggest, that lines 232cm-1 and 312cm-1 could be expected. Further measurements with improved signal-to-noise ratio should be done for more reliable identification of this lines and for verification of that suggestion.

In the case of Fig.2 Raman lines are more intense then in previous one due to resonance excitation. Nevertheless, again only lines 606 cm-1, 692 cm-1 and 752 cm-1 could be positively identified in the Raman spectra. Note, that intensity distribution obtained in resonance Raman spectra (Fig.2.) are essentially different from that observed in nonresonant case (Fig.1). This property may provide an important information about CuPc structure. There are only two lines in the low frequency region of Raman spectra:165 cm-1 and 238 cm-1 -which coud be suggested to appear in the case of improved signal-to-noise ratio measurements. Note, that strong line 94 cm-1 is spurious one since it originates from laser discharge radiation in the case of 647.1 nm argon-krypton laser regime.

Raman spectra for iron phthalocyanine L-B film

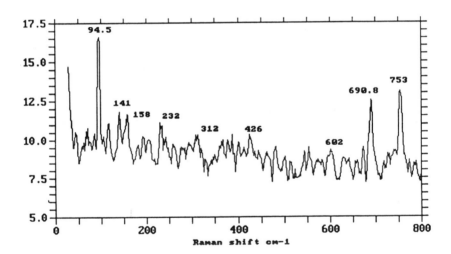

Fig. 3. Raman spectra for the 51 layer FePc LB film on the silica plate. Silica slide spectra have been subtracted. Excitation radiation wavelength is 647.1 nm (resonant excitation).

In the case of FePc the intensities of the lines in Raman spectra are also rather weak even for resonance excitation conditions. Nevertheless, lines 690 cm-1 and 753 cm-1 could be positively identified. There are several weak lines, which could be suggested to be expected for improved signal-to-noise ratio measurements. These lines could not be positively identified, since they are almost hidden in the noise signal of Raman spectrometer, but this data are listed below for completness: 141 cm-1, 158 cm-1, 232 cm-1, 312 cm-1, 426 cm-1, 602 cm-1. Strong spurious line 94 cm-1 is originated from laser discharge.

Raman spectra for cobalt phthalocyanine L-B film

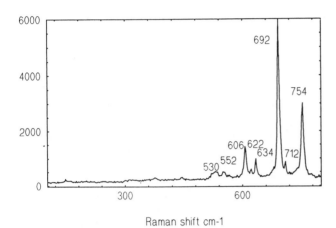

Fig. 4. Raman Spectra for the 50 layer CoPc LB film on the silica slide. Excitation radiation wavelength is 647.1 nm (resonant excitation). Silica slide spectra have been subtracted

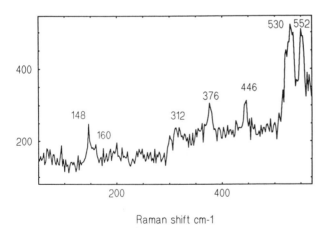

Fig.5. Low frequency part of the resonance Raman spectra presented on Fig. 4.

The most intense Raman lines with highest signal-to-noise ratio values have been obtained for CoPc LB films. There are several new lines, which could be positively identified both in the high frequency as well as in the low frequency parts of resonance Raman spectra. These lines are listed below: 148 cm-1, 160 cm-1, 312 cm-1, 376 cm-1, 446 cm-1, 530 cm-1, 552 cm-1, 606 cm-1, 622 cm-1, 634 cm-1, 692 cm-1, 712 cm-1, 754 cm-1. At least five lines in these spectra (146 cm-1, 160 cm-1, 312 cm-1, 376 cm-1 and 446 cm-1) could be used for normal mode analysis.

Discussion

Since phthalocyanine molecule have square-planar arrangement, their vibrational spectra should be analyzed in the frame of the D_{4h} symmetry model. The D_{4h} point group has the following vibrational representation:

$$\Gamma_{vib} = 14A_{1g} + 13A_{2g} + 14B_{1g} + 14B_{2g} + 13 E_g + 6A_{1u} + 8A_{2u} + 7B_{1u} + 7B_{2u} + 28 E_u$$

The A_{1g}, A_{2g}, B_{1g}, B_{2g} and E_g modes are Raman active from the first principles of vibration theory. Among Raman active modes only E_g modes correspond to out-of-plane vibrations, while A_{1g}, A_{2g}, B_{1g} and B_{2g} modes all correspond to in-plane vibrations.

Low frequency Raman active normal modes calculation have been done in [5] for FePc, but with restriction, that only in-plane vibrations have been accounted for. As a result only two low frequency Raman active normal modes related with Fe - N stretching and bending vibrations: A_{1g} (271 cm-1) and B_{2g} (175 cm-1) have been predicted.

Low frequency Raman spectra for CoPc, presented in this paper, suggest, that there are at least five low frequency Raman active lines (148 cm-1, 160 cm-1, 312 cm-1, 376 cm-1, 446 cm-1), which could be related with Co-nitrogen stretching and bending vibrations. So, one should expect, that out-of-plane vibrations also could contribute in the low frequency Raman spectra. Unfortunately, at present time there are not enough experimental data in this region of spectra for complete normal mode calculations with out-of-plane vibrations being accounted for. So, further more refined measurements of MePc low frequency Raman and IR spectra should be performed in order to understand more profoundly the role of metal-nitrogen vibrations. It was shown in present paper, that MePc LB films are perspective object for the study of the low frequency Raman spectra of the phthalocyanine molecule.

References

[1] *Phthalocyanines Properties and Applications*, ed. by C.C. Leznoff and B.P. Lever, (VCH Publisher Inc., UK, 1989).

[2] J. Simon and J.J. Andre. *Molecular Semiconductors: Photoelectrical Properties and Solar Cells*, ed. J.-M. Lehn, (Springer - Verlag, Berlin, 1985).

[3] D. Wohrle and D. Meissner, Advanced Materials, **30**, 129 (1991).

[4] A.L. Thomas: *Phthalocyanine Research and Applications*. (CRC Press, Boca Raton, 1990).

[5] C.A. Melendres and V.A. Maroni, Journal of Raman Spectroscopy, **15**, 319 (1984).

[6] B. Hutchinson, B. Spencer, R. Thompson and P. Neill, Spectrochimica Acta, **43A**, 631 (1987).

[7] C. Jennings, R. Aroca, A.-M. Hor and R.O. Loutfy, Journal of Raman Spectroscopy, **15**, 34 (1984).

[8] Y. Bai, Y. Zhao, L. Zhang, K. Tian, X. Tang and T. Li, Thin Solid Films, **180**, 249 (1989).

[9] D.P. Dilella, W.R. Barger, A.W. Snow and R.R. Smartzewsky, Thin Solid Films, **133**, 207 (1985).

[10] J.R. Fryer, R.A. Hann and B.L. Eyers, Nature (London), **313**, 382 (1985).

Materials Science Forum Vols. 173-174 (1995) pp. 237-242
© 1995 Trans Tech Publications, Switzerland

RAMAN AND NONLINEAR LIGHT SCATTERING FROM UNDERSURFACE LAYERS OF ION IMPLANTED SILICON CRYSTALS

V.S. Gorelik

P.N. Lebedev Physical Institute, Leninsky pr. 53, 117924 Moscow, Russia

Keywords: Silicon, Raman Scattering of Light, Nonlinear Light Scattering, Order-Disorder Phase Transition, Laser

Abstract. The results of Raman spectra investigations for ion implanted silicon monocrystals are presented. The effects of crystalline lattice succesively disordering have been observed with the help of Raman technique. Second harmonic generation (SHG) properties of the silicon samples, implanted by high and low energy ions, have been investigated. The increase of SHG-signal has been founded for high energy ions implanted samples; such effect was explained as a result of stress emerging in undersurface layers.

Introduction

After laser sources revealing the wide opportunities have been opened for laser Raman spectroscopy. The important objects for Raman investigation are the silicon monocrystals and connected inhomogeneous systems, having numerous applications in modern microelectronics.

Solid silicon is nontransparent in visible region. So with the help of laser sources of visible region from Raman investigations we can receive the information, connected with the thin undersurface layers, corresponding to penetration depth into the sample volume of exciting laser rays. Such depth depends from the laser wavelength and is in range 0,1 - 0,5 mkm.

From the laser Raman measurement we can receive the information concerning the mechanical stresse, the degree of crystalline lattice disordering, the parameters of ion implantation and annealing processes and a number of others important properties of undersurface layers of solid silicon samples.

Ion implanted silicon crystals are used in numerous modern semiconductor arrangements. For the succesful applications of such crystals it is necessary to know the physical properties of disturbed undersurface crystalline layers: the phase state, the concentration of free charges and anothers. In recent experimental works [1-3] there was established that the Raman scattering experiments have given a number of physical characteristics of ion implanted undersurface silicon layers. There was observed that at small implantation doses only the decrease of the Raman intensity had taken place (see work [2]). Such results was explained due to point defect, emerging in crystals after the ion implantation process.

When the ion implantation dose increases the additional Raman satellites have been observed. Such satellites were connected with the crystalline lattice disordering and correlated with the one-photon density of states pecularities. The dose value of the whole lattice disordering depended from the mass and kind of ions.

In the present work we report the new Raman results of the phase states implanted silicon investigations for the several types of implanted ions and for wide region of implantation doses. Besides that, we present the results on properties of second harmonic generation (SHG) for different types implanted samples.

Experimental technique

We have investigated the silicon crystals, implanted by ions B, P, As, Se, Sb with the energy E = 50 keV and doses from 10^{11} to 10^{16} cm $^{-2}$. Raman spectra have been received on experimental setup, described in work [4]. The excitation of spectra was made by argon laser (λ = 488,0 nm). We have used the reflection scattering geometry and the difference Raman technique for comparing of disturbed Raman spectra with the original silicon crystal spectrum. Second optical harmonics experiment have been fulfilled with the help of copper laser (λ = 510,5 and 578,2 nm; P_{max} = 10^4 W/cm^2 , P_{av} = 1 W).

Results and discuccions

In the Raman spectrum of nonimplanted silicon sample we have observed the known 520,5 cm $^{-1}$ fimdamemtal Raman maximum and the second order satellites, corresponding to hugh order scattering phonon processes (Fig.1).

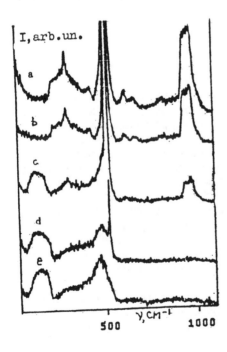

Fig.1. Raman spectra of (111) silicon before (a) and after P$^+$- implantation (E = 70 keV) with dose D = 1*10^{13} cm $^{-2}$ (b);6*10^{13}cm^{-2}(c); 3*10^{14} cm $^{-2}$ (d); 6*10^{14} cm $^{-2}$ (e).

After ion implantation the Raman intensity I of fundamental line 520,5 cm^{-1} decreases. Such result can be illustrared by Fig.1 and 2. At the Fig.2 there is the dependencies of $\xi = 1 - I/I_0$ value from dose of implantation; here I_0 - is the Raman intensity of original silicon sample. The parameter ξ characterizes the degree of lattice disordering and is equal to unity for whole disordered crystals (in this case $I = 0$). We can see that the phase transition to the disordered state takes place at different ions. We can consider that the linear part of $\xi = f(D)$ dependency corresponds to the point defect emerging; the nonlinear behavior of ξ value we explained as a result of sharp absorbtion of exciting light increase due to amorphous phase emerging. In this case the microcrystals of silicon also exist because of in the corresponding Raman spectra there are the

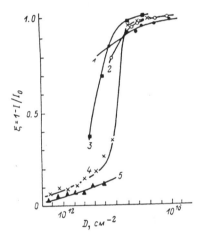

Fig.2. The dependencies of $\xi = 1 - I/I_0$ value from ion implantation dose for different ions (E = 50 keV); Sb (1), As (2), Se (3), P (4), B (5).

additional Raman maximum near the fundamental line due to the amorphous phase. At doses D ~ 10^{16} cm^{-2} the parameter ξ is close to the saturation value; in this case the whole amorphisation takes place. So the value ξ characterizes the type of phase state at different steps of ion implantation process.

For the greater fluences there was observed the intensity redistribution of Raman maxima, corresponding to spectra of amorphous phase. Fig.3 shows the evolution of the Raman spectrum of silicon, implanted with 50 keV Sb$^+$ for the fluences, ranging from 6*10^{13} to 3,6*10^{16} cm^{-2}. One can see that the intensity of wide band 470 cm^{-1} (TO - phonon) decreases while the intensity of 300 cm^{-1} - mode (LA - phonon) remains unchanged under the fluences greater then D = 3*10^{15} cm^{-2}. In the region 160 cm^{-1} (TA - phonon) the new feature appears with the frequency about 90 cm^{-1}. The intensity of this new maximum increasea as the fluence increases (Fig.3). At the fluence D = 3,6*10^{16} cm^{-2} (Sb) the intensities of TO - and LA - bands become comparable, while the lowerfrequency peak becomes sharper and more intensive.

When the silicon was implanted with P, B, As - ions the amorphous spectrum appeared beginning from fluences 3*10^{13}, 10^{15} and 10^{12} cm^{-2} respectively. The further increasing of fluence lead to the Raman spectrum resemble to the density of phonon states. For this ions no additional lines in low-frequency region and no intensity redistribution in amorphous spectrum for fluences up to $1*10^{16}$ cm $^{-2}$ were observed.

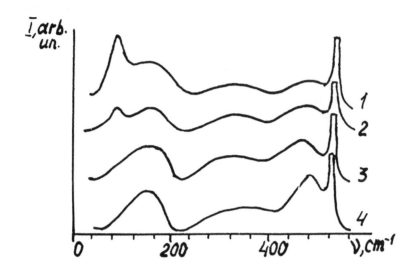

Fig.3. The shape of Raman spectra for different doses of Sb-implanted silicon; 1,2,3,4 - D = $4*10^{16}$, $1*10^{16}$, $6*10^{15}$ and $6*10^{13}$ cm $^{-2}$.

According to the work [5], the width and intensity of TO -peak are connected to the local order in amorphous silicon. In this work there is the proposal that the angular deviation of direction between silicon bonds from regular tetrahedra orientations should increase the width of discussed Raman maximum. If this consideration is valid, the observed widening of TO- maximum tells us about such type disordering. According to [5] the width Γ = 110 cm $^{-1}$ corresponds to $\delta\theta$ = 11°.

The appearing of the additional low-frequency maximum in Raman spectrum of Sb - implanted silicon with doses, greater then D = 10^{16} cm $^{-2}$, can be explained by the creation of small isolated silicon clasters. For such particles the main contribution to the density of phonon states is due to the acoustic resonator modes or so called boson peak. The frequency of such peak is v = v_s/L, where v_s- sound velocity, L - the cluster size. We can conclude, that the probability of the creation of atomic scale isolated clusters is greater for the larger ions (Sb). The boson peak width is connected with the clusters sizes dispersion. According to our calculations, based on Raman parameters of boson peak, the size of such cluster is about 10 Å.

The investigations of SHG effect have been fulfilled for sillicon plates, implanted by B and P with different doses and energies. Table 1 illustrates the changing of SHG signal for P-implanted silicon for different energies and doses.

From this table we can see the abrupt increase of SHG - signal of implanted silicon plate for high energy ions and small ion-implantation doses. Such effect could be understood as a result of stress emerging near the sample surface when high energy ions go into the crystal without essential destruction of surface layer.

Table 1. The dependence of SHG - intensity of silicon,
implanted by P - ion.

No	I, rel. un.	D, K/cm	E. keV
1	192	0	-
2	215	0,2	30
3	128	20	30
4	10,5	200	30
5	259	0,2	75
6	164	20	75
7	13,3	200	75
8	350	0,2	120
9	188	20	120
10	170	200	120
11	15,9	1200	120

Fig.4 shows the typical Raman spectra of P^+ - implanted silicon ($D = 2*10^{15}$ cm^{-2}) before and after irradiating with the monopulse of Q - switched ruby laser with W, ranging from 0,2 to 2,2 J/cm^2 . RS - spectrum of implanted silicon is amorphous - like. After irradiating with W, increasing from 2,0 to 0,4 J/cm^2, no indications of the crystalline phase appearance have been obtained.

Beginning from W = 0,4 J/cm^2 spectra demonstrate the superposition of a - Si wide bands and the novel narrow feature near 520 cm^{-1} . More detailed inspections of this feature reveals that it is a doublet. Its first line has the parameters, appropriate to c - Si, and arises from an unimplanted substrate. The second line, broadened by 7 cm^{-1} and down shifted by 6 cm^{-1} with respect to c - Si line, indicates the nanocrystalline - Si nucleating. The crystallites sizes, evaluated from RS -spectra, using phonon - confinement model, are of 50 A. As W increases the second line becomes narrower and moves toward to c - Si line, reflecting the nanocrystal sizes increase. Simultaneously the doublet intencity rises, while a - Si broad bands amplitudes decrease. It means that the volume fraction of crystallinity in the mixed structure increases.

At W = 1,5 - 2,2 J/cm^2 a - Si bands finally disappear and RS-spectrum becomes fairly similar to that of c - Si, indicating the recovering of monocrystalline structure.

Thus, for the recrystallization process we have revealed the threshold value for the variable parameter (the laser power density), below which there is no obvious changes in the local molecular structure. Above this level initially amorphous structure transforms into mixed amorphous-microcrystalline phase, and finally (at great power densities) into crystalline material.

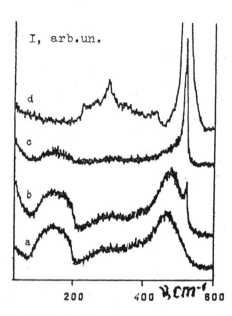

Fig.4. Raman spectra of P$^+$ - implanted silicon (D = 2*10^{15} cm , E = 70 keV) before (a) and after irradiating with monopulse of Q - switched ruby laser (t = 70 ns) with power density W = 0.4 J/cm^2 (b), 0,53 J/cm^2 (c), 2,2 J/cm^2 (d).

Conclusion

So in this work there was shown, that the successive increase of implantation dose results in the different type phase states of undersurface implanted silicon layers: 1) the crystalline matrics with point defects;   2)   the inhomogeneous phase, including the amorphous matrics and the microcrystal clasters;   3) the amorphous phase; 4) the inhomogeneous medium of regular amorphous phase and disturbed tetrahedra silicon clusters.

The result of SHG - investegation gives us the information concerning the stress, emerging in undersurface layers of silicon plates implanted high energy ions.

References

1. K.P.Jain, A.K.Shukla, R.Ashokan, S.C.Abbi. Phys. Rev. B. **32**, 6688 (1985)
2. A.K.Shukla, K.P.Jain. Phys. Rev. B. **35** , 9240 (1988)
3. L.P.Avakyants, E.D.Obraztsova, V.S.Gorelik. Kr. Soob. po Phys. (USSR) **8**, 7 (1988)
4. L.P.Avakyants, I.A.Kitov, A.V.Tsherviakov. PTE (USSR) **2**, 145 (1988)
5. J.Forther, J.S.Lannin. Phys. Rev. B. **37**, 10154 (1988)

Materials Science Forum Vols. 173-174 (1995) pp. 243-248
© 1995 Trans Tech Publications, Switzerland

RAMAN STUDY OF „BOSON PEAK" IN AMORPHOUS SILICON: DEPENDENCE ON HYDROGEN AND CARBON CONTENT

M. Ivanda[1,2], I. Hartmann[1], F. Duschek[1] and W. Kiefer[1]

[1] Institut für Physikalische Chemie, Universität Würzburg, Marcusstrasse 9-11,
D-97070 Würzburg, Germany

[2] Ruder Boskovic Institute, Bijenicka c. 54, 41000 Zagreb, Croatia

Keywords: Raman Scattering, Boson Peak, a-Si, Fractal Analysis

Abstract. A systematic Raman study of dependence of ."boson peak" on hydrogen and carbon content in amorphous silicon thin films is presented. With increasing the hydrogen content the spectral form of the "boson peak" changes while its intensity remains constant. The opposite behaviour was observed with increasing the carbon content. These observations were interpreted in the frame of fractal model. The temperature reduced Raman spectra were decomposed on to phonon-fracton curve and Gaussian shaped bands by fitting procedure. The increase of intensity of the "boson peak" (fractal component) with carbon content is qualitatively interpreted with appearence of nanometer-strained fractal regions which origin should be in homogeneous substitutional bonding of carbon atoms in a-Si network. The observed increase of fractal correlation length with hydrogen content is explained with reduction of internal strain influenced by hydrogen bonding. These findings were confirmed by Auger spectroscopy and atomic force microscopy at high magnifications where the homogeneous distribution of carbon atoms and nanometer-sized blobs of silicon atoms were observed.

Introduction

It is well known that fundamental difference in thermal and elastic properties between crystalline and amorphous solids exists [1]. Typical example is quartz. While the thermal conductivity of crystalline-phase has been explained by Debay theory at the beginning of the century, the basic understanding of the excitations responsible for the conductivity in glassy-phase is still lacking. Morever, the observed relatively low thermal conductivity, thermal plato, as well as temperature dependence have been found to be an inherent property of many amorphous solids, independent of chemical composition or bonding [2].

A great number of experimental results obtained recently show that thermal, optical and electronic properties of amorphous solids mainly depends on structural features at legth scale from atomic level to ~100 Å. These features also cause the "excess" of the vibrational density of states (VDOS) in comparison to the expected Debye value. In Raman spectra this "excess" of VDOS appears at low energies in the form of the so-called "boson peak" which has been observed in vitreous semiconductors like a-Se, a-As_2Se_3 [3] and, recently, in tetrahedral like a-Si, a-GaAs [4,5]. Due to simple structure, the discovery of the "boson peak" in tetrahedral amorphous semiconductors may shed a new light upon its origin. In that framework, here we presents the results of influence of internal parameters such as hydrogen and carbon content on the properties of "boson peak".

Experimental

Amorphous silicon thin films with variable hydrogen and carbon content were prepared by means of DC magnetron sputtering on silicon (111) substrate of 200 µm thickness at room temperature. Sputtering was performed in a gaseous mixture of Ar, H_2, and benzene vapor under the condition given in the Table 1.

Table 1: Experimental conditions for the preparation of the a-Si sample with different H and C concentrations.

Sample	pAr [Pa]	pH_2 [Pa]	pC_6H_6 [Pa]	I [mA]	V [V]
A	1.33	0.05	0.381	100	540
B	1.33	0.05	0.191	100	540
C	1.33	0.05	0	100	540
D	1.33	0.10	0	100	540
E	1.33	0.15	0	100	540
F	1.33	0.20	0	100	540

Deposition rate was ~ 100 Å/min and films thickness were ~ 1.3 µm. Samples A, B and C contain ~ 18, 10 and < 1 at% of bonded carbon measured by Auger spectroscopy; samples C, D, E, F 8.4, 10.8, 14.2 and 15.3 at% of bonded hydrogen measured by IR spectroscopy.

Raman spectra were performed with DILOR Z-24 triple monochromator spectrometer at 90° scattering geometry. The spectrometer slits were adjusted to 800 µm aperture. Signal accumulation time was 3 sec, and point-to-point distance was 3 cm⁻¹. Spectra were taken in a single scan. Laser power of 1.5 W from COHERENT INNOVA 100 argon ion laser was focused on the sample in the shape of elliptical spot with dimension of 100x400 µm.

Results and Discussion

Recently, it has been shown that the broad background signal observed in Raman spectra of a-Si:H and a-GaAs has the properties typical for the "boson peak" in glassy solids [4,5]. The fractal model has been successfully applied for its explanation as well. Here, we present new results of influence of hydrogen and carbon content on the properties of "boson peak".

Theory of Raman scattering from fractons [6,7] is based on dipole induced dipole (DID) mechanism that has been recognized as the major source of Raman scattering in dense systems. Considering the continuous transition from the phonon to the fracton scattering regime, the phonon-fracton curve of the temperature reduced Raman scattering intensity, $I^R(\omega)$, is given by:

$$I^R(\omega) \propto \omega^3 \left(\omega^2 + \omega_{col}^2\right)^{[\tilde{d}/D(\sigma+d-D)-5/2]} , \qquad (1)$$

where ω_{col} is the crossover frequency from phonon to fracton scattering regime, $\tilde{d}$ the spectral dimension, D fractal dimension, d space dimension and σ the scaling index describing the modulation of the density in the embedding space by the vibration.

Fig. 1 show phonon-fracton curves fited on Raman spectra in dependence on carbon, Fig. 1a, and hydrogen content, Fig. 1 b. Due to clearness the spectra were presented without other fited vibrational bands. Results of fitting are summarized in Table 2.

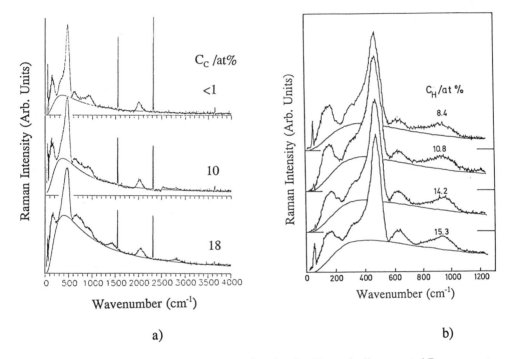

a)   b)

Figure 1. "Boson peak" fited on temperature reduced and radiometrically corrected Raman spectra of a-SiC:H and a-Si:H in dependence on carbon (a) and hydrogen content (b).

Table 2. Parameters from the fited phonon-fracton curves. Correlation length was calculated from the relation: $\xi = v/(\omega_{col}c)$, and spectral dimension with the assumption: D=2.5.

Sample	ω_{col} (cm^{-1})	$(\sigma+d-D)d/D$	ξ (Å)	$\tilde{d}$
A	285	0.30	5.1	0.47
B	295	0.32	5.0	0.50
C	280	0.28	5.2	0.44
D	269	0.35	5.5	0.55
E	250	0.42	5.9	0.65
F	230	0.49	6.4	0.77

From Fig. 1 and Table 1 it is evident that the intensity of the "boson peak" increases with the C-content while its spectral shape remains unchanged. In fractal model, it was assumed that amorphous or glassy solids, although homogeneous from the point of view of density down to the near atomic scale, could be self-similar fractals in their connectivity, i.e. one should expect the elastic modulus K to scale with length as $K \propto l^{-\alpha}$, where $\alpha = d - 2\sigma - D + 2D/\tilde{d}$ [8]. Therefore, in these systems fractals are expected in a strained micro- to nano-regions. In our case the substitutional bonding of C-atoms which is expected in a range of C-content till ~ 30 at% [9], exerts local-deformation of Si-network due to lower C-diameter. The number of a such strained nano-regions should increase with the C-content. Therefore, in our Raman spectra, intensity increase of the "boson peak" with C-content could be explained by this effect.

The behaviour of the "boson peak" with H-content is opposite - the spectral form changes while its intensity remains constant. In fractal model it means that the number of strained nano-regions are not changed, while their correlation length and spectral dimension, as presented in Tab. 1, increase. The increase of the length of strained regions should in principle reduce their internal strain. Therefore, higher concentration of SiH bonds in silicon network should reduce the network strain which is in accordance with the well known behaviour of the hydrogen bonding in amorphous silicon. The changes in d should be due to changes in morfology of nano-structural regions in materials.

To confirm the homogenous bonding of C-atoms and existance of nano-structures with correlation length of ≈ 5Å we have taken Auger deep profiles, Fig. 2, and AFM images, Fig. 3, of the sample with highest C-content (sample A). Fig. 2 shows that only at the surface the concentration of C slightly deviate from the value of 18 at%. The same AES profiles has been obtained at different positionen on the sample.

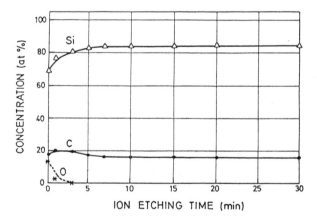

Figure 2. AES depth profiles of $Si_{0.82}C_{0.18}$:H film.

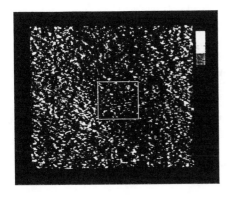 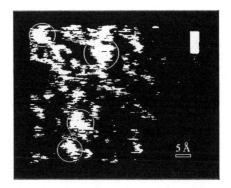

a)   b)

Figure 3. AFM images of the surface of $Si_{0.82}C_{0.18}$:H film. The scan sizes are
200x200 $Å^2$ (a), and 50x50 $Å^2$ (b). Square in fig. a) denotes the dimension
of fig. b). The vertical scales (black to white) are 5 Å.

Fig. 3a shows a 200x200 $Å^2$ large surface area of AFM image of sample A. It is covered with nanometer-sized structural correlations (blobs of Si-atoms). With larger magnification, Fig. 3b, it is evident that diameters of these blobs are between 5 and 10 Å. We exclude the possibility of some oxide origin of these blobs on behalf of low surface contamination with oxygen (< 16 at%) which should forms SiO or SiO_2 bonds following the morphology of surface. We also think that surface morphology is a fingerprint of a bulk morphology due to the fact that surface is frozen state of bulk formation.

Conclusion

Here, we have presented a systematically Raman study of "boson peak" in amorphous silicon with intention to check the validity of fractal model that has been used for its explanation. The fractal model qualitatively well explaines the behaviour of the "boson peak" in dependence on hydrogen and carbon content. Moreover, its quantitative prediction of existance of nanometer structural correlations have been confirmed by atomic force microscope images of the samples surface.

Acknowledgment: One of as (M.I.) thanks to Alexander von Humbolt Foundation for the Research Fellowship. The authors wish to thank to M. Ohnesorge for making AFM images, and S. Lugomer for supplying the Auger spectra.

References

[1] D. G. Cahill and R. O. Pohl, Ann. Rev. Phys. Chem. **39**, 93 (1988).

[2] R. O. Pohl, Phase Trans. **5**, 239 (1985).

[3] V. K. Malinovsky, V. N. Novikov and A. P. Sokolov, J. Non-Cryst. Solids **114**,61 (1989).

[4] M. Ivanda, Phys. Rev. B **46**, 14893 (1992).

[5] M. Ivanda, U. V. Desnica and T. E. Haynes, *Proceedings of 17th International Conference on Defects in Semiconductors*, Gmunden 1993, Materials Science Forum,TransTech Publication, Switzerland, 1993 (in press).

[6] S. Alexander, Phys. Rev. B **40**, 7953 (1990).

[7] E. Stoll, M. Kolb and E. Courtens, Phys. Rev. Lett. **68**, 2472 (1992)

[8] S. Alexander, Phys. Rev. B **40**, 7953 (1989).

[9] W. Y. Lee, J. Appl. Phys. **51**, 3365 (1980).

Materials Science Forum Vols. 173-174 (1995) pp. 249-254
© 1995 Trans Tech Publications, Switzerland

STRUCTURAL INVESTIGATION OF MICROCRYSTALLINE SILICON

P. Hapke, M. Luysberg, R. Carius, F. Finger and H. Wagner

Institut für Schicht und Ionentechnik, Forschungszentrum Jülich, D-52425 Jülich, Germany

Keywords: Raman Spectroscopy, Transmission Electron Microscopy, Microcrystallinity, Percolation

Abstract. Microcrystalline silicon was prepared by PECVD. The samples were grown using different silane to hydrogen ratios, which leads to a wide range of crystalline volume fractions. Annealing experiments were performed in order to study the influence of the changes of the microstructure on the electrical transport. The structural properties were investigated by Raman spectroscopy and by transmission electron microscopy. The correlation of the microstructure with the electrical properties gives evidence for percolation processes dominating the electrical transport in this material.

Introduction

Microcrystalline silicon (μc-Si:H) is prepared by Plasma Enhanced CVD processes from the mixture of silane and hydrogen. This material has the typical advantages of thin film technology, i.e. large area deposition ($0.5\text{-}1 \text{ m}^2$) and low deposition temperatures ($\approx200\ ^\circ$C). Doped μc-Si:H has orders of magnitude higher conductivities than it´s amorphous counterpart. Therefore, μc-Si:H is used for the fabrication of thin film devices like solar cells or thin film transistors. However, these applications require a optimization and understanding of the electronic and structural properties of this material.

Structural investigations have shown that μc-Si:H is a composite material of three phases: crystalline grains, grain boundaries, and amorphous phase. The electrical conductivity is strongly influenced by the microstructure of the films, i.e. the spatial arrangement of the three phases. Previously, we pointed out that the electrical transport is dominated by transport through percolation paths [1]. The knowledge of the crystalline volume fraction is of particular interest because of the strong dependence of percolation transport on this volume fraction. Since in the Raman spectra the transverse optical (TO) mode of c-Si (at 520 cm^{-1}) and the analogous mode in a-Si:H (at 480 cm^{-1}) are situated at different wave numbers, Raman spectroscopy is often used to distinguish between the two phases. From the integrated intensities of the two signals it is possible to calculate the crystalline volume fraction if the Raman scattering cross section and the dielectric function at the given excitation wavenumber is know. Usually these values are not easily obtained, therefore we determined the integrated intensities in a semiquantitative way. Annealing studies were performed to investigate the dependence of the transport properties on the microstructure of μc-Si:H.

Experimental

The μc-Si:H samples were prepared with PECVD at 70 MHz at substrate temperatures between 160 °C and 200 °C. The gas phase silane concentration was in general 1.5% with respect to hydrogen. As dopant gases phosphine and diborane were used. N- and p-type material and a sample prepared at a silane concentration of 4.5%, which characterizes the transition zone of the amorphous to microcrystalline formation, were investigated. The samples were annealed between the deposition temperature and 700 °C under high vacuum conditions for about 30 minutes. Raman spectra were

Raman spectra of sample A, <n>-type

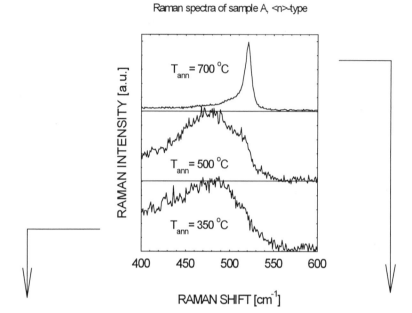

Fig. 1 Raman spectra of sample A at various annealing stages

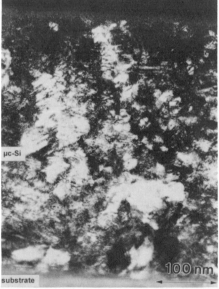

Fig. 2: TEM micrographs of sample A
a) in the as deposited state   b) after annealing at 700 °C

recorded with a double monochromator coupled with a cooled photomultiplier in the photon counting mode. The 514.5 nm line of a COHERENT INNOVA 100 Ar-Laser was used for excitation. Focusing of the laser beam was made by means of a cylindrical lens, in order to avoid a significant sample heating. The electrical transport properties were determined by conductivity measurements in the temperature range between 80 K and 300 K. In addition, the microstructure of the samples was characterized by transmission electron microscopy (TEM) using a JEOL 2000EX microscope operated at 200 kV and a JEOL 4000EX microscope at an electron energy of 400 keV. The TEM specimens were prepared as cross sections, which allows the observation of depth dependent structural modifications.

Results and Discussion

Fig. 1 shows the Raman spectra of sample A at different annealing stages. This sample is n-type and prepared in the transition zone of the amorphous to microcrystalline formation, i. e. at 4.5% silane diluted in hydrogen. The Raman spectra of this sample in the as deposited state and at annealing temperatures of 350 °C and 500 °C is dominated by a signal centered at 480 cm^{-1}, whereas at 520 cm^{-1} no significant contribution can be observed. This indicates, that the sample consists of an almost complete amorphous phase, which is not significantly changed upon annealing at these temperatures. Possibly a weak crystalline signal (at 520 cm^{-1}) is also seen in these spectra, but it is difficult to separate this phase from the overall spectrum because of the noise. In contrast the annealing at 700 °C leads to the formation of a considerable crystalline part, which can be deduced from the distinct signal at 520 cm^{-1}.

For the determination of the crystalline volume fraction one needs the knowledge of the Raman cross sections and the absorption coefficient of the different phases. At first the Raman intensities of the different phases have to be determined. This procedure is still controversially described in the literature. Some authors use the integrated intensities of the amorphous (at 480 cm^{-1}) and the crystalline (at 520 cm^{-1}) phase [2], others try to fit the Raman spectra by superposition of an a-Si:H and a c-Si spectra of different weight [3]. It is known that the Raman signal corresponding to the crystalline phase is asymmetric and shows a shift to lower energies due to a size effect of grains < 10nm [4,5,6]. Veprek *et al.* [6] found that the presence of an "amorphous" signal in films with small amorphous partition is due to an effect of the grain boundaries. It is also found, that the relative Raman cross section (which is directly related to the scattering intensity) is a strong function of the crystallite size [6,7]. For the calculation of the crystalline volume fraction many authors assume that the scattering cross section of the "amorphous" phase is the same as for a-Si:H [5,7]. This, however, is doubtful [5].

Because of these discrepancies the ratio of the integrated intensities $I_{520}/(I_{520}+I_{480})$ is used as a semiquantitative determination of the crystalline volume fraction. This ratio increases from zero to about 0.75 at annealing temperatures of 500 °C and 700 °C, respectively. The annealing above the crystallization temperature, such as the annealing at 700 °C, should result in a complete crystallization. Therefore the intensity ratio of 0.75 could in fact be considered to correspond to a crystalline volume fraction close to 100%. This is confirmed by the TEM studies, where a crystalline volume fraction of at least 90% was observed (see below).

Although there is no significant crystalline contribution detectable in the Raman spectra of sample A up to annealing temperatures of 500 °C, in the TEM dark field images crystallites can be observed even in the as deposited state (Fig. 2a). The white contrasts in Figure 2 are caused by crystallites of the same crystallographic orientation relative to the incident electron beam. In the as deposited state (Fig. 2a) the size of the crystallites varies from 5 nm to 10 nm. This result is in agreement with the work of Okada *et al.* [8] who compared TEM and Raman measurements for samples with an average particle size of $\approx$ 7 nm and also found that the Raman measurement shows only an amorphous signal although TEM shows the presence of microcrystallites. Large-angle tilt experiments reveal a sphere-like shape of the crystalline regions. In order to determine the crystalline

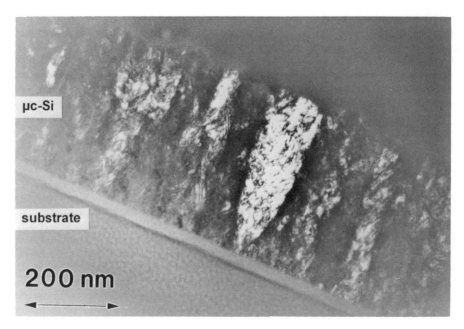

Fig. 3: TEM micrograph of sample B in the as deposited state.

volume fraction high-resolution TEM studies were performed. In these images the lattice planes of crystallites are clearly resolved resulting in a crystalline area of 20%.

Annealing at 700 °C induces the growth of the crystallites up to a size of 200 nm. Besides the large crystallites in the dark field image of this sample (Fig. 2b) also smaller crystallites of about 10 nm can be observed. Some of the crystallites reveal a fringe contrast. Whereas the fringes of regular distance can be attributed to Moiré contrasts, the fringes of irregular width may be caused by twinning. Additionally performed high-resolution studies reveal the existence of twinned crystallites. From the superposition of several dark field images and from high-resolution images only a very small amorphous volume fraction (<10%) is estimated.

Next we turn to the samples prepared with 1.5% silane diluted in hydrogen, sample B (n-type) and sample C (p-type). It is noticeable that the Raman spectra of sample B show only small changes upon annealing. Therefore, the intensity ratio of crystalline to amorphous phase remains approximately constant (0.75) upon annealing. Indeed, the TEM investigations of this sample in the as deposited state and at an annealing temperature of 700 °C show no significant changes of the structural properties. Fig. 3 shows a TEM micrograph of sample B in the as deposited state. The crystallites of the same crystallographic orientation (white contrast) form a columnar structure in direction of growth. Some of the columns show an increasing diameter from the glass substrate to the surface. The columns with diameters up to 70 nm and lengths up to 250 nm do not consist of one single crystal, but contain many small crystallites of the same orientation. The dark contrasts within one column correspond either to crystalline regions of different orientations or to amorphous regions. Again fringe contrasts can be observed, which may be caused by Moiré contrasts and/or twinned crystallites. The superposition of several dark field images and additionally performed high-resolution studies points to a very small amorphous volume fraction (<10%). Sample C is p-type and doped with boron. It is known, that boron hinders the crystalline growth [9]. This sample has a considerable portion of amorphous phase in the as deposited state. The Raman spectra (Fig. 4) show a continuous decrease of the amorphous signal with increasing annealing temperature. The analysis of the integrated intensities of the Raman spectra of this sample leads to a continuous increase from

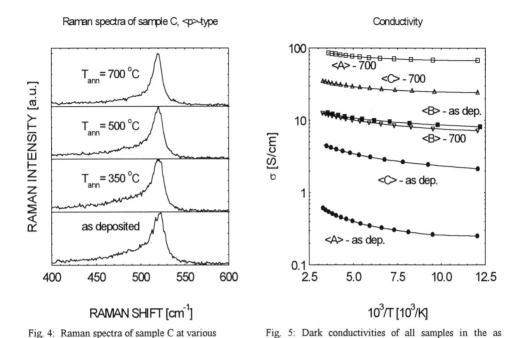

Fig. 4: Raman spectra of sample C at various annealing steps

Fig. 5: Dark conductivities of all samples in the as deposited state and at an annealing temperature of 700 °C.

0.5 to 0.75 of the crystalline phase with increasing annealing temperature. Despite the lower crystalline volume fraction in the as deposited state, the TEM micrographs show a microstructure very similiar to the results of sample B. Again a columnar growth of crystallites is observed.

In Fig. 5 an overview of the temperature dependent conductivity is shown for all samples in the as deposited state and at an annealing temperature of 700 °C. In all cases the temperature dependence of the electrical conductivity shows a strong deviation from a singly activated behaviour.

For sample A annealing up to 500 °C do not cause any changes of the conductivity. However, a drastic increase of the conductivity is observed when this sample is annealed above the crystallization temperature, i.e. at 700 °C. The weak temperature dependence and the high value of the conductivity in the as deposited state ($\sigma > 10^{-1}$ Scm^{-1}) are properties typical of microcrystalline silicon. The TEM measurements show that this sample has a considerable crystalline volume fraction (Fig. 2a). In contrast the Raman measurement is not sensitive enough and shows an almost pure amorphous phase. The high conductivity and the presence of microcrystallites for this sample in the as deposited state indicate a considerable contribution to transport by percolation. Indeed, theoretical and experimental work predicts a percolation threshold below 20% crystalline volume fraction [10], in line with the observed value for this sample. After annealing at 700 °C the conductivity increases by two orders of magnitude but the characteristic shape of the σ_D vs. 1/T curve does not change. It appears as if additional transport paths of the same type are opened up while no changes in the dominant transport mechanism is observed. The structural changes as indicated by the TEM measurements are correlated with the changes of the transport properties.

Sample B shows no changes of the temperature dependent conductivity over the whole annealing range (Fig. 5). This result is in agreement with the small changes in the structure as indicated by the Raman and TEM measurements. The crystallinity can not be improved by annealing and the crystallite sizes seem to be unaffected. The temperature dependence of the conductivity (compare previous section) implies that transport should be dominated by transport through percolation paths made of crystalline material.

The last sample, the highly conductive p-type sample C, shows a continous increase of the conductivity with increasing annealing temperature. This tendency is in agreement with the structural changes. After annealing and a considerable increase in crystallinity the characteristic shape of the σ_D vs. 1/T curve does not change, but only shifts to higher σ_D-values. It appears again (comp. sample A) that no change in the dominant transport mechanism is observed. In addition to the conductivity measurements we performed Hall effect measurements on the sample C. The Hall coefficient for this set of samples has always a positive sign, which means that the Hall effect is dominated by transport through the crystallites and the carrier density accompanies the increase of the conductivity. Evidently, annealing mainly affects the carrier density through dopant activation and crystallization but it is not clear which of both effects has the strongest influence on the increase of the conductivity.

Conclusions

The comparison of TEM and Raman measurements indicates that the Raman measurement is not sensitive enough if the crystalline volume fraction is below $\approx 20\%$ in μc-Si:H with small crystallites. TEM results clearly show the presence of microcrystallites although the Raman measurement indicates an almost pure amorphous phase. In agreement with the TEM measurements this class of samples show typical microcrystalline transport properties. Further, we note that the increase of the crystalline volume fraction increases the conductivity. The annealing of microcrystalline material has no influence on the dominant transport mechanism. Microcrystalline silicon prepared with low temperatur PECVD and a wide range of crystalline volume fractions shows a strong indication for electrical transport through percolation paths.

Acknowledgement

We thank the Institut de Microtechnique, Neuchâtel (Switzerland), for providing samples. This work was supported by the Bundesministerium für Forschung und Technologie (Germany).

References

[1] P. Hapke, F. Finger, R. Carius, H. Wagner, K. Prasad and R. Flückiger, J. Non-Cryst. Solids **164-166**, 981 (1993)

[2] J. Bandet, J. Frandon, F. Fabre and B. De Mauduit, Jpn. J. Appl. Phys. **32**, 1518-1522 (1993)

[3] R.J. Nemanich, E.C. Buehler, Y.M. Legrice, R.E. Shroeder, G.N. Parsons, C. Wang, G. Lucovsky and J.B. Boyce, J. Non-Cryst. Solids **114**, 813-815 (1989)

[4] H. Richter, Z.P. Wang and L. Ley, Solid State Comm. **39**, 625-629 (1981)

[5] J. Iqbal, S. Veprek, A.P. Webb and P. Capezzuto, Solid State Comm. **37**, 993-996 (1981)

[6] S. Veprek, F.-A. Sarott and Z. Iqbal, Physical Rev. B **36**, 3344-3350 (1987)

[7] E. Bustarret and M.A. Hachicha, Appl. Phys. Lett. **52**, 1675-1677 (1988)

[8] T. Okada, T. Iwaki, K. Yamamoto, H. Kasahara and K. Abe, Solid State Comm. **49**, 809-812 (1984)

[9] K. Prasad, U. Kroll, F. Finger, A. Shah, J.-L. Dorier, A. Howling, J. Baumann and M. Schubert, Mat. Res. Soc. Symp. Proc. **219**, 383 (1991)

[10] H. Scher and R. Zallen, J. Chem. Phys. **53**, 3759 (1979)

Materials Science Forum Vols. 173-174 (1995) pp. 255-258

RECOMBINATION LIFETIME IN SILICON FROM LASER MICROWAVE PHOTOCONDUCTANCE DECAY MEASUREMENT

C.H. Ling, H.K. Teoh, W.K. Choi, T.Q. Zhou and L.K. Ah

Department of Electrical Engineering, National University of Singapore, Kent Ridge, Singapore 0511

Keywords: Laser/Microwave Photoconductance, Recombination Lifetime in Silicon, Corona-Charged Silicon Oxide

Abstract. The effective minority carrier recombination lifetime in oxidized silicon wafers is measured using a laser microwave photoconductance technique. The effect of surface recombination is demonstrated through altering the silicon band bending at the oxide/silicon interface, through a non-contact injection of charges on the outer oxide surface by a corona discharge. The dependence of effective lifetime on wafer orientation and oxidation conditions is investigated.

Introduction

Minority carrier lifetime is a good indicator of the level of defects in a semiconductor material. Metallic species in silicon, for example Fe and Cr, common contaminants from processing chemicals, can seriously reduce lifetime and degrade device characteristics, such as the refresh time of dynamic random access memories (DRAM). A common and most widely used method of characterising metallic species in silicon is the deep level transient spectroscopy [1]. This method requires the fabrication of a good test device, and involves a number of processing steps. In silicon submicrometre technology, contaminants have to be controlled to a low level, and deep level transient spectroscopy may not be the most suitable tool for defect characterisation, because of the unintentional introduction of impurities during processing.

A non-contact, non-destructive technique based on the measurement of the photoconductive decay of the material was proposed [2] and has been extensively applied to the characterisation of minority carrier recombination lifetime in silicon [3-5]. Fig. 1 illustrates the experimental setup, which monitors the decay in a 9.6 GHz microwave power reflected off a silicon surface, as the electron-hole pairs, generated simultaneously by a 170 ns ($\lambda = 910nm$) laser pulse, recombine and the conductance of the silicon falls. Fig. 2 shows a decay spectrum of the reflected microwave power, taken to be proportional

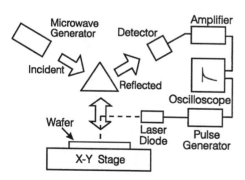

Fig. 1 Schematic of the laser microwave photoconductance technique

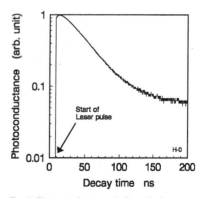

Fig. 2 Photoconductance (reflected microwave power) decay curve.

to the excess photoconductance, on a log-linear plot. From the linear part of the spectrum, an effective carrier lifetime is obtained.

The effective lifetime τ_e is due to the contribution from the bulk of the silicon wafer and the surfaces, and may be expressed as follows

$$\frac{1}{\tau_e} = \frac{1}{\tau_s} + \frac{1}{\tau_b},\tag{1}$$

where subscripts b, s denote bulk and surface. Bulk lifetime τ_b is the quantity of interest to wafer manufacturers and various algorithms [4,5] exist for its extraction from Eqn (1).

Surface lifetime τ_s depends on the recombination of carriers at trapping centres located at the silicon wafer surfaces, and often dominates the bulk lifetime. The recombination kinetics can be modelled by the Shockley-Hall-Read statistics [6]. A maximum rate of surface recombination occurs for those traps whose energy levels E_t are near the midgap E_i, and for a surface having approximately equal electron and hole concentrations. For a surface that is in strong accumulation or in strong inversion, this condition is not satisfied and surface lifetime contribution to Eqn (1) is expected to be greatly diminished. For surface traps that are distributed in energy across the band gap, recombination can also be expected to be reduced for surface in strong accumulation or inversion.

In this paper, we present preliminary experimental data obtained for oxidized silicon wafers, showing that effective lifetime can be varied by varying the silicon surface band bending, using a non-contact technique.

Experimental

Two inch silicon wafers were cleaned in standard RCA solutions before oxidation in dry oxygen at $900°C$ or steam at $1050°C$. A corona-charging apparatus, consisting of a sharp electrode biased at $\pm(5{\sim}10)$ kV, a grid electrode at ±500 V, and the silicon wafer at ground potential was used to charge the wafers up to a maximum of ±500 V, depending on the oxide thickness [7]. Charges were trapped in a thin layer at the outer surface of the oxide. The surface charge resulted in silicon band bending at the oxide/silicon interface and a surface voltage, which could be measured using a Trek 366 electrostatic voltmeter. The surface charge σ_s is related to the surface voltage V_o approximately through the oxide capacitance C_{ox} [8], thus

$$\sigma_s \sim C_{ox} V_o.\tag{2}$$

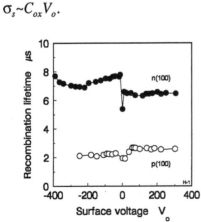

Fig. 3 Recombination lifetime versus surface voltage for n(100) and p(100) wafers, grown with 1 μm wet oxide.

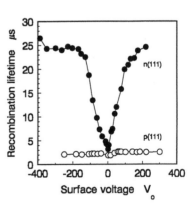

Fig. 4 Recombination lifetime versus surface voltage for n(111) and p(111) wafers, grown with 1μm wet oxide.

In a dry ambient, the surface charge can be retained over long periods of time, extending into many months. However, under normal laboratory conditions, surface charge leaks away due to humidity in the air, in a matter of days, causing the band bending to change and thus allowing the effective minority carrier lifetime to be observed as a function of band bending. This field-effect is realized without the physical presence of a metal or polysilicon gate.

It is to be pointed out that the wafers used in this work are low-grade silicon and consequently, the observed lifetime is considerably lower than those used in wafers in today's industry.

Results and Discussion

Fig. 3 and 4 show the recombination lifetime as a function of surface voltage for n- and p-silicon wafers with both (100) and (111) orientations, and having 1 μm wet oxide. A minimum in the effective lifetime is observed for surface voltage $V_o \sim 0$, corresponding to the maximum surface recombination condition. Lifetime increases for large surface voltages. This result is in agreement with the surface recombination model, namely, in silicon surface accumulation or inversion, the recombination rate is reduced. Because the wafers are sourced from non-standard suppliers with doubtful quality control, it is not possible to compare the results for the various wafers. However, it is instructive to note that for the n(111) wafer, changing the surface band bending results in an increase in the lifetime by a factor of 28. The corresponding change for n(100) wafer is much smaller. One of the reasons could be due to the larger interface trap density in (111) wafers.

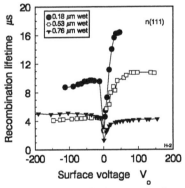

Fig. 5 Recombination lifetime versus surface voltage for n(111) wafers, grown with oxide of different thickness.

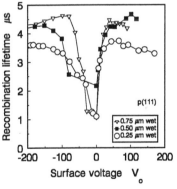

Fig. 6 Recombination lifetime versus surface voltage for p(111) wafers, grown with oxide of different thickness.

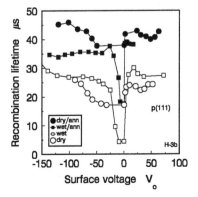

Fig. 7 Recombination lifetime versus surface voltage for p(111) wafers, with 0.1 μm oxide grown under different conditions

Shown in Fig. 5 and 6, are lifetime data for n(111) and p(111) wafers for a number of oxide thickness. For n-type silicon, larger lifetime increase in accumulation or inversion is observed for thinnner oxides. But for p-type silicon, larger lifetime increase is observed for thicker oxides. The reasons for this difference are not clear, and this result is under investigation.

Finally, we show in Fig. 7, lifetime data for p(111) wafers with 0.1 μm wet oxide, with and without post-oxidation anneal. For the condition that corresponds to maximum surface recombination, dry oxide wafers have higher lifetime than wet oxide wafers. However, little difference is noted when the wafer surfaces are in accumulation or inversion. For the annealed wafers, dry oxidation wafers again have higher lifetime over the wet oxidation wafers, in particular at zero surface voltage. This is in agreement with established results that dry oxidation produces an interface with fewer interface traps, and that interface trap density decreases with anneal.

Conclusion

Minority carrier recombination lifetime in silicon wafer is measured using the laser microwave photoconductance decay technique. By injecting charges on to the outer surface of the oxide layer to control the silicon surface band bending, the effect of surface recombination on the effective lifetime has been demonstrated. This non-contact field effect thus allows the extraction of bulk lifetime from the observed effective lifetime.

Acknowledgment

The authors acknowledge funding support under AAECP Microelectronics Project.

References

[1]   D.V. Lang, J. Appl. Phys. **45**, 3023 (1974).
[2]   Y. Mada, Jpn. J. Appl. Phys. **18**, 2171 (1979).
[3]   F. Shimura, T. Okui and T. Kusama, J. Appl. Phys. **67**, 7168 (1990).
[4]   A. Buczkowski, Z. L. Radzimski, G. A. Rozgonyi and F. Shimura, J. Appl. Phys. **69**, 6495 (1991).
[5]   A. Buczkowski, Z. L. Radzimski, G. A. Rozgonyi and F. Shimura, J. Appl. Phys. **72**, 2873 (1992).
[6]   A. S. Grove: *Physics and Technology of Semiconductor Devices* (Wiley, New York, 1967), p. 136.
[7]   P. Gunther, IEEE Trans. Electrical Insulation **26**, 42 (1991).
[8]   G. M. Sessler: *Topics in Applied Physics Vol. 33: Electrets* (Springer Verlag, Berlin, 1987), p. 13.

Materials Science Forum Vols. 173-174 (1995) pp. 259-264
© 1995 Trans Tech Publications, Switzerland

INVESTIGATION OF ELECTRONIC TRANSPORT IN SEMICONDUCTOR JUNCTIONS BY PHOTOINDUCED LASER BEAM DEFLECTION

P. Grunow[1], R. Schieck[2] and M. Kunst[2]

[1] PEMM/COPPE/UFRJ, 68505 Rio de Janeiro, Brazil

[2] HMI, Bereich CS, Glienicker Str. 100, D-14109 Berlin, Germany

Keywords: PDS, Photovoltaic

Abstract. The investigation of the electronic transport in bare silicon wafers and pn-solar cells by the method of the photoinduced laser beam deflection is presented. The dependence of the measured signals on frequency, applied external voltage and additional bias illumination is successfully described theoretically.

Introduction

Photoinduced laser beam deflection (PD) measures the deflection of an infrared light beam inside an illuminated semiconductor due to the induced refraction coefficient n(x) change [1]. The gradient dn/dx is proportional to the deflection angle φ and consists of a thermal and an electronic contribution. The PD technique is used to investigate the electronic transport in silicon single crystals and pn silicon solar cells. The dependence of the deflection signal on the modulation frequency is demonstrated for a bare silicon wafer. Dependences of the PD-signal on external bias voltage in the closed circuit and on additional bias illumination in the open circuit case are simulated successfully for a pn-junction by a semi-microscopic approach using the expressions for the current-voltage characteristics obtained from the diode theory. The experimental results on crystalline silicon solar cells demonstrate the ability of the method for a contactless all-optical characterisation of photovoltaic interfaces.

Experimental

Sample preparation. The measurements presented are carried out on a commercial p^+n-solar cell (Siemens, test cell, emitter: $p^+ = 10^{17}$ cm^{-3}, 1 μm; base: $n = 10^{15}$ cm^{-3}, 500 μm) and on a n-Si wafer. All samples are polished at the edges with a diamond paste (0.7 μm) in order to allow an unperturbed transmission of the probing laser beam.

Set-up. The sample is excited by an Ar$^+$ - ion laser beam modulated with the frequency f by an acousto/optical-modulator. The gradient of the refractive index induced by the excitation light deflects the infrared (IR)-probe beam (HeNe 1152 nm) inside the sample with the deflection angle φ. This deflection of the probe beam is detected by a two segment Ge - diode. The difference of the two photo currents is measured with a resolution of about 10^{-5} rad. The use of a two-phase lock-in gives the modulus and the phase of the PD - signal with respect to the reference signal. It is possible to apply a potential to the sample via the potentiostat or in a contactless manner via the bias illumination.

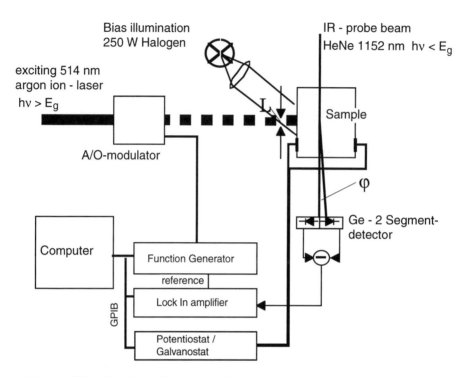

Figure 1: Schematic picture of the experimental set-up.

Theory

The deflection angle φ is determined by the excess charge carrier pair distribution $N(x,t)$ and the temperature distribution $T(x,t)$. These quantities were calculated from the respective continuity equations. The appropriate boundary conditions include the surface recombination at the front- and backside of the sample.

The deflection angle can be written as the sum of the derivatives of N and that of T with respect to the spatial variable x with the weighting factors dispersion volume $\partial n/\partial N$ and thermo-optical coefficient $\partial n/\partial T$ according to:

$$\varphi \sim \frac{dn}{dx} = \frac{\partial n}{\partial N}\frac{\partial N}{\partial x} + \frac{\partial n}{\partial T}\frac{\partial T}{\partial x} \quad . \tag{1}$$

For linear surface recombination, the solution in the considered case of strong absorption is given as

$$PD \sim Ae^{-x/\lambda_{th}} + Be^{x/\lambda_{th}} + Ce^{-x/\lambda_{el}} + De^{x/\lambda_{el}} \tag{2}$$

with the effective thermal diffusion length λ_{th} and the effective electronic diffusion length λ_{el}. They are distinguished by their different frequency behavior. As a consequence, the thermal contribution dominates the signal in the low frequency range, whereas the electronic dominates in the high frequency range. This means one can examine the recombination processes appearing either as a consequence of heat production or as a change of the excess charge carrier gradient. An increasing surface recombination for example gives rise to the heat production on the surface of the sample and therefore increases the thermal gradient. In addition carriers diffuse towards the surface and diminishes the electronic gradient in the bulk.

As a consequence the thermal contribution would be observed in the signal only and a lower limit for the surface recombination velocity can be given.

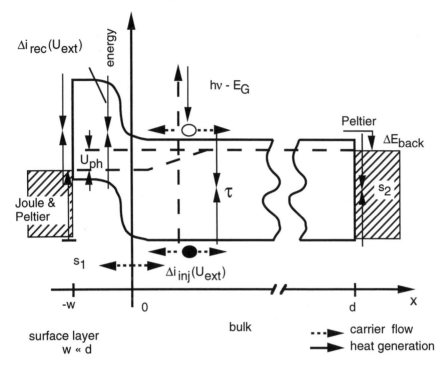

figure 2: The photovoltaic interface in the open circuit case shows the main features of the carrier flow and heat generation mechanism. After the band edge relaxation, the excess charge carrier pairs diffuse to the space charge region and the back contact. Heat is generated due to the band edge relaxation, recombination in the bulk, at the front and back contact and in the space charge region. In addition, Peltier and Joule heats at the rear and front contacts have to be considered.

The parameters of interest, i.e. surface recombination velocities s_1 and s_2, bulk-lifetime τ, recombination current Δi_{rec}, diffusion constant D_{el}, the externally applied voltage U_{ext} and the dc bias illumination, are included in the coefficients A, B, C and D. The unknown transport parameters are determined by comparison of the experimental PD results to the values calculated with Eq. (2).

The main feature in describing the pn-junction is the separation of the collection of carriers by a constant collection velocity [2] and the re-injection of carriers into the bulk due to the photovoltage. This change of the injection current Δi_{inj} and the change of the recombination current Δi_{rec} are derived from the current-voltage characteristics of the pn-junction, which is written as

$$i_{ext} = i_{inj}(U_{ext}) + i_{rec}(U_{ext}) + i_R - i_{ph} \qquad (3)$$

with i_{ext} measured in the external circuit, the recombination i_{rec} and injection i_{inj} current, the photocurrent i_{ph} and the leakage current i_R due to a parallel Resistor.

The small signal modulation with ΔU_{ph} gives Δi_{inj} and Δi_{rec} at the working point around U_{ext}. Thus, one gets as an additional parameter of interest: the recombination current i_{rec}, while i_{inj} is determined theoretically [3].

Results

Single crystalline silicon. PD-measurements on single crystalline n-type silicon are shown in Fig. 3. The frequency dependence of the signal (modulus and phase) are successfully simulated with the parameters given in the caption of the figure. For low frequencies the thermal contribution dominates the signal and a linear dependence on the square root of the modulation frequency is observed. The slope is proportional to the square root of the thermal diffusion constant D_{th} and the position x of the probing laser beam. For high frequencies the electronic contribution is dominating. The minima in the curves are due to the different phases of the thermal and electronic contributions. The value of the frequency at which the transition occurs depends on the volume lifetime τ. At even higher frequencies the slope of the curves becomes linear again where the values of the slopes are proportional to the square root of the electric diffusion constant D_{el} and the x-positions of the detection laser.

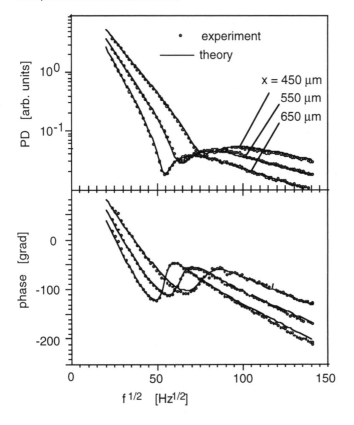

Figure 3: Determination of the electric transport parameters of an n-type silicon wafer sample with the PD-technique. The calculated curves fit well with the parameters from the literature (heat conductivity k=1.41W/cm, hν=2.41eV, $\partial n/\partial T$=1.28·10^{-4} K^{-1}, E_g=1.12eV, D_{th}=0.9cm^2/s) and the fit parameters s=1600cm/s, D_{el}=13cm^2/s, τ=360 µs and $\partial n/\partial N$=-2·10^{-23}cm^3.

pn-junction. The possibility of a contactless characterisation of photovoltaic devices is shown in Fig. 4,5 for a p$^+$n-silicon solar cell. The PD-signal is measured as a function of the square root of the modulation frequency for different intensities of the additional white light illumination ($\approx$AM 1.5). Due to the high lifetime τ, the signal shows no thermal contribution. The increase of the electronic contribution caused by the increasing additional white light illumination intensities is observed clearly and is successfully calculated from the theory. This means that the induced photovoltage works as a external forward biasing voltage. The deviation from theory at zero bias light could be explained by the photovoltage caused by the relatively low absorption of the probing laser beam.

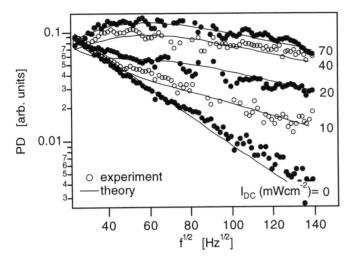

figure 4: PD-signal as a function of the square root of the modulation frequency for different white light illumination intensities I_{DC}.

Figure 5 shows in addition the calculated recombination and injection current changes. The increasing contribution of the injection current to the internal current for increasing bias photovoltages leads to the increase of the PD-signal. The higher sensitivity of the injection current to the bias voltage than the recombination current increases the carrier gradient at the edge of the space charge region and therefore the PD-signal.

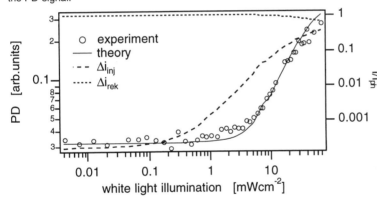

figure 5: Calculated injection and re-combination cur-rent change for different addi-tional white light illumination in-tensities and the respective PD-signal change at 9.4 kHz.

Conclusion

The presented new application of the laser beam deflection technique has been proven to be a useful tool for substrate and device characterisation. Its sensitivity to the different recombination processes gives promising results for the understanding of electronic transport in photovoltaic interfaces.

References

[1] A. C. Boccara, D. Fournier, J. Badoz, Appl. Phys. Lett. **36**, 130 (1980)
[2] H. C. Card, J. Appl. Phys. **47**, 4964 (1976)
[3] C. T. Sah, R. N. Noyce, W. Shockley, Proc. IRE **45**, 1228 (1957)

Materials Science Forum Vols. 173-174 (1995) pp. 265-272
© 1995 Trans Tech Publications, Switzerland

NONDESTRUCTIVE AND CONTACTLESS EVALUATION OF ELECTRICAL AND THERMAL PROPERTIES OF THIN SEMICONDUCTING LAYERS

H.-D. Geiler

Jena Wave Engn. and Consult., Friedrich-Schelling-Str. 11, D-07745 Jena, Germany

Keywords: Nondestructive Evaluation, Photothermal Analysis, Surface Recombination Velocity, Semiconductors

Abstract. Laser assisted response techniques for nondestructive evaluation of material properties are especially useful for contactless measurements in small structures of some µm dimension. The single beam double modulation technique of photothermal response detection is used to determine surface recombination velocities in small silicon structures produced by focus ion beam technology. Thin films of silicon can be analyzed to detect the influence of defect creating radiation and surface passivation accuracy. Subsurface defect recognition is demonstrated by mapping the carrier capture area of a dislocation network.

1. Introduction

Advanced laser and ion beam assisted technologies allow the processing in small size structures with the aim of establishing defined physical properties [1]. This processing needs nondestructive and at least contactless analytical tools to control the performance of the structure-property relations [2]. The paper presents the method of linear response measurement in µm-structures of silicon with the performance of laser assisted microscopy. An intensity modulated laser beam excites the sample by energy deposition in a deposition region which is determined by the spot radius w on the surface and the absorption depth $1/\alpha$ (see Fig.1). The energy dissipation creates a time dependend response field the so called thermal wave which follows the exitation

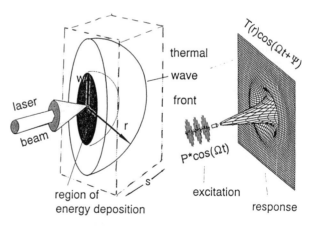

Fig.1: Scheme of the creation of response waves $T(\vec{r},t)$ in a solid of size s

frequency with an magnitude and phase shift Ψ both characterising the thermal material properties. The thermal wave can be detected by the influence of the temperature on the dielectric function of the material [3]. The main feature of this analytical technique is the possibility to change the excitation frequency in a wide range (frequency sweep). The penetration depth l_T of the thermal wave for instance is governed by the relationship $l_T = \sqrt{(2\kappa/\Omega)}$ including the modulation frequency Ω and the thermal diffusivity κ of the material under consideration. So the high frequency response (10 MHz) allows the probing of the thermal properties in a scale s of 1 µm (see Fig.1). The complex response can be plotted in the complex plane versus frequency which gives a circle plot in analogy to the Smith diagrams of the frequency analysis of electrical networks [4]. This circle plot is uniquely defined by the set of material parameters probed by the response wave in the dimension s. In the following this technique is applied to the determination of the surface recombination velocity in small silicon structures.

2. Analytical method of single beam double modulation response technique

To detect the response wave the double modulation technique [5] is applied to the measurement in small structures. The Gaussian laser beam is modulated by 2 frequencies Ω_1, Ω_2 which are separated by a small difference Ω_{12}. The modulation of the dielectric function of the material by both frequencies causes in the sample a mixing creating among all intermodulation products this differential frequency Ω_{12} in the scattered beam, which is detected by a photodiode in connection with sensitive lock-in amplification. Realizing a wide range frequency sweep under the condition of the synchronious change of both excitation frequencies the response can be extracted at the constant differential frequency with high sensitivity (photothermal heterodyn [6]).

Because the differential frequency is not contained in the incident laser beam, no reflectivity but in analogy to the second harmonic generation a conversion coefficient K can be defined.   This conversion coefficient of photothermal   response is   a   physical   quantity describing   that   part   of incident power which is converted into the differential frequency due to the mixing caused by a normalized   response

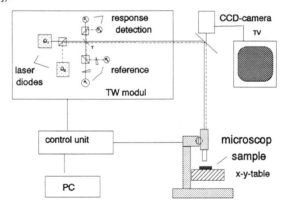

Fig.2: Scheme of the laser assisted response microscope

magnitude. Because the response field of Gaussian laser beams (spot radius w, excitation power P) scales with the relation P/w, the dimension of K is [m/W]. The experimental arrangement is shown in Fig.2. The double modulated laser beam is created by two current modulated laser diodes ($\lambda = 785$ nm). The beam is fitted into a microscop and the backscattered light is detected by a photodiode. To define the conversion coefficient the incident power has to be measured carefully by a reference diode.

3. Superposition of carrier wave and thermal wave response in semiconductors

The carrier and thermal wave analysis bases on the following model [7,8]: The sample is excited by the near surface absorption (primary energy deposition) of a continously intensity modulated laser beam of modulation frequency Ω (see Fig.1). The energy deposited into the electron gas dissipates via different radiationless channels (secondary energy deposition) [9]. Because of the near surface absorption the channel of surface recombination of the excess carriers plays an important role. The laser excited electron-hole concentration is determined by the following electrical material parameters: the excess carrier diffusion governed by the ambipolar diffusivity D, the bulk excess carrier recombination [10] determined by the lifetime τ and the surface recombination expressed by the recombination velocity s. The modulated part of the carrier concentration the so called carrier or plasma wave $\hat{c}$ then follows [7]:

$$\Delta \hat{c} - p^2 \hat{c} = -\frac{\hat{v}}{\hbar \omega D} \qquad (1)$$

In Equ.(1) the carrier wave number p is introduced:

$$p^2 = \frac{1+i\Omega\tau}{D\tau} \qquad (2)$$

The energy deposition function $\hat{v}$ can be deduced from the divergence of the Poynting vector of the electromagnetic laser field with the optical frequency ω [9].
The silicon surface gives rise to a boundary condition which is governed by the surface recombination velocity s:

$$D\nabla\hat{c}(0) = s \cdot \hat{c}(0) \qquad (3)$$

The energy stored in the electronic system dissipates due to radiationless recombination and causes a periodically time dependent temperature field, the so called thermal wave $\hat{T}$ [8] :

$$\Delta\hat{T} - q^2\hat{T} = -\frac{\hbar\omega - E_g}{\hbar\omega} \cdot \frac{\hat{v}}{k} - \frac{E_g}{k\tau}\hat{c} \qquad (4)$$

The two terms of the right side of Equ.(4) describe the direct heating by intraband relaxation and the energy deposition by bulk recombination, respectively [9]. The heat dissipates via heat conduction governed by the thermal conductivity k. Using the relation to the heat diffusivity κ the complex thermal wave number q is introduced:

$$q = \sqrt{\frac{\Omega}{2\kappa}} \cdot (1+i) \qquad (5)$$

Both p and q depend on the modulation frequency Ω which gives the possibility of the thermal wave analysis by the so called wide range frequency sweep (from 100kHz to 10MHz). Now the surface recombination channel causes a heat source at the surface expressed by the corresponding boundary condition of the heat flow $k\nabla T$ at the surface:

$$-k\nabla\hat{T}(0) = sE_g\cdot\hat{c}(0) \tag{6}$$

The response of the sample due to the periodical laser excitation results in the complex quantities $\hat{c}$ and $\hat{T}$ (amplitude and phase). The knowledge of this response of the sample depending on the modulation frequency Ω gives the possibility to calculate the surface recombination velocity s by use of Equ. (1) to (6).

The measured conversion coefficient K is proportional to the response quantities $\hat{c}$ snd $\hat{T}$ under consideration. For abbreviation we introduce the dimensionless carrier and thermal response functions $f_c(\Omega)$ and $f_T(\Omega)$, respectively [4]. Both quantities do not further depend on the excitation power:

$$K = \frac{1}{R}\frac{\partial R}{\partial c}\cdot\frac{1}{\hbar\omega D}\cdot f_c(\Omega) + \frac{1}{R}\frac{\partial R}{\partial T}\cdot\frac{1}{k}\cdot f_T(\Omega) \tag{7}$$

So it is evident that optical, thermal and electrical properties of the material are involved into the conversion coefficient.

The separation should be possible by changing the optical wavelength (photothermal spectroscopy [3]) and the modulation frequency related to the thermal wavelength by $\sqrt{(2\kappa/\Omega)}$ (thermal wave analysis).To understand the behaviour of the conversion coefficient during the frequency sweep one has to add the two complex numbers in Equ.(7) as it is shown schematically in Fig.3. Note that in the case of silicon the modulated thermoreflectivity dR/dT is positive whereas the Drude model of free carrier absorption predicts a negative sign for the term dR/dc. So conversion coefficent measured in the experiment results from the vector sum (see Fig.3) and its phase angle is very sensitive to

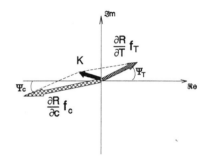

Fig.3: Vectorial superposition of the carrier and thermal response f_c and f_T, respectively, to construct the conversion coefficient K in the complex plane

changes of one or both components of the response. A small amount of surface recombination decreases the carrier wave vector and in the same manner increases the thermal wave vector. The resultant changes from the third to the second quadrant. That behaviour can be monitored by measuring the real and imaginary parts of the conversion coefficient versus the frequency and plot the values in the Gaussian plan as a circle diagram. The conformal transformation to the pure carrier wave diagram finally is used to determine the surface recombination velocity.

4. Surface recombination velocity in nanostructured silicon layers

The sensitivity of the response analysis with respect to radiation damage in thin surface layers is demonstrated using Ga$^+$implantation (100)-oriented Si wafer (5Ωcm). The surface was passivated by an HF-dip immadiately before processing. Stripes were implanted by the direct writing technique with different doses of 30 keV Ga$^+$-ions extracted from a liquid metal ion source. Two sets of 10 µm long and 150 nm broad strips separated 4µm and 1 µm, respectively, are created for each dose value. The resulting structure inclusively the 50 µm surrounding is analysed by a TWIN-thermal wave inspection system of the company

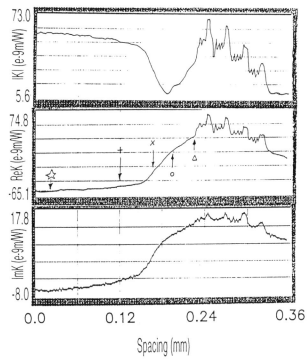

Fig.4: Magnitude, real and imaginary part of K for the linescan across the implanted structure

Jenoptik with an effective spot size of the laser beam w=2.9µm. Fig.4 shows the linescan versus 4 sets of stripes implanted with different ion doses.

The 4 µm spaced stripes were resolved laterally whereas the 1µm spaced stripes stay unresolved because of the convolution with the laser spot size. The lines capture the excess carriers due to the radiationless recombination channels caused by ion implantation. More in detail the halo of enhanced recombination of more than 100µm radius around the structure was investigated. Fig.5 shows the circle plots of the complex response gained by the frequency sweep in defined regions signed by arrows in Fig.4. Each point plotted in the complex plane of Fig.5 corresponds to the measured complex conversion coefficient for one specified modulation frequency. These circle plots can be understood by looking at the cooperation between carrier wave and thermal wave

shown in Fig.3.

Plots in the third and second quadrant means dominating carrier waves with small surface recombination and the location os the circle pot in the first and second quadrant expresses a dominant thermal wave generated by dissipation processes due to surface recombination. The whole circle plot can be understood from the solution of the coupled equ.(1) and (4). As can be shown [4] among the electrical material properties the surface recombination velocity plays a dominant role because of the near surface investigation with high modulation frequencies. So with the knowledge of D and κ by fitting with the analytical solution

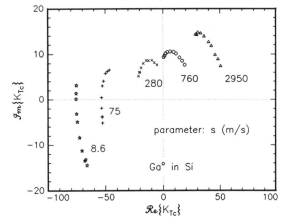

Fig.5: Circle plots of the conversion coefficient measured in the areas of scattering radiation marked in Fig.4

of Equ. (1) to (7) the surface recombination velocities marked in Fig.5 were extracted. From these data the influence of neutral and scattering irradiation in the neighbourhood of the focused ion beam can be evaluated and the radiation effect expressed by the enhanced surface recombination velocity. The circle plot representing 8 m/s correlates nearly with the value of with native oxid passivated Si. For gate oxide coverered Si this value decreases to 1 m/s [4].

5. Subsurface defect recognition

An analogous effect on the excess carrier concentration causes a dislocation network schematically shown in Fig.6. The response inspection by mapping of |K| makes the influence of the subsurface defect visible by a reduced carrier life time (see Fig.7). More information can be gained by a linescan across the buried dislocation network with different modulation frequencies. Fig.8 displays the sensitive imaginary part of K versus the scan distance for two modulation frequencies. The response of the higher

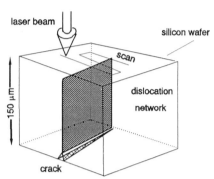

Fig.6: Scheme of the front side inspection of a dislocation network caused by a back side crack in a silicon wafer

frequency reveals the electrical fine structure of the dislocation concerning the mobility and life time.

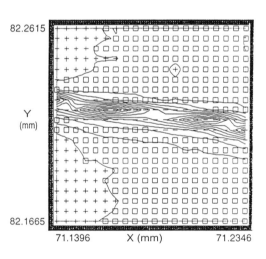

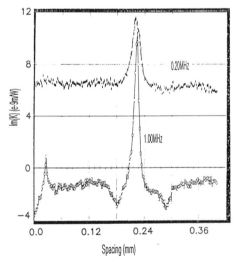

Fig.7: Mapping of the magnitude of K at 1 MHz (- symbolizes decreased carrier life time)

Fig.8: Imaginary part of K across the dislocation network for two modulation frequencies

6. Conclusion

The combination of lateral imaging with the wide range frequency sweep is a poweful tool to analyse small structures of semiconducting material and gives the possibility to calculate from the cicle plots the electical material parameter s, D and τ and the thermal diffusivity κ.

Acknowledgement

The author thanks Dr. R. Mühle and the ETH Zürich for the preparation of the demonstration samples.

References

[1]   Y. Horiike, R. Yoshida, H. Okano, M. Nakase, T. Takigawa: in *Science and Technologie of Microfabrication,* ed. by R. E. Howard, E. L. Hu, S. Namba, S. W. Pang, Mat. Res. Soc. Proc. 76,39(1987)

[2]   G. Busse, H. G. Walther: in *Principles and Perspectives of Photothermal and Photoacoustic Phenomena,* ed. by A. Mandelis, (Elsevier New York, 1992), p.205

[3]   B. Batz: in *Semiconductors and Semimetals,* Vol.9 ,edited by R. K. Willardson and A. C. Beer, (Academic Press, New York, 1972), p.315

[4]   S. Käpplinger, F. Buchmann, H. D. Geiler, Proc. 8.ITMP^3, Guadeloupe 21.-24.January 1994, in press.

[5]   M. Wagner, H. D. Geiler, Meas. Sci. Technol. 2,1088(1991)

[6]   M. Wagner, H. D. Geiler, P. Kowalski, Laser&Optoelektronik 26,1(1994)63

[7]   A. C. Boccara, D. Fournier: in *Photoacoustic and Thermal Wave Phenomena in Semiconductors,* ed. by A. Mandelis, (North Holland, New York, 1987), p.288

[8]   A.Rosencwaig: in *Photoacoustic and Thermal Wave Phenomena in Semiconductors,* ed.by A. Mandelis, (North Holland ,New York, 1987), p.98

[9]   H. D. Geiler, Nucl. Instr. & Meth. B65,9(1992)

[10] H. D. Geiler, S. Krügel, J. Nützel, E. Frieß, G. Abstreiter, Appl. Surf. Sci. 63,260(1993)

Materials Science Forum Vols. 173-174 (1995) pp. 273-278
© 1995 Trans Tech Publications, Switzerland

DIAGNOSTICS OF SEMICONDUCTOR SURFACE BY LASER-INDUCED PHOTOVOLTAGE

B. Pohoryles and A. Morawski

Instytut Fizyki PAN, Al. Lotnikow 32, PL-02668 Warszawa, Poland

Keywords: Dislocation, Surface Photovoltage, Semiconductor

Abstract. The (110) surface perpendicular to the dislocation lines in plastically deformed Ge was investigated by the SPV technique. Pronounced qualitative effects of dislocations on the surface charge and its kinetics under laser pulse were observed and explained invoking charge transfer from the surface via conducting dislocation segments to the bulk of the sample.

Introduction

Since early '70th it is known that surface photovoltage (SPV), i.e. the change of surface potential barrier under illumination, from photo-stimulated population and/or depopulation of surface states (SS) provides an effective means for determining the energy and the dynamic parameters of surface states [1].

We have implemented the SPV technique to characterise the (110) surface of plastically deformed Ge and Si. The further described deformation procedure is known to produce the homogenous (with respect to the direction, density and the length of undisturbed dislocation segments), high density dislocation set [2]. The investigated surfaces were perpendicular to the <110> direction of the prevailing number of dislocations introduced by uniaxial compression.

When studying SPV on highly dislocated samples we have discovered an unexpected phenomenon that never occurred neither in the reference, undeformed samples nor on the $(1\bar{1}1)$ glide planes of the deformed samples. The first illumination of the (110) surfaces of the deformed samples after cooling them in the darkness down to T≤240K induces the SPV signal of the opposite polarity to all the subsequent illuminations. This "first illumination puzzle" dominated our interests.

Presentation of the preliminary results of our investigations will be confined to Ge. Diagnostics of the (110) surface in plastically deformed Si will be given elsewhere. We shall start up with reminding an experimental set-up commonly used for investigation of the contact potential difference (cpd) and principles of the SPV measurements. In the subsequent section experimental results will be given followed by their discussion. Final remarks will speculate on a possible exploration of the new phenomenon.

Experimental procedure

Sample preparation. Experiments were carried out on P-doped Ge single crystals of resistivity 14Ωcm and the as-grown dislocation density 2.4×10^3cm^{-2}. The samples with this dislocation density will be further referred to as "dislocation-free". Additional dislocations were introduced into the samples by the method of two stages plastic compression along the <123> axis. The pre deformation took place at 550°C with the low shear stress of 20MPa. The final length reduction was 3%. After such pre deformation the Ge ingot was annealed at 550°C for 30min. and deformed again, but this time at low temperature (T=220°C) with the high shear stress of 300MPa. Cooling down to 50°C was performed under the load used for the main deformation. Dislocation density after the deformation exceeded $10^8 cm^{-2}$. The Ge ingots were kindly deformed by the group of Prof. H. Alexander of Cologne University. The (110) surfaces were investigated in undeformed and deformed samples. Additionally, also the ($1\bar{1}1$) glide plane of the deformed sample was investigated.

Experimental principles and set-up. The surface charge is compensated by the charge in the surface space charge region in order to maintain electrical neutrality in the system. The corresponding surface potential barrier is a function of surface charge, of free electrons and holes concentrations and of the concentrations of electrons and holes trapped in localised bulk levels. A change of any of these quantities under illumination leads to a change in the surface barrier, i.e., to surface photovoltage.

The SPV was generated by 20 mW He-Ne laser. The change in steady state cpd was measured with respect to gold reference electrode. This was a small vibrating-boss electrode which enabled point-like detection of the SPV signal and scanning throughout the plane of the wafer. Single harmonic detection of the signal, utilising unity-gain FET preamplifier and PAR Lock-in, permitted off-null linear measurements of cpd with an accuracy of better then 1mV. The final signal was continuously registered either by a recorder or the IBM computer.

The sample was placed in LNT cryostat and was electrically isolated from the grounded cold finger. The pick-up gold electrode, the sample and the cold finger, although not directly connected, were all capacitively coupled what enabled the non-contact SPV measurements.

High intensity, long lasting, laser light was used in order to assure flat band condition under illumination and filling up the traps.

Experimental results

The first striking difference between the (110) surfaces of the deformed and the dislocation-free Ge sample showed up already at room temperature: only the latter was photosensitive, i.e. gave the measurable SPV signal. This signal increased with decreasing temperature, keeping qualitatively the same shape throughout the whole temperature range. The time constant involved in the processes leading up to steady state conditions increased at 77K by several orders of magnitude. Typical SPV signal vs. time characteristic for dislocation-free Ge is shown in Fig. 1 for two different temperatures. Similar behaviour of the SPV signal was observed on the ($1\bar{1}1$) surface of the deformed sample.

In the following we present the characteristic features of the SPV signal from the (110) surface of the deformed Ge. No SPV signal could be measured from this surface at room temperature. A typical transient behaviour of the (110) surface charge induced by the long-lasting He-Ne laser pulses is illustrated schematically in Fig. 2 for a sample cooled in the darkness down to 170K. The SPV transients were induced by the following sequence of illuminations:

1) First illumination induces a slow monotonic increase of the SPV signal (solid line). The sign of the SPV indicates an increase of the negative surface charge in Ge.

2) Switching the light off results in an overshoot-type transient: after initial, fast increase of the SPV signal a marked, slow decrease of it is observed.

3) Renewed application of illumination decreases the signal, again in an overshoot manner, as shown in Fig. 2.

4) Subsequent switching the light off and on results in sequential increase and decrease of the SPV signal, always by the overshoot-like process.

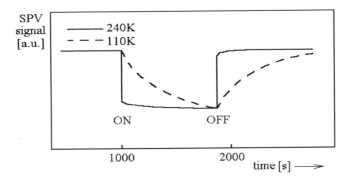

Fig.1. SPV transients for undeformed Ge at 110 and 240K

The investigations carried out lately on another set of the deformed samples revealed that also the first illumination induces an overshoot-like capacitance transient (see Fig. 2, dotted line). The dislocation density as measured by the density of etch pits was in both sets the same. Moving the vibrating-boss electrode throughout the surface we have noticed that the difference in the transient behaviour is due to the local fluctuation of the dislocation and/or surface charge density.

This type of behaviour was never observed neither in dislocation-free samples nor on the ($1\bar{1}1$) surface of the deformed sample. Therein each illumination decreased the surface charge and switching it off always resulted in an increase of this charge.

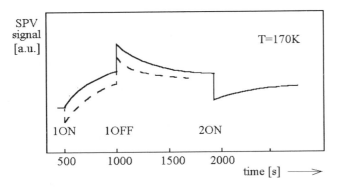

Fig.2. SPV transients for deformed Ge. They correspond to different
 sites on (110) surface

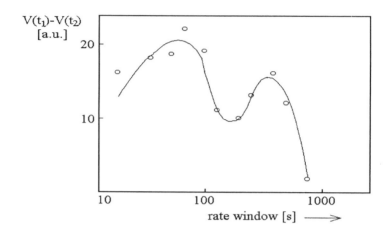

Fig.3. Isothermal SPV transient spectrum taken at 113K

A close inspection of the low temperature kinetics measured on the deformed Ge revealed that at least two different surface traps are involved in the slow relaxation processes. We have implemented the rate window concept [3] to analyse the SPV relaxation transient after switching off the laser at 113K. After the illumination pulse the SPV signal was sampled at two times t_1 and t_2 chosen so that $t_2/t_1=2$. The $V(t_1) - V(t_2)$ values as plotted against the rate window - the inverse of some preselected time constant of the transient decay - show two maximums corresponding to different surface traps characterised by different emission time constants τ. For one trap $\tau{\cong}50s$; for another $\tau{\cong}400s$. The plot is shown in Fig. 3.

Discussion

When acceptor-like SS are present below the Fermi level, as it is usually the case in n-type Ge, they will not be in equilibrium with the energy bands as long as they remain unoccupied. Those of the electrons in the CB that fall into the SS make the surface negatively charged while a positive space-charge layer form below it. Consequently, the energy bands at the surface bend upwards thus building up a barrier for electrons in the CB. Any change of the height of this potential barrier under illumination constitutes the SPV signal.

In this context, characteristics shown in Fig. 1 for dislocation-free sample and the ($1\bar{1}1$) surface of the deformed sample may be readily explained. At the onset of the laser pulse, ($h\nu>E_g$), a strong band-to-band electron excitation flattens the surface potential barrier abruptly giving rise to the SPV signal of the analogues character. Thereafter some electrons are trapped by the SS thus trying to stop the barrier drop and slowing down the last stage of the relaxation of the SPV signal. At the termination of the pulse, photovoltage decays rapidly with a time constant dictated by the minority carriers lifetime while the potential barrier is rebuild. The subsequent thermal release of additional, non equilibrium electrons from the SS slows down the last stage of relaxation of the SPV signal towards its new steady-state value. The trapping and retrapping components of the SPV signal change with time constant strongly dependent on temperature because they reflect thermally enhanced charge transfer from and to the localised SS. The lower is the temperature the slower are the transients.

Starting up with our experiments we have expected only quantitative differences between the results obtained from the (110) surfaces of dislocation-free and deformed samples. However, two striking effects showed up in the deformed sample: 1) the lack of the SPV signal at room temperature and 2) the before mentioned "first illumination puzzle" at the temperatures below 240K.

The lack of the SPV signal means the lack of the potential barrier at the (110) surface what, in turn, means the lack of the surface charge. After standard processing of the Ge surface (polishing and etching in the CP4) this is very unusual. The more so that at low temperatures and on the ($\bar{1}\bar{1}1$) surface a strong SPV signal was observed. How might that be possible?

The effect is observed in the deformed sample only and there, only on the surface perpendicular to the direction of dislocation lines. It is thus natural to prescribe the lack of potential barrier, strictly connected with the lack of a surface charge, to the charge transfer from the SS via conducting dislocation segments to the bulk of the sample.

The upper dislocation band distinguished inside the band gap of plastically deformed Ge is situated in the vicinity of the Fermi level [4]. The barrier SBD separating electrons in the SS from those in the dislocation upper band is thus most likely small and may easily be overflowed at room temperature. At 300K also a thermally assisted tunnelling from the dislocation band to the CB (previously observed in the deformed Si [5]) does not seem to be an obstacle. The resulting electron transfer from the SS to the CB reduces the potential barrier height at the surface. Low potential barrier means low SPV signal. In heavily dislocated samples this signal drops below the detectability level.

Electrons that drop into the SS from the CB are immediately transferred back via dislocations. Lowering the temperature cuts off the influx of electrons from the CB to the SS as well as the outflux of electrons from the SS to the CB thus freezing the electrical configuration reached at 300K. This time, however, the SPV signal appears due to the build up of the surface barrier - laser emitted electrons from the VB and excited over the low surface barrier in the CB drop into the empty SS. Now, in the considered temperatures, an outflux of the SS electrons over the SBD is negligible and the steady-state value of the SPV signal is determined by the balance between the process of electron trapping at the SS, their slow, thermal re-emission to the VB and their quick re-emission by the same laser pulse to the CB. The latter re-emission stops after ceasing the illumination and thus the additional build-up of the barrier at the termination of the first laser pulse is observed.

The difference between the SPV response to the first and to the subsequent laser pulses results from different initial conditions. Initially almost empty SS increase their occupation under the first illumination pulse. All the subsequent laser pulses meet already occupied SS. The initial, "empty" steady state may be reached only by warming the sample up to the room temperature and cooling it again in the darkness.

The overshoot-like character of the transients taken from the (110) surface of the deformed sample is due to the fact that the SPV signal usually consists of the two components. At the onset (termination) of the laser pulse ($h\nu>E_g$) the surface potential barrier drops (builds up) abruptly due to the strong band-to-band electron excitation (recombination) and due to a rapid photo-emission of electrons from the SS into the CB (drop of electrons from the flat CB edge to the SS). Thereafter the slow component begins to play a role. Due to the barrier drop (build up), some electrons are retrapped by the SS (released from the SS to the VB), slowing down the last stage of the relaxation of the SPV signal towards its steady-state value.

The SPV response to all the laser pulses is qualitatively the same. However, at some sites where the initial occupation of the SS is low enough the slow component may completely mask the abrupt one under the first illumination onset. In that case only a slow increase of a barrier (resulting from an increase of the SS occupation) is observed. This is the solution of the "first illumination puzzle".

Final remarks

We have started to investigate the deformed and dislocation-free Ge in order to reveal the SS possibly introduced during the plastic deformation. However, at its early stage, our work was dominated by the effect that we called the "first illumination puzzle" and by the unexpected lack of the SPV signal from the (110) surface of the deformed samples. The model just presented explains consistently both peculiarities. Although it is no doubt an oversimplification of the actual situation it has the important advantage manifesting itself in its general character. No particular type, origin or density of the SS was assumed.

The final remark is on the possible application of the effect of the reduction in barrier height resulting from the charge transfer from the SS via dislocations to the CB. The surface barrier, as a rule prevents the achievement of a good ohmic contact between a metal and a semiconductor, particularly at low temperatures. Might the generation of dislocations, e.g. by micro-indentation improve the situation? In view of our results this may well be possible.

ACKNOWLEDGEMENTS. The creative technical assistance of A.świątek is appreciated. One of the authors (B.P.) gratefully acknowledges the financial support of The Alexander von Humboldt Foundation during the Ist International Symposium on "Semiconductor Processing and Characterisation with Lasers", 18-20 April, 1994, Stuttgart, Germany.

References

[1] J. Lagowski, C.Balestra and C.Gatos, Surf. Sci. 29, 203 (1972)
[2] E. R. Weber and H. Alexander, J. Phys. C4-319 (1983)
[3] D. V. Lang, J. Appl. Phys. 45, 3022 (1974)
[4] B. Pohoryles, phys. stat. sol.(a) 116, 349 (1989)
[5] R. Nitecki and B. Pohoryles, Appl. Phys. A, 36, 55 (1985)

Materials Science Forum Vols. 173-174 (1995) pp. 279-284

CHARACTERIZATION OF THE M.O.S. STRUCTURE BY THE SURFACE PHOTOELECTRICAL VOLTAGE METHOD

B. Akkal[1], Z. Benamara[1], M. Chellali[1], H. Sehil[1] and B. Gruzza[2]

[1] Laboratoire de Microélectronique, Université de Sidi Bel Abbès,
22000 Sidi Bel Abbès, Algérie

[2] Laboratoire de Matériaux pour l'Electronique et d'Automatique, Université Blaise Pascal
(Clermont Ferrand II), U.A CNRS 1793, F-63177 Aubière Cedex,France

Keywords: MOS Structure, Photovoltage Method, Characterization, Laser

Abstract. In this paper a new method for characterization of the MOS structure, based on the measurements of the surface photoelectrical voltage (SPV) is proposed. Under a pulsed laser illumination, the variation of the photovoltage as a function of the gate tension is recorded using an adapted electronic circuit. Concerning the theoretical aspect, we have assumed:
- any MOS capacitor to be composed of a large number of elemental capacitors having a well defined surface voltage (SPV);
- the surface potential fluctuations having a Gaussian distribution.
Next, the interface states density in the gap and the standard deviation were determined by fitting the experimental curves with the theoretical ones. The results confirm that the surface photovoltage is an efficient method of characterization. Analysis can be performed also in the depletion mode. Moreover this method allows us the determination of the oxide charge density through direct reading of the flatband voltage.

Introduction

The stability and the reliability of semiconductor devices are dependent on the surface states and on the work function changes. They can disturb physical and electrical properties of the electronic elements, indeed the characteristics of MOS structures are influenced by states at the semiconductor-oxide interface.

These states can be precised with the knowledge of the relation between the photovoltaic signal and the V_G gate voltage, as well as the $C^*_{MOS}(V_G)$ capacitance values for high frequencies and under light illumination.

In order to study the work function variations and to estimate the surface photovoltage we have elaborated numerical programmes based on the macrocapacity model [2].

One can expect that the total capacity is composed of microcapacities with well known surface potentials and having a Gaussian distribution with a σ_u scatter.

First we have resolved the Poisson's equation to calculate the total charge Q_{sc}^* at the semiconductor surface for different generation rates. For different values of the polarization, changes of the photovoltaic voltage $U_{ff}(V_G)$ can be obtained by this method.

The interface states density N_{ss} and the standard deviation σ_u are determined after the adjustment of the theoretical curve with the experimental one.

Experimental part

The home-made experimental system, connected to a C(V) measuring unit, allows us to characterize the oxide-semiconductor interfaces. The photoelectric voltage $U_{ff}(V_G)$ and the value of the MOS capacitance can be recorded versus the bias polarity.

The schematic diagram of this unit is shown in the figure 1. We have used a helium-neon laser (P = 1mW , λ = 632,8 nm) [1] and the frequency of the optical signal was fixed to 160Hz using a rotative chopper. Then the light reflected by a mirror above the sample is impinging on the gate of the structure, the V_G and U_{ff} voltages are measured using the output of the differential unit ($S_0 - S_4$) and the output of the amplifier (S_3).

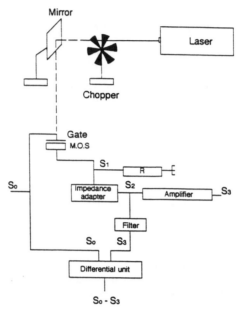

Figure 1: Schematic diagram of the experimental system

Model

The surface photovoltage U_{ff} [2] is defined as the change in gate voltage V_G required to keep the total semiconductor surface charge Q_{sc} constant after turning on the illumination.

$$Q_{sc}(\psi_s) = Q_{sc}^*(\psi_s^*) \tag{1}$$
$$U_{ff} = |V_G^* - V_G| \tag{2}$$

where $Q_{sc}^* = -\varepsilon_s kT.F[U_F^*,U^*(x)]/q.L_D^*$ with $F[U_F^*,U^*(x)]$ is the Kingston function for the quasi-equilibrium condition [3] depending on the carrier generation rate ξ.

After turning on the illumination, the surface potential decreases where as the voltage drops across the dielectric and the flat-band voltage remains unchanged. When the illumination is turned off the equilibrium state of the system is restored. This is illustrated in figure 2 which shows the change in the total surface charge as a function of the surface potential for various values of the parameter ξ corresponding to different levels of illumination.

In the depletion region for low excess carrier injection level and for surface states density higher than 10^{10} eV^{-1}.cm^{-2} the surface photovoltage is approximately proportional to exp($2U_s$), where U_s is the surface potential. On the other hand the surface photovoltage reaches a constant value in the strong inversion and accumulation regions.

For the theoretical description we have used the so-called macro-capacitor model in which the MOS capacitor is divided into small area capacitors each with a defined surface potentiel SPV. The fluctuation of the surface potential caused by an inhomogeneous distribution of oxide charge at the semiconductor-oxide interface is assumed to be of a Gaussian distribution. The probability to have the surface potential Us on an elementary area [2,3] is given by:

$$P(U_s) = [2\pi\sigma_u^2]^{-1/2}.\exp[-(U_s - \bar{U}_s)^2/2\sigma_u^2] \tag{3}$$

where $\bar{U}_s$ is the mean surface potentiel defined as:

$$\bar{U}_s = \int_{-\infty}^{+\infty} U_s.P(U_s).dU_s$$

σ_u is the standard deviation.

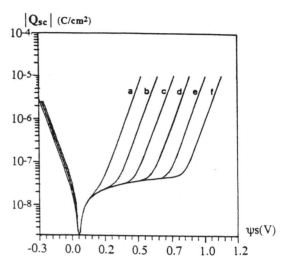

Figure 2: Total surface charge in semiconductor as a function of surface potential for various levels of light-induced rate of carrier generation:

a) $\xi = 10^4$

b) $\xi = 10^2$

c) $\xi = 1$

d) $\xi = 10^{-2}$

e) $\xi = 10^{-4}$

f) $\xi = 10^6$

In the analysis of the surface photoelectrical voltage for every macrocapacitor (SPV) as a function of potential U_s we must take into consideration these observations get from the figure 2. Then, we note the following informations:
- varying the flat band voltage V_{FB}, the characteristic SPV(V_G) is shifting to the left, toward a higher negative gate voltage (figure 3).
- considering two kinds of the states density distribution (constant or parabolic) the induced effects on the surface voltage variation SPV(V_G) can be shown in figure 4 .

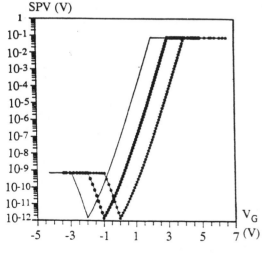

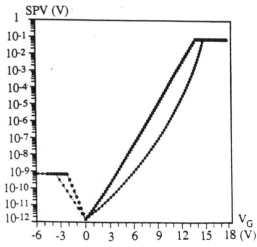

Figure 3: The variation of surface photovoltage versus the gate voltage V_G for different values of the flat band voltage (with $\xi = 5.10^{-4}$):

  *** $V_{FB} = 0V$

  ooo $V_{FB} = -1V$

  —— $V_{FB} = -2V$

Figure 4: The variation of surface photovoltage versus the gate voltage VG for the different surface states densities (with $\xi = 5.10^{-4}$):

  ●●●●●● N_{ss} constant

  ●●●●●● N_{ss} parabolic

The previous theory refers to a single elementary macrocapacitor. For the whole capacitor of the MOS structure the total surface photovoltage U_{ff} can be expressed as:

$$U_{ff} = \frac{\int_{-\infty}^{+\infty} SPV(U_s).C^*_{MOS}(U_s).P(U_s).dU_s}{\int_{-\infty}^{+\infty} C^*_{MOS}(U_s).P(U_s).dU_s} \qquad (4)$$

$SPV(U_s)$ is the surface photovoltage for each elementary macrocapacitor.

The numerator represents the variation of the interface charge, the demoninator is the average value of the C_{MOS} capacitance including space charge region, interface traps and oxide layer capacitances.

Figure 5 shows the surface photovoltage U_{ff} of a MOS structure as a function of the tension voltage V_G for various values of the standard deviation σ_u.

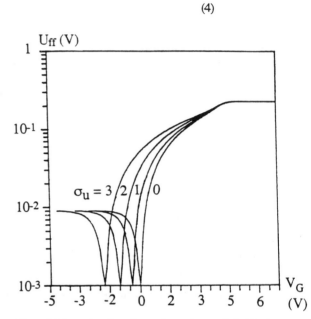

Figure 5: The total surface photovoltage characteristics U_{ff} the gate voltage for different values of σ_u.

Results

Measurements have been performed on silicon MOS structures (Al-SiO2-Si). The substrates were p type [111] Si, doped with boron atoms, with a resistivity close to 1,6 Ω.cm ($N_a = 10^{16}$cm^{-3}). The thickness of the SiO2 film, thermally deposited, was about 1000Å. Aluminium was evaporated in a 10^{-5} Torr vacuum and condensed on the sample at 250°C. The film thickness was 400Å and the area of the metallic gates were about 5.10^{-3} cm^2.

The figure 6 shows the capacitance-voltage characteristics $C_{MOS}(V_G)$ obtained at 1MHz frequency, under and without laser illumination.

In the figure 7, we have reported the theoretical and experimental surface photovoltage U_{ff} for various values of the gate tension.

The $U_{ff}(V_G)$ and $C_{MOS}(V_G)$ lines remain constant in the inversion and in the accumulation regions. Moreover in the depletion region, the photovoltage decreases exponentialy,that is in agreement with the model.

As we can see from the $U_{ff}(V_G)$ curve, the flat band voltage is close to -2,3 eV for this test structure. So we can conclude that charges are present in oxide as well as the interface with the semiconductor, and their amount is estimated to 8.10^{11}charges per cm^2.

Next, the $U_{ff}(V_G)$ experimental curve is adjusted to the theoretical one by changing both the σ_u scatter and the parameters of the density of states, which were supposed to obey to a parabolic statistical distribution.

Results are reported in the figures 7 and 8. The adjusement between the lines corresponding to U_{ff} is indicated in the figure 7. The deduced N_{ss} values, in the gap of semiconductor, are indicated in the figure 8.

The numerical expressions obtained for σ_u and N_{ss} are:

$$\sigma_u = \alpha(kT/q) \qquad \text{where } \alpha = 2/3$$

$$N_{ss} = [3.10^{12}(\psi_s - \psi_F)^2 + 6.10^{11}] \, eV^{-1}.cm^{-2}$$

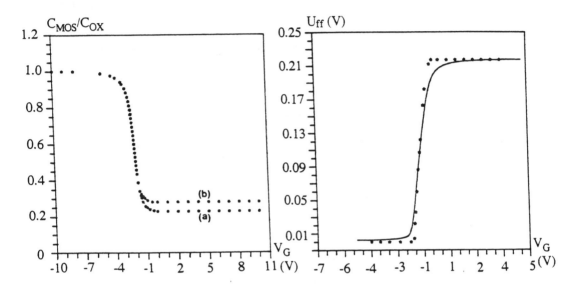

Figure 6: The capacitance-voltage characteristics measured at 1MHz:
(a) laser lighting off
(b) laser lighting on

Figure 7: The variation of the total surface photovoltage characteristics U_{ff} as a function of gate voltage V_G:
—— theoretical characteristic
* * measured values

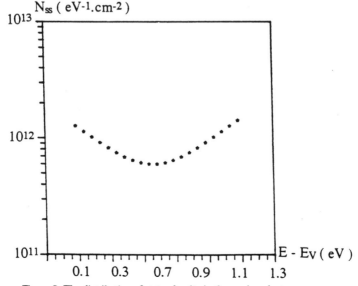

Figure 8: The distribution of states density in the semiconductor gap

The present photovoltage procedure seems convenient to carry out the states density in the gap and the flat band voltage. With this method the charges density and the variation of the surface potential can be also estimated.

Conclusion

This paper deals with characterization of a MOS structure using the photovoltage method. With the experimental set up described in the text, one get the U_{ff} tension and gives the variation of the surface potential ψ_s in the depletion region when the laser is switched on. That cannot be determined using conventional methods as for example capacitance-polarization voltage measurements.

The theoretical part is based on the macro-capacity model; the comparison between experimental results and calculations gives the following parameters of the MOS sample:
- the amount of charges in the oxide (8.10^{11} charges per cm^2)
- the states density (6.10^{11} $eV^{-1}.cm^{-2}$ in the midgap)
- the standard deviation ($\sigma_u = (2/3)(kT/q)$)

The used technique can be applied to any MIS structures, junctions and Schottky diodes [5] elaborated on others materials (InP, GaAs, Ge ...). This experimental system can be also used for the characterization of structures on polycrystalline silicon [6].

References

[1] O.Engström, A.Carlsson, J. Appl. Phys. **54**, 5245 (1983)

[2] S.Krawczyk, A.Jakubowski, Physica. Status. Solidi. **72**, 49 (1982)

[3] S.Gourrier, P.Friedel, Acta. Electronica. **25**, 3 (1983)

[4] Reinhard B.M. Girisch, Robert P.Mertens, Roger F. De Keersmaecker, IEEE Trans. Electr. Dev. **ED-35**, 203 (1988)

[5] S.R. Dhariwal, B.M.Deoraj, Semicond. Sci. Technol. **8**, 372 (1993)

[6] Arnold R. Moore, J. Appl. Phys. **54**, 222 (1983)

Materials Science Forum Vols. 173-174 (1995) pp. 285-290
© *1995 Trans Tech Publications, Switzerland*

OPTICAL SECOND-HARMONIC GENERATION (SHG) ON SEMICONDUCTOR ELECTRODES BY MEANS OF FEMTOSECOND AND NANOSECOND-PULSE LASERS

W. Kautek, N. Sorg and J. Krüger

Laboratory for Laser and Chemical Thin Film Technology, Federal Institute for Materials Research and Testing, D-12200 Berlin, Germany

Keywords: Femtosecond-Pulse Laser, Sub-Picosecond-Pulse Laser, Semiconductors, Silicon, Second-Harmonic Generation, SHG, Nonlinear Electroreflectance, Etch Process, Oxide, Interfacial Electronic States, Fermi-Level Pinning, MOS Diode

Abstract. *In situ* optical second-harmonic generation (SHG) on centrosymmetric crystalline semiconductor electrodes opens up a new field of *in situ* investigations of *buried solid state interfaces* and *metal front contacts* relevant to electronic and photovoltaic devices, which are rarely accessible by other methods.

The use of *in situ* SHG was restricted to metal electrodes so far. Photoelectrochemical nanosecond- and femtosecond-pulse laser investigations of silicon(111) electrodes show that *in situ* SHG is feasible in such complex interfacial systems. In a p-p polarization configuration, the azimuthal dependence of the SHG from oxide-covered and bare n-Si(111) electrodes, with and without nickel contact deposits, have been studied. Etching and regrowth of silicon oxides as well as buried interfacial electric field distributions (e.g. in Si/SiO$_x$) were monitored. *In situ* SHG is shown to be extremely sensitive to *trapped interfacial charge*, *crystal misorientations* and *surface step arrays*. Conventional nanosecond-laser pulses require an operation near or beyond the damage threshold of semiconductors. An advantage of *femtosecond-pulses* is the fact that illumination fluences well *below* the *damage threshold*, but still with sufficient power density, can be applied.

Introduction

Techniques for materials surface analysis *in vacuo* are relatively common, but it is the recent emergence of *in situ* interfacial probing techniques like radiotracer, fluorescence, infrared, Raman and neutron scattering, that enable direct physico-chemical process control of surfaces in gaseous and liquid environment. However, the restrictions placed on these techniques, such as interference from the bulk, engendered search for new viable surface and interfacial tools.

Modern microelectronic computer integrated manufacturing (CIM) systems have to include real-time control capabilities by effective *in situ* sensors. There exists a number of sensor technologies for dry plasma etch processes [1], but no dependable real-time *in situ* endpoint control in wet etching technology: conventional linear reflectance is insensitive to the presence or the removal of oxide layers on silicon [2]. SHG investigations of *thermally oxidized Si* represent promising steps to clarify this technologically important thin film system [2-4]. A chemical etching SHG experiment [3] showed a major signal drop when the innermost 1-nm-region of the SiO$_x$ layer was etched. It was concluded that only this region contributes to the SHG assuming a completely constant and homogeneous etch rate and absolute atomic flatness. Real systems, however, exhibit roughnesses of the order of nanometers and inhomogeneous etching leading to oxide islands and free Si surface patches, before the oxide has been entirely removed [5,6].

Early SHG studies of the Si-electrolyte interface gave some insight into the electric polarization dependence of the SHG process with 690 nm [7] and 1060 nm laser irradiation [8]. The use of a poorly inert counter electrode and the lack of potential control puts some uncertainty to the reliability of these data. It was concluded that when a strong field $E^{(0)}$ (10^4 - 10^7 V cm^{-1}) is applied, the nonlinear polarizability $P^{(2\omega)}$ changes on account of the appearance of an additional term which contains the cubic susceptibility of the semiconductor. Setting the product of the static field $E^{(0)}$ and the extension of the space charge region W approximately equal to the potential drop in the space charge, V_{sc}, one gets a parabolic relation of the SH intensity and the field $E^{(0)}$: $I^{(2\omega)} \propto [E^{(0)}W]^2$.

Our recent studies with ns laser pulses showed that even semiconductor electrodes can be probed with SHG when the duration of the light pulses (τ) is faster than electrode processes of interest, like metal plating ($\tau \ll$ ms) [9,10], or electron transfer and nucleation ($\tau <$ ps) [11]. Therefore, laser pulses of ~300 fs duration have, for the first time, been used to excite SHG at semiconductor electrodes [11]. The laser pulse interaction, and hence the SHG is finished (~300×10^{-15} s) before the fastest electron transfer processes (> 10^{-12} s) can take place. Moreover, the key point is that SHG exhibit efficiencies of less than 10^{-10}, and that signals are proportional to the square of the light intensity. It is thus necessary to apply a high peak intensity laser to produce signals whilst maintaining a low average power to avoid sample damage. Subpicosecond pulses allow much higher power densities necessary to generate sufficiently detectable SHG without depositing high fluences which can cause thermal damages in the substrate [12]. It is demonstrated that optical second harmonic generation (SHG) on semiconductor electrodes opens up a new field of *in situ* monitoring of electronically relevant *buried solid state interfaces* which are rarely accessible by other methods.

Experimental

n-Si(111) electrodes (P doping, ~10^{18} cm^{-3}, miscut 3°, Werk für Fernsehelektronik, Berlin) with evaporated gold back contacts were mounted onto brass cylinders using a conducting epoxy. They were isolated by silicon rubber leaving only the crystal surface of ~0.3 cm^2 exposed to the electrolyte. A specially designed electrochemical cell could be rotated by 360° around the normal of the electrode surface (azimuthal variation; Fig. 1). The cell was equipped with a quartz window, a gold wire counter electrode (~ 1.5 cm^2), and a saturated calomel electrode (SCE) as a reference connected via a Luggin capillary. 0.5 M K$_2$SO$_4$ prepared from analytical grade reagents and bi-distilled water was used without deaeration. Potential control was achieved by a potentiostat (HEKA). For experiments with oxide-free electrodes, the oxide was dissolved by etching in 2 M NH$_4$F (pH $\approx$ 4.5 by addition of H$_2$SO$_4$).

Nanosecond-pulse laser experiments were performed with p-polarized frequency-doubled 532 nm radiation of a Nd:YAG laser (repetition rate 10 Hz, pulse width 7 ns, average power 40 mW, intensity $3 \cdot 10^6$ W·cm^{-2}) [9,10]. The p-polarized second harmonic specular reflex at 266 nm was spectrally filtered from the fundamental, recorded by a photomultiplier tube, and normalized with respect to a reference SH signal (KDP). Alternatively, a colliding pulse mode-locked (CPM) ring laser (BESTEC, Berlin) delivered pulses shorter than 100 fs (1 fs = 10^{-15} s) at 612 nm wavelength with a repetition rate of 80 MHz and an average power of 5 - 10 mW (single pulse energies 60 - 120 pJ) [11,12]. In a four-stage dye amplifier pumped by 308 nm radiation from a XeCl-excimer laser (EMG 150, Lambda Physik, Göttingen) at a repetition rate of less than 10 Hz, they were amplified to single pulse energies of up to 200 µJ with a duration of ~300 fs. Amplified spontaneous emission (ASE) at the output of the amplifier was always less than 3%. Pulse energies were measured by a pyroelectric detector. The laser light was softly focused with a lens (f = 45 cm), and the electrode was located 8 cm out of focus. The p-polarized laser light hit the Si surface at an angle of 45° (Fig. 1). In the p-polarized specular reflex the second harmonic at 306 nm was separated from the 612 nm fundamental by means of colour glass filters (UG11). The SH radiation was detected in a photomultiplier tube (1P28A, Burle, Lancaster) the signal of which was fed to a boxcar amplifier (SR250, Stanford

Research Systems, Sunnyvale) and normalized with respect to the bulk SH generated in a KDP crystal. The time characteristics of the photomultiplier tube (PMT) current is determined by the transition function of the PMT and not by the pulse duration. Therefore, it is of the order of a few nanoseconds. The gate widths (signal and reference) of the boxcar were in the order of 100 ns so that the jitter in the PMT signal did not affect the results. Since the wavelengths of the second harmonic and the pumping radiation from the excimer laser are so close to each other, a very thorough optical shielding of the SHG set-up was necessary. To compensate the stray light background from the excimer laser, the signal without fs pulses was averaged over 200 laser shots and subtracted from the fs pulse signal averaged over 800 laser shots. The absence of damage at the Si surface after the SHG experiments was confirmed by optical microscopy.

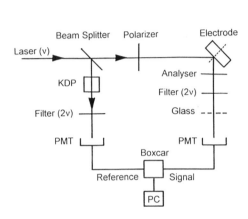

Fig. 1 Principle of the SHG-setup

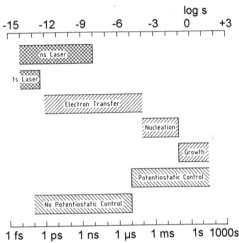

Fig. 2 Time scales of processes at semiconductor electrodes illuminated with laser pulses

Results and Discussion

Kinetics of laser-photochemical processes. Photons with an energy above the bandgap create electron-hole-pairs. One can expect relaxation, spacial redistribution of the initially hot carriers, and flattening of the electronic band edges on a time scale of picoseconds. This situation will persist till bulk recombination after τ_{rec} (> 100 ms in common Si wafers) sets in, and the bands return to their equilibrium position [11]. During this comparatively long period (τ_{rec}), electron transfer processes with rate constants of up to 10^{12} s^{-1} [13,14] can take place over both the conduction and the valence band within time domains of picoseconds and longer at open circuit (Fig. 2). At n-Si in aqueous NiSO$_4$ contact, Ni nucleation is observed within milliseconds followed by growth of Ni deposits under such conditions [10]. The photogenerated holes in the valence band obviously do not totally reoxidise the Ni-deposits. They are scavenged at the semiconductor substrate to cause either oxide growth or dissolution processes. Anodic potentiostatic control becomes effective only after the time $\tau_{pot} \approx 10$ µs. One can assume that a finite uncontrollable amount of reduced Ni species exist on the surface within and after the laser pulse duration, $\tau_{pulse} = 7$ ns, up to $\tau_{pot} \approx 10$ µs. Then, the majority carriers, the electrons, are withdrawn from the interface, and the relatively slow Ni nucleation ($\tau_{nucl} >$ 1 ms) cannot proceed.

Azimuthal SHG characteristics and Surface morphology. Bare Si(111) surfaces in contact with F$^-$-containing electrolytes as well as oxide-covered surfaces with and without Ni deposited on top of the oxide exhibit a threefold SH symmetry [9-11]. The orientational information contained in the azimuthal SH anisotropy reflects the Si surface either adjacent to the more or less disordered SiO$_2$ phase, or, in the case of bare Si, the Si/electrolyte interface. On the partly Ni covered oxidized surfaces the SH signal originates at the Si/SiO$_2$ interface [9,10].

The azimuthal dependence of the femtosecond-SHG at 612 nm in Fig. 3 shows six maxima according to the angle dependence of the intensity [11]

$$I_{pp}^{(2\omega)} \cong \left[a + b \cdot \sin(\phi) + c \cdot \cos(3\phi) \right]^2 \cdot \left(I_p^{(\omega)} \right)^2. \tag{1}$$

This is at variance with an analogous measurement with 7 ns pulses of 532 nm where three maxima were observed [9,10]. Irrespective of the wavelength and pulse duration, the amplitude b can be traced back to an ordered step structure along [$\bar{1}10$] (Fig. 4). Both the ns and the fs SHG result point out the threefold structure on the (111) terraces of the Si/SiO$_x$ interface. Amplitude b and c are only weakly affected by the different wavelengths. However, coefficient a representing the isotropic z-contributions normal to the interface, is reduced relative to the anisotropic coefficient b and c in the femtosecond-experiments at 612 nm. It is known that z-SHG components are particularly sensitive to resonant light coupling with surface states [15].

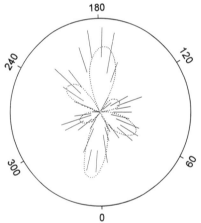

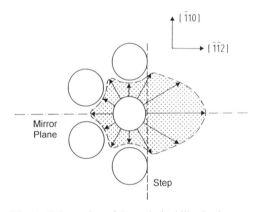

Fig. 3 Azimuthal dependence of the fs pulse laser SH signal from Si(111)/SiO$_2$; Fit parameters a=0, b=7, c=23 (Eq. 1)

Fig. 4 Schematics of the polarizability in the Si(111) plane with a step

SHG-monitoring of metallization. The nanosecond-SH signal from an oxide-free electrode in contact with 0.5 M NiSO$_4$+0.2 M NH$_4$F is by an order of magnitude smaller than that from an oxide-covered surface [10]. The azimuthal distribution from the oxide-free electrode exhibits a clear structure (Fig. 5a) that vanishes after the surface has been covered by *in situ* electrodeposited Ni before the optical measurement (Fig. 5b). The SH signal from the Ni-covered surface, however, exhibits completely different characteristics: a SH signal increase is observed at ~90°.

SHG-monitoring of oxide growth. The SHG development during regrowth of the Si oxide in humid air suggests relatively slow stepwise layer-by-layer growth. After 1000 min, a complete mono-layer of SiO$_x$ has been formed (~0.3 nm). After less than 3000 min air exposure, the SHG approxi-mately doubles. This can be related to the growth of the second SiO$_x$ layer which is known to be completed after about 10000 min (~1 week) [16]. The SHG increase can be qualitatively explained according to the electric field induced second harmonic generation (EFI-SHG) mechanism [7,8,11]. This relates the SH intensity to the square of an interfacial strong field, e.g. between the semiconduc-tor space charge and the fixed oxide charge. Each additional nonstoichiometric layer in the growing SiO$_x$ increases this fixed charge, and consequently, the induced static field. SHG should therefore keep increasing with each further monolayer till the SiO$_x$ transition layer is completed. SiO$_2$ growth after the completion of this transition layer should not affect the SH generation process any more.

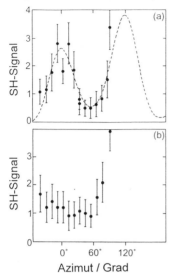

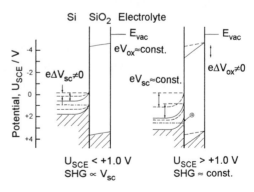

Fig. 5 Azimuthal SHG dependence of an oxide-free n-Si(111) electrode without (a) and with (b) electroplated Ni in 0.5 M NiSO$_4$ + 0.2 M NH$_4$F at -0.38 V$_{SCE}$ [10]

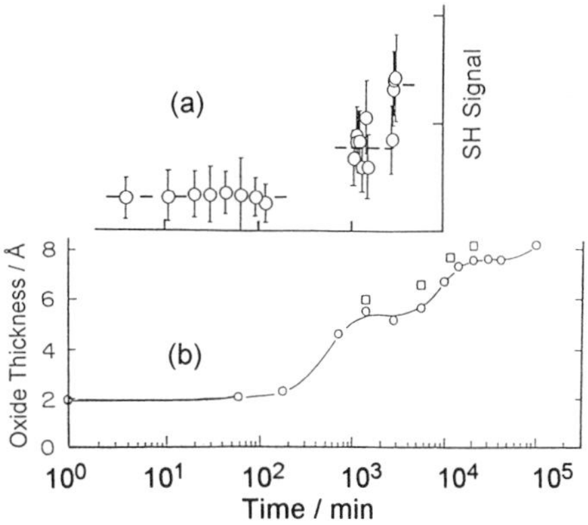

Fig. 6 Oxide growth on n-Si in air contact: (a) fs pulse laser SH signal, n-Si(111), N$_D \approx$ 10^{18} cm^{-3}; (b) XPS results, Si(100), (O) N$_D \approx$ 10^{15} cm^{-3}; ($\square$) N$_A \approx$ 10^{20} cm^{-3} [16]

Electric field dependence of SHG. The SH intensity clearly increases when the potential is moved from the conduction band edge to the valence band edge, thus inducing an increasing depletion (Fig. 7). The monotonous SHG increase (indicated by the dashed line) might be represented by a parabolic branch according to the above mentioned EFI-SHG models [7,8]. Once the electrode potential reaches the valence band edge, either Fermi level pinning by interfacial states energetically positioned near the valence band, or inversion takes place (Fig. 8). This latter case would render

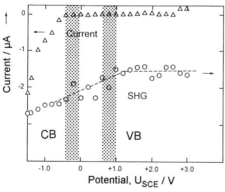

Fig. 7 Electrode potential (field) dependence of femtosecond-pulse laser SHG on Si/SiO$_x$/0.5M K$_2$SO$_4$ electrolyte [11]

Fig. 8 Scheme of the SHG field dependence in the Si/SiO$_2$/electrolyte system [11]

a p-type interfacial layer between the oxide and the semiconductor bulk which is electronically blocked by the forbidden gap from the n-type bulk. Actually, negligible anodic currents were observed (Fig. 7). Polarizing more positively, in any case, unpins the bands, while the barrier height as well as the SH intensity keep their maximum value constant. Potential changes now occur across the the oxide layer, and do not affect the SHG process. We can conclude that the height of the Schottky barrier (V_{sc}) exerts a strong influence on the SHG process. In other words, *in situ* SHG can serve to directly monitor electric field changes at the Si/SiO$_x$ interface.

Acknowledgement

This work was partially supported by the DFG (German Science Foundation).

References

[1] G.G. Barna, L.M. Loewenstein, S.A. Henck, P. Chapados, K.J. Brankner, R.J. Gale, P.K. Mozumder, S.W. Butler, and J.A. Stefani, Solid State Technol. **37**, 47 (1994).

[2] L.L. Kulyuk, D.A. Shutov, E.E. Strumban, and O.A. Aktsipetrov, J. Opt. Soc. Am. B **8**, 1766 (1991).

[3] I.V. Kravetsky, L.L. Kulyuk, A.V. Micu, D.A. Shutov, E.E. Strumban, C. Cobianu, and D. Dascalu, Appl. Surf. Sci. **63**, 269 (1993).

[4] G. Lüpke, D.J. Bottomley, and H.M. van Driel, Phys. Rev. B **47**, 10389 (1994).

[5] W. Kautek, N. Sorg, and W. Paatsch, Electrochim. Acta **36**, 1803 (1991).

[6] N. Sorg, W. Kautek, and W. Paatsch, Ber. Bunsenges. Phys. Chem. **11**, 1501 (1991).

[7] C.H. Lee, R.K. Chang, and N. Bloembergen, Phys. Rev. Lett. **18**, 167 (1967).

[8] O.A. Aktsipetrov and E.D. Mishina, Sov. Phys. Dokl. **29**, 37 (1984).

[9] J. Krüger, N. Sorg, J. Reif, and W. Kautek, Appl. Surf. Sci. **69**, 388 (1993).

[10] N. Sorg, J. Krüger, W. Kautek, and J. Reif, Ber. Bunsenges. Phys. Chem. **97**, 402 (1993).

[11] W. Kautek, N. Sorg, and J. Krüger, Electrochim. Acta **39**, (1994), in print.

[12] W. Kautek and J. Krüger, in *Laser Materials Processing: Industrial and Microelectronics Applications*, SPIE Proceedings Vol. **2207** (1994), in print.

[13] F. Willig, in *Advances in Electrochemistry and Electrochemical Engineering*, Vol. **12** (eds. C. Tobias and H. Gerischer) Wiley, New York, (1981), p. 1.

[14] W. Kautek and F. Willig, Electrochim. Acta **26**, 1709 (1981).

[15] P.V. Kelly, J.D. O'Mahony, J.F. McGilp, and Th. Rasing, Appl. Surf. Sci. **56-58**, 453 (1992).

[16] T. Ohmi, T. Isagawa, M. Kogure, and T. Imaoka, J. Electrochem. Soc. **140**, 804 (1993).

Materials Science Forum Vols. 173-174 (1995) pp. 291-296
© 1995 Trans Tech Publications, Switzerland

OPTICAL AND ELECTRICAL PROPERTIES OF SEMICONDUCTING THIN FILMS NEAR THE DEPOSITION SUBSTRATE

J. Ebothe

Laboratoire de Microscopie Eléctronique, G.R.S.M./E.M.E.T., UFR Sciences, Université de Reims Champagne Ardenne, 21 rue clément Ader, F-51100 Reims, France

Keywords: Thin Films, Semiconductor, Electro-Optics

Abstract. The investigation of sprayed CdS film optical transmision and reflection at 510 nm wavelength is here reported. It is shown that the two properties are both insensitive to the fim thickness at this irradiation frequency close to the material band gap. From these results, it is analytically demonstrated that the optical absorption coefficient behaves as an hyperbolic function versus film thickness near the deposition substrate. A relationship is established between this optical behaviour and the film carrier density whose value is supposed to vary due to the migration of some metallic elements during the deposition process from the bulk to the surface of the film substrate.

INTRODUCTION

Semiconducting thin films play an important role in modern technologies, what explains the intensive research activity devoted to this material group. In photovoltaïc area, there is some restriction as regards the film thickness required which must be delimited between 5 and 50 µm [1]. The consequence is that little attention has been paid to thinnest samples in this research activity for a long time. Most of physical properties attribuated to thin films were those of the corresponding bulk material. In fact, it seems difficult to avoid the influence of the substrate on the film properties at the initial stage of its formation. This influence has been mostly investigated in structural aspects and reported in terms of memory effect [2,3]. Some other film properties also can be considered as quite free from this influence only when a certain thickness threshold is crossed.

The present work concerns CdS film transmission and reflection properties respectively measured at different irradiation frequencies increasing the film thickness. The results are analytically examined and correlated to the film absorption coefficient and carrier density.

EXPERIMENTAL

CdS films used in the present work have been prepared according to the experimental details described previously [4]. The two steps of the spray pyrolysis film deposition technique were as follows:

i) Formation of the rough CdS films on a highly transparent and conductive substrate material consisting of SnO_2- coated CORNING 7056 glass maintained at 380° C from a spraying solution made of $CdCl_2$+ thiourea reactive media,

ii) Annealing of the films as deposited at 420°C to get rid of volatile impurities.The film structure obtained is the hexagonal type.

The film thickness value is directly dependent to the spraying time. It is calculated here by weighing from the material density (4.82 g cm^{-3}). An agreement is obtained with thickness values measured with a RANK-TAYLOR-HOBSON Talystep profiling stylus.

A BECKMAN UV 5270 spectrophotometer was used for the optical measurements with a normal incident illumination of the film surface.

RESULTS AND DISCUSSION

Transmission and reflection measurements. In this section, each of the two film properties are respectively investigated at three irradiation frequencies selected below, near and above the band gap energy (2.42 eV). Of course, T and R are both known to be characteristics of the irradiation frequency but, it seems worth examining their respective behaviour versus film thickness d, in the different cases.

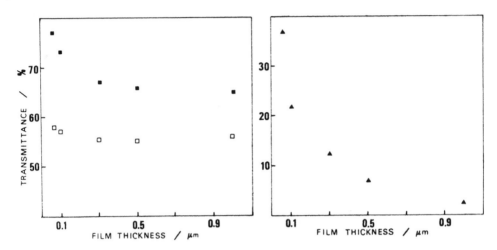

Fig. 1 Behaviour of the CdS transmittance with an increasing of the film thickness
Experimental points: ▲ , at 480 nm;   □ , at 510 nm;   ■ , at 600 nm.

In Fig. 1, a decrease of T versus d is remarkable at 480 nm wavelength, leading to a quasi exponential shape. This decrease is attenuated at 600 nm and practically inexistent at 510 nm.

Evaluation of this behaviour can be given in terms of speed of decrease $\Delta T/\Delta d$ in the investigated film thickness range.

In Table 1, the lowest slope is obtained at 510 nm when the irradiation photon energy is close to the material band gap. Eight fold increase over this value is given for 480 nm and only threetimes at 600 nm.

Table 1 - Comparative characteristics of sprayed CdS film optical properies -

wavelength (nm)	480	510	600
$\Delta T/\Delta d$	38	5	14
ΔR (%)	2	2	6

As regards the behaviour of R versus d, it appears in Fig. 2 that a similar curve profile is obtained at the three selected frequencies. An increase in R values is observed between 0.1 and 0.6 μm thickness with a slight maximun near the last value. A certain discrepency is observed in the experimental points at 600 nm so that one has a variation ΔR of about 6% in the investigated film thickness range. This variation is reduced to 2% at 510 and 480 nm wavelengths.

It comes out of the results that T and R properties are relatively sensitive to the film thickness at 480 and 600 nm and both practically remain insensitive at 510nm. Let us examine the consequence of these results on the behaviour of the film absorption coefficient.

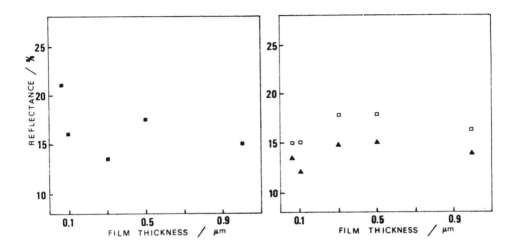

Fig. 2 Behaviour of the CdS reflectance with an increasing of the film thickness
Experimental points: ▲ , at 480 nm; □ , at 510 nm; ■ , at 600 nm.

Analytical expression of the film absorption coefficient. The behaviour of a semiconducting material under illumination mainly depends upon the optical absorption coefficient α, which has a precise value for a monochromatic irradiation frequency. A direct determination of this constant is a difficult task. It is usually obtained from the transmission and reflection measurements performed at specific experimental conditions.
If one assumes that the performance of the transmission and reflection measurements fulfilled the conditions of:
i) High transparency of the film deposition substrate,
ii) high absorptivity of the material,
ii) Normality of the irradiation to the film surface,
T and R are simply related in the optical model proposed by CHOPRA [1] as

$$T = (1-R)^2 \, e^{-\alpha d} \qquad (1),$$

with α, the film absorption coefficient. α can thus be deduced from Eq. (1) to give

$$\alpha = d^{-1} \ln[(1-R)^2 \, T^{-1}] \qquad (2a).$$

In the hypothesis of a great variation of T and R, taken individually or both versus d, Eq. (2a) corresponds to a complex relationship between α and d. However, if T and R are insensitive to d value, the relationship between α and d obeys an hyperbolic law

$$\alpha = K d^{-1} \qquad (2b),$$

with K a proportionality constant including T and R constant values. Eq. (2b) clearly shows that thinner the film, the higher is α , which practically evoluates towards a constance from a certain film thickness.

The conditions applied in the transmission and reflection measurements of sprayed CdS films are those ones of the model proposed. In actual fact, SnO_2 substrate is highly transparent, the film absorptivity is relatively high at the selected wavelengths and the photon irradition is always kept normal to the film surface. The experimental results obtained in Figs. (1) and (2) lead to consider that Eq. (2a) is valid only at 510 nm wavelength since T and R are insensitive to d value. This analytical conclusion should be corroborated by experimental results.

Carrier density in Sprayed CdS films. An intimate relationship between α and N is described in the MAXWELL and DRUDE- LORENTZ theories where a proportionality is well established between the two parameters. It comes out of this fact that a certain similariry can be expected in the behaviour of each of them versus d. Experimentally, we reported this similarity with sprayed CdS films elsewhere [7]. It was precisely observed that α and N both decrease with an increasing of d

value. This behaviour, at least as regards α, seems to obey Eq. (2b). The result concerning N necessitates an examination of the related literature.

The dependence of N to CdS film thickness was studied by AMITH with evaporated samples who found an increasing of N with an increase of d value due to S vacancies or Cd excess [3]. One can conclude that this deviation from stoichiometry is thus sensitive to d parameter. A comparable relationship between N and d has been reported also with sprayed CdS samples by KWOK et al [8]. It clearly appears here that this behaviour can be considered as a general tendency of the CdS material with no connection with the last electromagnetic theories. In actual fact, no link has been evidently established between the non stoichiometry, the oxygen chimisorption in sptrayed CdS films and the material optical absorption coefficient.

We think that the origin of the similarity in the behaviour of α and N versus d mentioned above is explained by the fact that the film properties are investigated from bare substrate, increasing gradually its thickness. one knows that the experimental conditions applied in the film deposition process are sufficient to give rise to a migration of metallic elements from the inner substrate towards its surface. Thinner CdS films can thus be considered as submitted to the predominance of these elements which can exist in different oxidation states as shown elsewhere [9]. Their carrier density can thus be very

high and their optical absorption coefficient the result of several processes. α and N both will decrease with the increasing of the film thickness since one is more and more away from the back interface and the influence of these metallic elements.

Since the role of S vacancies in the film electrical properties is accentuated with the increase of d, as shown in the literature, we think that the predominance of migrated elements only precedes that one of S vacancies in a more marked manner.

CONCLUSION

The influence of SnO_2 substrate through the migrated elements in the CdS film properties presented here can be considered as a general problem whenever the deposition process of the semiconducting film is performed at a certain high temperature. The use of a sublayer between the film and the substrate material attenuates the phenomenon but cannot suppress it completely. We have shown that this phenomenon can modify the film optical absorption coefficient. This modification is practically inavoidable and must be taken into account when one is dealing with very thin samples.

REFERENCES

[1] K. L. CHOPRA and S. R. DAS in *Thin film Solar Cells* (New York, Plenum, 1983)p.10.
[2] E. FATAS, R. DUO, P. HERRASTI, F. ARJONA and E. GARCIA-CAMARERO, J. Electrochem. Soc. **138**, 2246 (1984).

[3] A. AMITH, J. Vac. Sci. Technol, **1 5**, 353 (1978)

[4] P. CHARTIER, B. BA, J. EBOTHE, N ALONSO VANTE and H. NGUYEN CONG,
J. Electroanal. chem. **138**, 381(1982).

[5] J. EBOTHE, Thèse de Doctorat d'Etat ès Sciences Physiques, Université Louis Pasteur de
Strasbourg, France (1987).

[6] M. PEROTIN, J. BOUGNOT, OUDEACOUMAR, J. MARRUCHI, M. MARJAN and
M. SAVELLI, Revue Phys. Appl. **1 5**, 585(1980).

[7] J. EBOTHE, P. CHARTIER and H. NGUYEN CONG, Thin Solid Films, **138**,1(1986).

[8] K. L. KWOK and W. C. SIU, Thin Solid Films, **6 1**, 249 (1979).

[9] J. EBOTHE, Materials Letters, **4**, 85 (1986).

Materials Science Forum Vols. 173-174 (1995) pp. 297-310

CRYSTALLINE SILICON MATERIALS AND SOLAR CELLS

J. Knobloch and A. Eyer

Fraunhofer-Institute for Solar Energy Systems, FhG-ISE, D-79100 Freiburg, Germany

Keywords: Monocrystalline Silicon, Multicrystalline Silicon, Silicon Sheets, Ribbons, Crystalline Thin Film Silicon, Solar Cell of High Efficiency, Structured Surface, Double-Step Emitter, Passivation of Front and Rear Side, MIS-Cell, Back Surface Field

Abstract. Crystalline silicon is the most important solar cell material for power applications. About half of these cells are made from monocrystalline silicon wafers fabricated by the Czochralski-method (Cz-Si). As an alternative, multicrystalline silicon (mc-Si) produced by directional solidification of silicon melts in large crucibles is already widely used in industrial solar cell fabrication. The conversion efficiency is only about 2 - 3 % absolute lower compared to Cz-Si, but costs are reduced substantially. In order to eliminate the material losses by slicing, several techniques were developed to produce sheets or ribbons of adequate size directly from the melt. But all of these processes are very complicated and most of them were only studied on a laboratory level. An even higher cost reduction potential offer the so-called crystalline thin film silicon solar cells on low cost substrates.

In production, the conversion efficiency of solar cells manufactured from mono- or multicrystalline silicon has to be improved and production costs have to be lowered. The efficiency is limited by optical and electrical energy losses. Various structures have been developed to minimize these losses. Inverted pyramid structures on the surface of the cell, performed by anisotropic etching, allow effective light trapping. The front surface is passivated with a SiO_2-layer, which acts simultaneously as an antireflective coating. The emitter is a two step emitter. A back surface field (BSF) structure has been introduced at the rear side of the cell. An even better rear side passivation is a SiO_2-layer, combined with a local BSF. The metallisation of the rear side acts simultaneously as a mirror for the long wavelength light. For the moment cost limitations allow only some of these features to be realized in industrial solar cells, but they will be developed further to make them commercially applicable.

Introduction

Crystalline silicon is still the most widely used material for the fabrication of solar cells for terrestrial power applications. Amorphous silicon is only playing a role in consumer electronics but nevertheless in the total peak power produced it has a market share of about 25 %. Gallium arsenide which shows a substantially larger conversion efficiency than even the best silicon is only used for space applications due to its high fabrication cost. Other promising solar cell materials like $CuInSe_2$ or CdTe are still on a research level but they show promising results and might be important for future applications.

For these reasons crystalline silicon will remain the dominating material for at least the near future. In 1993 solar modules of about 60 MW were produced and in total about 400 MW have been installed world wide in the last decade. The solar cell market is growing (about 10 % per year during the last few years) but it is slowly growing due to the high cost per KWh electrical energy compared to conventional energy production.

The aims of solar cell research and development activities therefore are:
- to improve the conversion efficiency and to simplify the solar cell processes
- to reduce the cost of the silicon substrate or wafer as well as of the solar cell process

On the substrate side the following fabrication techniques for crystalline silicon exist and are in competition to each other. It is still an open question which technology will win the race.

Monocrystalline Silicon Materials

Monocrystals or single crystals of silicon grown from the melt and cut into wafers are the classical semiconductor material for integrated circuits (IC) as well as for solar cells. The float-zone (FZ) technique that leads to the most perfect silicon crystals is too expensive for solar cell material. Therefore, monocrystalline solar silicon is produced by the Czochralski (Cz) technique exclusively which is also for IC silicon still by far the dominating technique. The principle is well described in many books.

The produced crystals up to 200 mm in diameter and about 1.5 m in length are dislocation free but contaminated by about 10^{17} Atoms of oxygen per cm^3 due to pulling from an SiO_2 crucible. The oxygen content reduces the solar cell efficiency compared to FZ material but the cost advantages of Cz are dominating. Cz grown crystals (as well as FZ crystals) show the disadvantage of having a circular cross section and heads and tails of lower diameters. Therefore the cylindrical rods are cut into square rods before slicing. Modern slicing by multi-wire saws reduced the wafer thickness from classically 0.4 - 0.5 mm to 0.2 mm or even less. For solar cell applications 125 or 150 mm diameter crystals are used to produce standard 100 mm square or quasi-square wafers. But even with the wire saws the kerf loss is in the order of 0.2 - 0.3 mm i.e. it is as thick as the final wafer itself. After slicing a surface layer of about 50 μm thickness has to be etched away chemically before the solar cell process because the as-cut surface is highly damaged and full of impurities. Due to all these losses from the original Cz crystal only about 30 % are used as final wafers.

In spite of this low yield and in spite of the slow and complicated crystal pulling technique Cz wafers are still dominating the solar cell market because they show higher solar cell efficiencies than multicrystalline wafers described in the next section and because for the moment the selling price is far below the real cost due to lots of surplus or reject material from the IC market. In order to reduce the real wafer production cost and to be prepared for a growing solar cell market which can probably not be served from reject IC material, R & D activities are in the following areas of the Cz technology:
- increase of the pulling speed
- increase of the crystal length by continuous replenishment of the melt
- use of slightly dislocated crystals [1]

Multicrystalline Silicon Materials

Cast ingots. In order to reduce the cost of the crystal growth process, techniques from metallurgy were applied to silicon. They are fairly simple but the required semiconductor quality imposes some restrictions. The techniques which are used for industrial production of multicrystalline silicon (mc-Si) can be divided into 2 categories:

- melting and solidifying the silicon material in the same container
- melting in one container (called crucible) and pouring the melt into one or several moulds for solidification (the principle is sketched in Fig. 1)

Solidification is initiated by lowering the heater power or by moving the crucible out of the furnace.

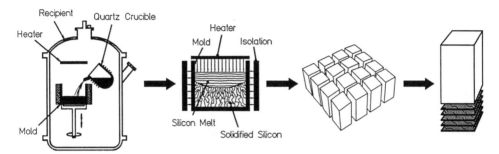

Fig. 1   Principle of the Wacker ingot casting and cutting technique

The main problem is the high reactivity of liquid silicon with any crucible material and the introduction of impurities by this reaction. In the first method liquid silicon is in contact with very hot crucible walls for a long time (several hours) but the configuration of the furnace is more simple. In the second method the mould can be kept at a temperature below the melting point of silicon during pouring. The melt solidifies immediately at the "cold" walls and forms a silicon container in which the rest of the melt can solidify slowly and without gathering impurities. The big disadvantage of the "cold" wall is that the solidification front is very concave, the grain structure of the growing ingot is inhomogeneous and all impurities are segregating to the top center of the ingot. As a consequence, the electrical homogeneity of such ingots is not good.

Fig. 1a Multicrystalline Si-Wafer 10 x 10 cm^2

Today all producers of mc-Si material are using hot walls in order to get a horizontal solidification front, a homogeneous columnar grain structure and a homogeneous electrical quality. The major activities are given in Table 1. The crucible material is still a fundamental problem. Some use quartz crucibles which are the best choice from the purity point of view but which crack during solidification and thus are a severe cost factor. Others use graphite crucibles with special and highly secret coatings which can be used many times but which introduce impurities. The products get more and more similar. The solidified ingots are in the weight range of 50 - 150 kg, square shaped with dimensions to allow for 4, 9, 16 or even 25 columns of 10 x 10 cm^2 cross section (see Fig. 1). These columns are sliced like Cz crystals. A typical mc-Si wafer of 10 x 10 cm^2 is shown in Fig. 1a. The main differences are the impurity levels, the thermal stresses correlated with the dislocation level and the grain size. The interaction of all these defects and their influence on the solar cell parameters are intensively investigated worldwide but little is fully understood.

Industrially produced mc-Si and industrially fabricated solar cells on these wafers are always 2 - 3 % absolute behind Cz based solar cells. On the laboratory level, conversion efficiencies go as high as 18 % only slightly below the records with Cz (20 %). A good and up to date overview is given in [2]. The paper also contains a large reference list.

Table 1 Directional Solidification Techniques for Multicrystalline Silicon

Name Company	Process	Crucible / Mould Re-use	Block Size (mm^3)	Characteristics	R + D Production
SILSO Wacker (Kyocera)	casting, solidification, 2 containers RF-, C-heater	quartz crucible, graphite mould + coating n x	430 x 430 x 280 120 kg	low pressure Ar, quasi columnar structure	R + D P
POLIX Photowatt France	direct solidification, 1 container, RF-heater	graphite mould + liquid encapsulation n x	410 x 410 x 130 60 kg	columnar structure	R + D P
Crystalox England (Italsolar, Bayer)	direct solidification, 1 container, RF-heater	slip cast quartz c. + Si$_3$N$_4$-coating 1 x	440 x 440 x 170 75 kg	1 atm Ar columnar structure	R + D (P)
SEMIX Solarex USA	dropping, direct solidification, 2 containers, C-heaters	ceramic crucible, ceramic mould, stapled 1 x	230 x 230 x 180 30 kg	> 1 atm Ar columnar structure	R + D P
SUMITOMO SITIX Japan	direct solidification, 1 container, RF-heater	graphite mould + Si$_3$N$_4$-coating n x	330 x 330 x 200 60 kg		R + D P
HEM Crystal Systems USA (BP Solar)	direct solidification, 1 container, C-heater	slip cast quartz c. + Si$_3$N$_4$-coating 1 x	440 x 440 x 140 80 kg ? (100 kg)	vacuum columnar structure	R + D (P)
SUMITOMO SITIX Japan	continuous casting and solidification RF-heater	cold crucible n x	117 x 117 x 200	1 atm Ar, 1-4mm/min, inhomogeneous structure	R + D

Among these ingot solidification techniques is one crucible free technique [3] which has to be mentioned. At the top of an almost square silicon rod (up to about 200 x 200 mm^2 cross-section) quite a large volume of liquid silicon (about 1 kg) is kept in position and shape by the electromagnetic forces of the high frequency heater. Silicon granules are fed into this melt continuously thus leading to a continuous growth of the rod by moving either the heater or the rod. High quality material is being produced by this technique on a laboratory level.

Sheets and ribbons. Slicing of Cz-Si crystals or mc-Si ingots leads to high quality wafers with perfect geometry and desired thickness. The main disadvantages, however, are the high cost and the material loss of the slicing step. Therefore, many techniques were developed during the past 15 years to grow directly silicon sheets or ribbons from the melt. Table 2 gives an overview of several methods which were fully evaluated. Many others were abandoned at an earlier stage either for economic or technical reasons. A review paper of all these techniques is given in [4].

The basic problem is that the final geometry of the sheet or ribbon has to be formed directly during growth from the melt. Later shaping by rolling e. g. is not possible. For shaping of the melt there are several possibilities, but in most cases a foreign material is used as a shaping element.

The melt can be sucked into a narrow slit of a die by capillary forces or by additional pressure onto the melt as shown in Fig. 2. At the upper end of the slit a ribbon can be solidified. The EFG technique at Mobil Solar e. g. used this kind of shaping. The slit was formed like an octagon and octagonal tubes of 800 mm circumference and several meters length have been crystallized [5].

Table 2 Silicon Sheet and Ribbon Technologies

Name	Company	Method	Product	Crystal structure	Output (cm^2/min)	Efficiency (%)	Problems
Cast Ribbon	Hoxan Japan	growth from melt, shaping by a flat C-cavity	ribbons 100 mm wide	columnar grains of some mm^2	200	≤10	
D-Web	Westinghouse USA	growth from melt, shaping by two Si-dendrites	ribbons 50 mm wide	monocrystalline (twin)	10	≤17	high skill needed
EFG	Mobil Solar USA	growth from melt, shaping by C-die	octogonal tubes 800 mm cf.	large elongated grains of some cm^2, dislocations 10^4 [cm^{-2}]	160	≤15	carbon, SiC, stress
RAFT	Wacker Germany	growth from melt, reusable ramps	square sheets 50 x 50 mm^2	columnar grains of ~1 mm^2, dislocations 10^7	20.000	≤10	dislocation density, ramp reuse
RGS	Bayer Germany	growth from melt, reusable substrate	ribbons 100 mm wide	columnar grains of ~1 mm^2, dislocations 10^4	6.000	not rep.	grain size, defects
Spin Cast	Hoxan Japan	spinning melt into flat C-cavities	square sheets 100 x 100 mm^2	columnar grains of < 1 cm^2	400	≤12	oxygen, dendritic structure
SSP	FhG-ISE Germany	zone melting of powder layer by optical heating	ribbons 100 mm wide	large elongated grains of some cm^2, disloc. 10^3-10^6	20	≤13	bowing, width, growth rate
S-Web	Siemens Germany	growth from melt on a C-net	ribbons 100 mm wide	columnar grains of some mm^2, disloc. 10^4-10^6	1.000	≤12	flatness, impurities
TSE	Academy Sci. USSR	growth from melt on a graphite foil through a feeder	ribbons				
For refence: CZ	different sources	growth from melt	rods up to 200 mm diameter	monocrystalline	200	≤ 18	slicing

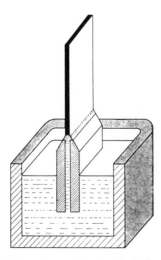

Fig. 2 Schematic graph of egde-defined film-fed growth (EFG)

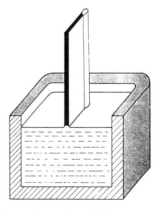

Fig. 3 Schematic graph of edge-supported pulling method

The principle of edge supported pulling is shown in Fig. 3. Quartz or graphite filaments were used as supporting elements. A very sophisticated version of this method used two silicon dendrites as supporting filaments [6]. High quality monocrystalline ribbons were grown. Instead of two filaments also graphite foils or ceramic sheets were pulled through the melt and silicon solidified on both sides of such substrates.

A different approach which uses no foreign material in contact with the melt is the SSP (silicon sheets from powder) method developed at our institute [7]. The principle is shown in Fig. 4. In the first step silicon powder is poured onto supporting plates made of quartz. The shape of the powder layer determines the shape of the final ribbon. The supporting plates are transported into the reaction chamber by quartz rollers. The second step is surface melting by focussed light. Liquid silicon penetrates into the interspaces of the powder particles but does not reach the supporting plates. Thus a fine-grained self supporting pre-ribbon is formed. The supporting plates are removed afterwards and the ribbon is moved onwards on lateral quartz rails. In the third step a liquid zone is formed which penetrates the ribbon thickness completely and very large grains are grown out of the liquid as shown in Fig. 4.

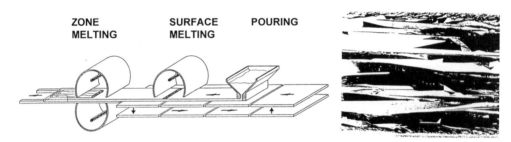

Fig. 4　Principle of the SSP technique (Silicon Sheets from Powder) and a typical grain structure

In all these techniques the growth direction is opposite to the pulling direction and the solidification interface is perpendicular to the pulling direction. The dissipation of the very high latent heat of silicon can occur only via the narrow ribbon cross-section and restricts the pulling velocity to values in the range of 10 - 20 mm/min. Such growth rates are very low from the economical point of view because, depending on the lateral extension, area growth rates of only about 100 cm^2/min can be achieved. On the other hand large grains and low defect densities can be achieved with some of these techniques. Solar cells reached values in the 14 - 16 % range.

Much higher throughput can be obtained when an extended wedge-shaped solidification front is used as shown in Fig. 5. Growth direction and pulling direction are perpendicular to each other. Heat dissipation occurs through a very large area. Area growth rates of some m^2/min can be achieved. One possibility is to grow sheets or ribbons on substrates which separate again after solidification and can be reused like in the RGS method [8]. Another possibility is to grow on graphite nets which are incorporated and lost like in the S-Web method [9]. In the case of moderate throughput (below 1 m^2/min like S-Web) fairly good grain and defect structures were achieved and solar cells were at least in the 12 - 13 % region. Highest throughputs only gave small grains and high defects structures and therefore poor solar cell properties.

For all sheet and ribbon technologies mechanical and geometrical properties are of serious concern. High temperature gradients are applied for crystallization and lead to stress, bowing and brittleness. In addition inaccurate lateral dimensions and roughness of the surface may be deleterious to solar cell processing. All these problems and the very early stage of development for

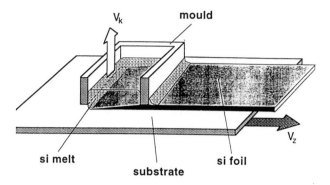

Fig. 5   Principle of the Bayer RGS technique (v_k growth velocity, v_z pulling velocity)

most of these techniques make a real economic analysis still very difficult. Meanwhile all techniques except the Bayer RGS are abandoned. Only the Mobil Solar EFG technique was developed to the stage of a large pilot production of some hundred kW solar cell power per year. But nevertheless it was stopped in 1993 due to economical considerations.

Crystalline Thin Film Silicon

In all silicon materials discussed above the wafer has two functions. On one hand the photovoltaic active material and its thickness of 0.2 - 0.3 mm is necessary for complete absorption of the sunlight but on the other hand it is also the supporting structure for the photovoltaic device.

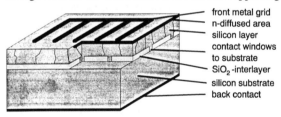

front metal grid
n-diffused area
silicon layer
contact windows
to substrate
SiO_2-interlayer
silicon substrate
back contact

Fig. 6   Diagram of a crystalline silicon thin film solar cell

In order to reduce the volume of expensive high quality silicon substantially, there is a new approach that needs only a thin layer of 20 - 50 μm supported by an inexpensive substrate (Fig. 6). The substrate might be either a low quality low cost silicon sheet produced by one of the high throughput techniques of the previous chapter or a foreign substrate. The active photovoltaic silicon layer can be deposited from the gasphase. This would be the most direct way because all semiconductor silicon originates from the gasphase ($SiHCl_3$ in most cases). High temperature deposition is necessary to achieve economic deposition rates. But that means that the substrate has to withstand high temperatures without contaminating the silicon layer. Some ceramics like Al_2O_3, SiC or Si_3N_4 would be suitable but might be too expensive. Low temperature approaches with glass substrates would be much more favorable but a sufficient silicon layer quality is probably not achievable. As a 20 - 50 μm silicon layer is not sufficient to get the red and infrared sunlight absorbed completely, special structures at the rear side of the silicon layer are necessary to reflect the light and achieve a second pass or even more passes due to total reflection at the front side.

Furthermore, an electrical reflector (electrical confinement) and/or passivation at the rear side is needed to avoid recombination of the charge carriers. If a non conductive substrate (like ceramic) is used both electrical contacts have to be realized at the front side. All these remarks make clear that crystalline thin film silicon solar cells require very complicated structures. It is certainly not easy to make full use of the economic advantages that the thin film approach offers.

Many groups are working worldwide in this new field of silicon solar cells [10], [11], [12], [13], [14], [15],[16]. Efficiencies of more than 10 % even in the 15 % range were reported. Most of these approaches are using silicon substrates as a first step but intend to change over to low cost substrates in the second step.

Diffused Junction Solar Cells

Basic considerations. In principle, a solar cell is a large area diode. Fig.7 shows a shallow n-conducting layer, the emitter, and a substantially thicker p-conducting layer, the base. N-type regions have large electron densities but small hole densities. Exactly the opposite is true for p-type regions. These concentration gradients will cause electrons to flow from regions of high concentration (n-type side) to regions of low concentration (p-type side) and similarly for holes. However, electrons leaving the n-type side will create a charge imbalance in this side by exposing ionized donors (positive charge). Similarly, holes leaving the p-type side will expose negative charge. These exposed charges will set up a space charge layer and an electric field from n- to p- region. This electric field, which is always present, is the essential factor for the operation of a solar cell.

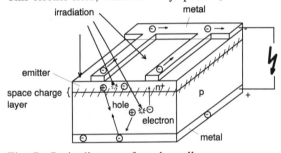

Fig. 7 Basic diagram of a solar cell

Under irradiation, photons with sufficient energy (higher than the bandgap) are falling on the surface of the solar cell, pass through emitter and space charge layer, in order to be absorbed in the base, there creating hole-electron pairs. Holes are majority charge carriers in the base and already existing in large quantities. Electrons are minority charge carriers and diffuse within the base until they reach the space charge layer. There, the strong electric field accelerates and transfers them to the emitter side. A separation of the generated charge carriers has taken place, the electric field acting as separation medium. Requirements for the separation are long diffusion lengths of the electrons, which enables them to reach the space charge layer, otherwise recombination between electrons and holes would have taken place, resulting in loss of energy of the photons. Analogous processes occur when photons are being absorbed in the emitter. Here, holes are minority charge carriers. In case of high diffusion lengths, the holes will reach the space charge layer, and the electric field will move them to the base side. Absorption within the space charge layer, leads to a immediate separation of electrons and holes by the electric field. Irradiation with light results in an increased concentration of electrons in the emitter and of holes in the base. In consequence an electric voltage has been built up, a current is flowing in an outer circuit as long as the irradiation continuous. Light energy has been transformed in electric energy.

Besides this photovoltaic effect, namely the absorption of light, the generation of charge carriers, the separation of charge carriers, there are processes which result in losses. The minimization of these losses is the main task in developing efficient solar cells.

The most important loss processes. Light will get lost at the cell surface by reflection or by shadowing of the metal contacts. The absorption within the cell can be incomplete and a part of the light will escape. The generated mobile charge carriers have only a very short lifetime, some will recombine and are lost for the current generation. Furthermore, there are ohmic losses in the grid

structure, in the contact layers and in the bulk of solar cell.
Table 3 summarizes the losses in a solar cell.

Table 3 Losses in solar cells

Optical losses	Electrical losses
Shadowing by the grid structure	Recombination losses of the generated carriers in the bulk and at then surfaces
Reflection of light at the upper cell surface	
Incomplete absorption of light with longer wavelength	Ohmic losses in the grid structure, in the emitter and in the base

Measures to reduce losses. The most important parameter is a low recombination, particularly in the base. That means a very high lifetime of the generated charge carriers, synonymous with a high diffusion length. All further measures to increase the efficiency are doomed to failure without a high diffusion length,. For high current this diffusion length should be at least double and for high voltage at least 3-5 times the cell thickness, particularly in combination with a good rear side passivation (see Fig. 9a/b).

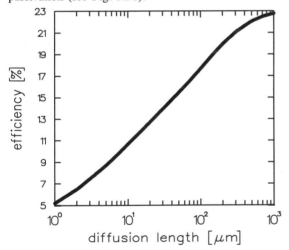

Fig. 8   Efficiency of a solar cell as function of the diffusion length in the base

Fig. 8 shows (under certain assumptions) the correlation between efficiency and diffusion length in the base. In order to increase the efficiency from a value of 10 % to above 22 %, the diffusion length must be increased by a factor of about 70 - 100, that means the recombination centers - impurities - have to be reduced by a factor of about 5 - 10.000. That requires a very pure monocrystalline silicon to start with, and very sophisticated laboratory equipment and production processes to maintain this low level of impurities during the manufacturing of the cell.

Shadowing losses can be reduced by the application of very fine grid structures, produced with photolithographic processes. Alternative techniques are the buried contacts cell, or the point contacts cell, with emitter and base contacts at the rear side.

Reflection losses at the front surface can be lowered with one or two layer antireflective coatings, and / or structurized surfaces (inverted or randomized pyramids or grooves).

Transmission losses due to weak absorption for light of longer wavelength can be reduced by a highly reflective rear Aluminum layer, increasing the effective path length. In combination with a structurized front surface, light once penetrated into the cell has practically no chance anymore to escape (light trapping, light confinement).

Limited ohmic losses in the metal and semiconductor layers require a careful engineering of the dimensions and of the doping, forcing always to compromises in base resistivity and contact structures.

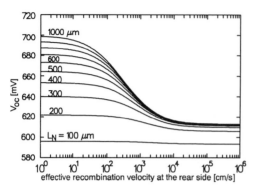
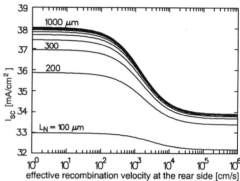

Fig. 9a V_{oc} as a function of the recombination at the rear side . Parameter is the diffusion length in the base

Fig. 9b I_{sc} as a function of the recombination at the rear side. Parameter is again the diffusion length in the base

Fig 9a and 9b show the tremendous influence of the recombination at the rear side on open circuit voltage (V_{oc}) and short circuit current (I_{sc}). (Assumptions: base resistivity 1 Ωcm, cell thickness 200 μm, ideal emitter, 8% reflection + shadowing losses, AM1,5). Suppression of the recombination will improve the open circuit voltage by about 70 mV, and the short circuit current by about 3-4 mA/cm^2. The curves further indicate that for current saturation recombination velocities below 1000 cm/s are sufficient, but voltage saturation calls for values below 100 cm/s. High voltage is more difficult to obtain than high current. So, the open circuit voltage is a good indicator for the quality of a solar cell.

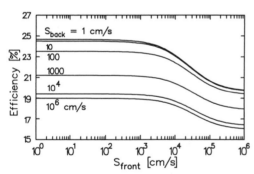

Fig. 10 Efficiency of a solar cell as a function of recombination velocity at the front and at the rear side

Naturally, the reduction of the surface recombination velocity at the emitter side has a positive influence on V_{oc} and I_{sc}, but the conditions on S_{front} are less severe than on S_{back}. Fig. 10 demonstrates that in an impressive manner. There, the theoretically obtainable efficiency is plotted as function of S_{back} and S_{front}. (For the calculation a cell thickness of 200 μm and a diffusion length in the base several times the cell thickness has been assumed). In order to obtain very high efficiencies the effective recombination velocity at the front side must be below 1000 cm/s, but below 100 cm/s at the rear side.

Technology for solar cells of high-efficiency. Fig. 11 shows the principle structure of a solar cell, incorporating the above mentioned features to reduce losses and increase the efficiency [17],[18]. As an indirect semiconductor, silicon absorbs sun light only weakly, therefore the active part of a solar cell includes front and rear side. A striking feature is the structurized surface to reduce reflection. These are inverted pyramides with base lengths of 10-20 μm, produced by anisotropic etching. Photons, not penetrating into the silicon when striking the surface for the first time get a second chance to be absorbed. Such structures, particularly in combination with AR-coating, reduce reflection losses down to 3-4%. Surfaces of this kind look black to the viewer.

The metal grid on the side of the cell exposed to sun light, is in finger number and finger width a careful compromise between shadowing and ohmic losses. Common finger widths are in the range of 10-20 μm, these structures are produced by photoresist techniques, lift-off processes and evaporation of titanium, palladium and silver, the latter intensified by electroplating.

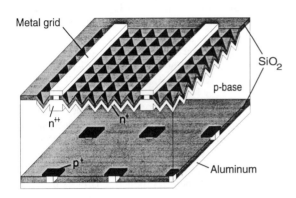

Fig. 11 Cross section of a solar cell for high efficiency

But the most important step is an almost complete coverage of the emitter surface with an excellent, thermally grown oxide to passivate the surface thus keeping charge carrier recombination as low as possible. The emitter surface is the area with the largest charge carrier concentration, because there the light intensity is highest. The thickness of the oxide is chosen to serve simultaneously as a one layer anti-reflective coating. For further reduction of the carrier recombination the contact points between semiconductor and metall are kept as small as possible, and the semiconductor is highly doped (n^{++}) at these points. Such a structure is known as two-step-emitter.

As already discussed, the behavior of the rear side is of tremendous importance. By far the largest part of the current is generated in the base and without special measures a large part of the charge carriers will be lost by recombining at the back side. Again, the best passivation is obtained with a thermally grown oxide layer. At the contact points (semiconductor-metal), a locally high-low-junction p^+ (local-back-surface field) is diffused in. The whole rear side is covered with an evaporated 2 μm Al-layer, serving as electrical contact and as a mirror for long wavelength photons. In this way, an effective recombination velocity below 100 cm/s for the whole rear side is obtained.

Production of solar cells of high efficiency. Without any doubts, a highly efficient silicon solar cell is a high-tech-product. The manufacturing of such a solar cell incorporates five main steps:
1) Structuring of the front side
2) Diffusion of the local back surface field at the rear side
3) Diffusion of the two step emitter on the front side
4) Oxide passivation of front and rear side
5) Metallisation of front and rear side + annealing of the contacts

To 1) A masking oxide layer is grown. By photolithographic means the structure for the inverted pyramids is opened, and the structure is etched anisotropically in hot alkaline solution. Afterwards, the remaining rest of the oxide layer is removed by etching in HF.

To 2) A new masking oxide layer is grown. The windows for the p^+- points are opened, a boron predeposition with BBr_3 follows, a deep in-diffusion of the p^+-area under inert atmosphere is added. Finally, all oxide is completely removed.

To 3) A new masking oxide layer is grown. The windows for the n^{++}-emitter structure are opened, followed by an n^{++}-predeposition with $POCl_3$ and n^{++}-drive in. The large windows for the n^+-area are opened, n^+- predeposition with $POCl_3$ is added. Finally, all oxide is completely etched away. This finishes the emitter structure, but the final junction depth has not yet been obtained.

To 4) An oxide layer of 105 nm thickness is grown, serving as passivation layer on front and rear side, and simultaneously as one layer antireflective coating on the emitter side.

To 5) At the rear side, the electrical contact holes at the p^+- points are opened, and the whole side is covered with $2\mu m$ of evaporated aluminum. On the front side, the windows for the grid structure are opened, followed by evaporation of the metals titanium, palladium and silver. A lift-off process removes the surplus metal, and the thickness of the silver top layer is increased by electroplating up to a few μm. An anneal in forming gas at about 400 ˚C for 20 minutes finishes the production process for such a cell.

Cells of this type have given the highest efficiency under one sun AM1,5:

- Fz-Si: 22-23 % [University of New South Wales, Australia; FhG-ISE, Germany] for areas of 4 cm^2, and 21-22 % for areas about 25 cm^2;
- Cz-Si: 20% [FhG-ISE] for 4 cm^2 area.

Other high efficiency si - solar cell concepts. Fig. 12 shows the LGBC (laser grooved buried contact) solar cell [19]. The special characteristic are grid contacts buried in grooves cut by laser. That saves a photolithographic step and results in low ohmic losses in the grid and low shadowing. Extension to larger areas (10x10 cm) cells is relatively easy. A second photo resist step can be avoided using a surface structure with random pyramids. The highest efficiencies of such cells, using Fz-Si, are in the range 19-20 % Cells of this type are already in production [BP Solar, UK].

Fig.13 shows the so called point contact cell, mainly used for concentrated light, but also showing excellent efficiencies under one sun illumination [20]. The p-n-junction consists only of n^+- p^+- dots on the rear side of the cell. There too, both contact structures are realized inter-

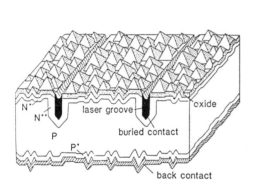

Fig. 12 Diagram of a buried contact cell   Fig. 13 Diagram of a point contact cell

locking like two combs. The front side has no shadowing losses at all. Requirement for high efficiency is extremely long diffusion lengths of the charge carriers and extraordinaire good passivation of the front surface because of the high carrier concentration and the long distance to the p-n-junction. Efficiencies are in the range 21-22 % for Fz-Si and a cell area around 20 cm^2 [Sun Power, USA]

Induced Junction Solar Cells

MIS (metal-insulator-silicon) solar cells. Unlike conventional silicon solar cells, a MIS cell contains no p-n-junction. Fig.14 shows schematically the construction of such a cell. The most important features are: rear side metal contact, p-doped base, inversion layer, silicon tunnel oxide, front side grid (simultaneously MIS contact), silicon nitride layer [21].

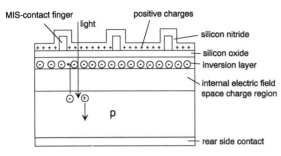

Fig. 14 Diagram of a MIS cell

The silicon nitrid layer contains a high density of fixed positve charges on the side directed toward the silicon. These positve charges induce in the p-doped silicon a very thin (~ 30 nm) layer with high electron concentration, named inversion layer because the conductivity type of the base material has been inverted from p- to n- type, followed by a space charge region. The internal electric field of the space charge zone separates the charge carriers created by the absorption of photons. The separated holes diffuse to the rear side, but the electrons migrate within the well conducting inversion layer to MIS contacts, tunnel through the very thin silicon oxide and reach the metal contacts.

An advantage of MIS - solar cells is given by the production technology. No high temperature processes are required and the number of steps is relatively small. Si-material which does not allow high temperature treatment can be used, but of course long diffusion lengths are required as well. The SiO$_2$-layer (1,3 nm) is grown at about 500 °C, the silicon nitride layer which serves simultaneously as AR coating is deposited at about 270 °C. A disadvantage is the required fine grid structure, due to the relatively low conductivity of the inversion layer. Furthermore, to avoid degradation of the efficiency the cell has to be covered with glass, containing an UV-absorbing layer. The obtained efficiencies are around 15%.

Summary

Monocrystalline silicon wafers from Czochralski crystals and multicrystalline silicon wafers from cast ingots are the dominating materials in industrial solar cell production. With increasing demand the multicrystalline wafers might win due to lower production cost. Sheet or ribbon materials as well as crystalline thin films are still far away from production, but they might be the material for a future large scale photovoltaic market.

In production, the conversion efficiency of solar cells manufactured from mono- or multicrystalline silicon has to be improved and production costs have to be lowered. As has been shown, the measures for efficiency increase are well known and relevant technologies are well established. For the moment cost limitations allow only some of these features to be realized in industrial solar cells, but they will be developed further to make them commercially applicable.

References

[1] K. L. Pauls, K. W. Mitchell, W. Chesarek: In *Proc. of 23rd IEEE PV Spec. Conf.*, (IEEE, New York, USA 1993) p. 209

[2] H. Watanabe: In *Proc. PVSEC-6*, (IREE, New Delhi, India, 1992) p. 745

[3] K. Kaneko, T. Misawa, K. Tabata: In *Proc. of 21st IEEE PV Spec. Conf.*, (IEEE, New York, USA, 1990) p. 674

[4] A. Eyer, A. Räuber, A. Goetzberger, Optoelectronics, Dev. and Techn. Vol **5**, No 2, 239 (1990)

[5] F. V. Wald: In *Crystals: Growth, Properties, and Applications*, Vol.**5**, ed. by J. G. Grabmaier (Springer Verlag, Berlin, 1981) p. 149

[6] R. G. Seidensticker: In *Crystals: Growth, Properties, and Applications*, Vol.**8**, ed. by J. G. Grabmaier (Springer Verlag, Berlin, 1982) p. 144

[7] A. Eyer, N. Schillinger, J. Reis, A. Räuber, J. Cryst. Growth, **104**, 119 (1990)

[8] H. Lange and I. A. Schwirtlich, J. Cryst. Growth, **104**, 108 (1990)

[9] J. G. Grabmaier and R. Falckenberg, J. Cryst. Growth, **104**, 191 (1990)

[10] B. F. Wagner, F. Friedrich, N. Schillinger, A. Eyer: In *Proc. of 11th EC PV Solar Energy Conf.*, (Kluwer, Dordrecht, The Netherlands, 1992) p. 217

[11] E. Demesmaeker, M. Caymax, R. Mertens, Le Quang Nam, M. Rodot, Int. J. Solar Energy **11**, 37 (1992)

[12] S. Kolodinski, J. H. Werner, U. Rau, J. K. Arch, E. Bauser: In *Proc. of 11th EC PV Solar Energy Conf.*, (Kluwer, Dordrecht, The Netherlands, 1992) p. 53

[13] B. F. Wagner, Ch. Schetter, O. V. Sulima, A. Bett: In *Proc. of 23rd IEEE PV Spec. Conf.*, (IEEE, New York, USA 1993) p. 356

[14] T. Matsuyama, M. Sasaki, M. Tanaka, K. Wakisaka, S. Nakano, Y. Kishi, Y. Kuwano: In *Proc. PVSEC-6*, (IREE, New Delhi, India, 1992) p. 753

[15] A. M. Barnett, R. B. Hall, J. A. Rand, W. R. Bottenberg: In *Proc. PVSEC-6*, (IREE, New Delhi, India, 1992) p. 737

[16] A. Takami, S. Arimoto, H. Morikawa, S. Hamamoto, T. Ishihara, H. Kumabe, T. Murotani: In *Proc. of 12th EC PV Solar Energy Conf.*, (Kluwer, Dordrecht, The Netherlands, 1994) in press

[17] M.A. Green: In *Proc. of 10th EC PV Solar Energy Conf.*, (Kluwer, Dordrecht, The Netherlands, 1991) p. 250

[18] J. Knobloch, A. Noel, E. Schäffer, U. Schubert, F. J. Kamerewerd, S. Klußmann, W. Wettling: In *Proc. of 23rd IEEE PV Spec. Conf.*, (IEEE, New York, USA 1993) p. 271

[19] A.W. Blakers, J. Zhao, A. Wang, A.M. Milne, X. Dai, M.A. Green: In *Proc. of 9th EC PV Solar Energy Conf.*, (Kluwer, Dordrecht, The Netherlands, 1989) p. 301

[20] R.M. Swanson, S.K. Beckwith, R.A. Crane, W.D. Eades, Y.K. Kwark, R.A. Sinton, S.E. Swirhun, IEEE-TED, Volume ED-31, 1984, p. 661

[21] R. Hezel, Elektronik Applikation Nr. 9, 1988, p. 38

Materials Science Forum Vols. 173-174 (1995) pp. 311-318
© *1995 Trans Tech Publications, Switzerland*

THE USE OF LASERS IN THE FABRICATION OF HIGH-EFFICIENCY SILICON SOLAR CELLS

Ch.B. Honsberg and M.A. Green

Centre for Photovoltaic Devices and Systems, University of New South Wales, Kensington, 2033, Australia

Keywords: Silicon Solar Cells, Buried Contact Cells, Thin Film Silicon Cells, Multijunction Solar Cells, High Efficiency Solar Cells

Abstract: The laser grooved, buried contact approach appears to be the only practical way yet suggested of transferring recent improvements in bulk silicon laboratory cells into commercial practice. Close to 1 MW of high efficiency modules have now been produced using this improved technology. To apply the technology to polycrystalline silicon wafers, some form of surface texturing is desirable. Laser texturing has been investigated with good results. However, mechanical texturing would appear to have economic advantages. A rear junction structure also has advantages for polycrystalline substrates in cases where the diffusion length is only about the half the substrate thickness. Such structures have also been successfully implemented using the laser grooved approach.

More recently, the laser grooved, buried contact approach has been adapted for use with thin film silicon layers deposited onto supporting substrates or superstrates. The strength of this approach lies in its compatibility with a parallel multijunction cell design.

Introduction

The last decade has seen quite substantial increases in the energy conversion efficiency of silicon cells (Figure 1). The most recent result has been the demonstration of 23.5% efficiency [1] with the PERL cell (Passivated Emitter, Rear Locally-diffused cell). These laboratory cells require microelectronics-based photolithographic sequences to achieve these high efficiencies. To transfer such efficiency improvements into commercial production, more rugged, higher throughput sequences are required.

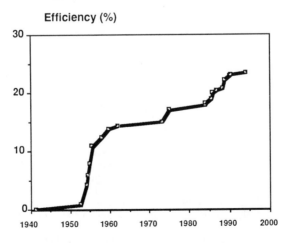

Figure 1: Evolution of silicon laboratory cell efficiency.

The approach developed at the University of New South Wales (UNSW) for such transfer is based on the use of lasers to form deep grooves in the surface of the cell midway during processing. These grooves are then used to selectively confine the effects of subsequent processing steps to the grooved areas. The application of this approach to bulk devices has formed the basis of a technology now licensed to many of the major silicon cell manufacturers. Pilot production experience with the technology has been described by these licensees on several previous occasions [2-5]. More recently, the approach has been applied to thin film silicon devices with possibly an even greater potential for future impact.

Generation I Bulk Devices

The structure of the first generation device, developed at the same time as the first 20% efficiency cell in 1984-5, is shown in Figure 2. After saw damage removal and surface texturing of the starting wafer, a light top surface diffusion is completed followed by surface oxidation. Deep grooves are then cut into the top surface, through the oxide and diffusion, using a neodymium-YAG laser. This forms grooves that can be very deep (> 60 μm), while being very narrow (20 μm or so). After etching to remove residues from the groove, the wafer undergoes a second, much heavier diffusion. Due to the presence of protective oxide on most of the cell surface, this second diffusion is restricted to grooved areas. Aluminium is then deposited on the rear of the cell by evaporation, sputtering, screen printing, or plasma arc and, after deposition, is fired appropriately. The cell is then metallized by an electroless plating method. This is very simple in practice, requiring immersion of the wafers, in plastic cassettes, in the appropriate plating baths. Plating consists of depositing a thin layer of nickel, after which the cell is briefly sintered, followed by deposition of copper, and finally a thin layer of silver. Both top and rear contacts are deposited simultaneously with this approach.

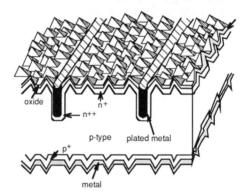

Figure 2: Laser grooved, buried contact solar cell.

This structure incorporates many of the features responsible for the high performance of the first 20% efficient laboratory cells. These include a lightly diffused cell emitter, oxide surface passivation, heavy doping of contact areas and a back surface field of acceptable quality. In pilot production, cells fabricated with similar techniques have substantially out-performed the normal commercial screen-printing sequence, due to the inability of the latter to accommodate such high efficiency features. As an example, Figure 3 shows results published by one licensee in pilot production [6]. Starting with the same quality substrate material, this figure compares the output of the normal screen-printed product produced by this licensee with that produced in pilot production using the buried contact sequence. An efficiency advantage of over 30% is obtained.

Economic studies suggest that the processing cost per unit area are very similar to that of the screen printed product, in similar production volumes. Given the increased power output per unit area, this results in substantial cost savings per watt of output as well as a marketing advantage arising from the higher product efficiency.

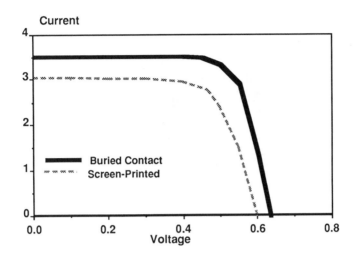

Figure 3: Comparison of output curves of screen printed and buried contact cells on identical starting substrate material [6].

Given these advantages, this technology seems destined to make a large impact on the commercial market in the near term. Production volumes have expanded greatly over recent years with the first large system (24 kW peak) installed in 1992. A much larger 550 kW system based on this technology was installed in Toledo, Spain in 1993. Both systems are the highest efficiency photovoltaic systems of these sizes yet installed.

Generation II Bulk Devices

Since the buried contact cell technology was developed in 1984-5, there have been further substantial improvements in laboratory cell efficiency [1]. These have arisen largely as a result of improved passivation of the rear surface of the cells. Accordingly, there has been considerable developmental work at UNSW aimed at simply incorporating similar benefits into the buried contact cell sequence. This has resulted in a second generation device which is already outperforming generation I devices despite the remaining potential for further substantial performance improvement.

The structure of the second generation device is shown in Figure 4. The difference from Figure 2 is the use of a grooved contacting technique on the rear surface and the use of a "floating" junction to provide high quality rear surface passivation.

The processing approach is similar to that used in first generation devices, apart from the need for a third diffusion (boron). Highlights with this process have been the extremely high open-circuit voltages demonstrated (up to 685 mV), much higher than with first generation devices. The major developmental problem has resulted from the higher susceptibility of the floating rear junction to shunting than for the top junction. This is most fundamentally due to the fact that more photogenerated current is available for biasing the front junction shunt than for the rear shunt by about a factor of 10. This means that the rear junction shunt resistance has to be 10 times higher than that of the front junction. Having now recognized this requirement, rapid progress is expected in achieving the full performance potential from this device. When fully developed, 20% cell efficiency upon Czochralski wafers is expected.

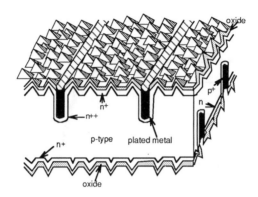

Figure 4: Bifacial buried contact solar cell.

Multicrystalline Bulk Cells

To extract the full potential from multicrystalline substrates, some form of surface texturing is essential. The UNSW group has pioneered the use of both laser and mechanical texturing of such surfaces [7,8].

Additional improvements are also possible for multicrystalline cells in which diffusion lengths are generally much smaller than the substrate thickness. In this case, performance improvement is possible by using a double junction structure, with a collecting junction also along the rear of the cell. The structure of Figure 5 has been under development at UNSW.

Studies show that this structure gives most advantage over single junction structures when diffusion lengths are about half the wafer thickness, a common situation with multicrystalline substrates. Multicrystalline cells have demonstrated output current densities comparable to those seen on high quality floatzone wafers, with a small loss in open-circuit voltage as a penalty [9]. However, the processing of the structure of Figure 5 is quite demanding. Processing simplification is being sought by removing the need for contacting the top junction of the cell.

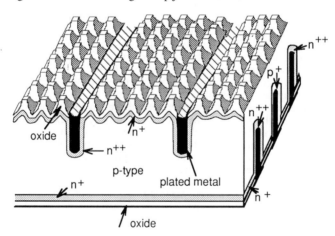

Figure 5: Double junction, multicrystalline buried contact cell.

Thin Film Buried Contact Cells

Since 1989, the UNSW group has also been developing its capabilities in the thin film silicon cell area. Recently, it established a new efficiency record of 15.6% for a thin film cell grown by liquid phase epitaxy upon an electronically "inert" single crystal silicon substrate [1], surpassing the highest efficiency previously reported with this approach. This cell was quite rudimentary in that it included virtually no features to trap light into the cell. The outstanding performance parameter of the cell was its open-circuit voltage with values above 650 mV demonstrated. However, if thin film technology is to be successfully developed, a less expensive substrate will be essential. Accordingly, the UNSW group has been experimenting with the deposition of thin film silicon on glass and other inexpensive substrates or superstrates. At the same time, new device structures have been developed to allow cell fabrication into such deposited films and automatic partitioning and interconnection of cells from material deposited uniformly onto large area substrates. Several patents have been filed in this area by the group[10].

The most promising approach is the buried contact multijunction approach shown in Figure 6. This structure combines at least three innovative elements. It uses the buried contact approach to provide electrical connection to layers of the same doping type within the multijunction stack. This approach is then developed to allow series interconnection between successive cells on the same substrate. Finally, the design of the multijunction cell allows the active collecting volume of the solar cell to be extended to an indefinite thickness, regardless of material quality.

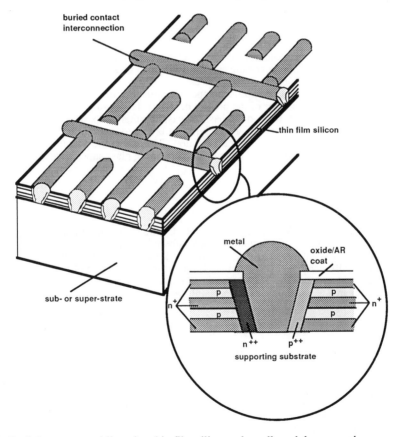

Figure 6: Buried contact, multijunction thin film silicon solar cell module approach.

The ultimate processing approach would involve deposition of successive multilayers uniformly over a treated glass superstrate. This would be followed by laser grooving to define grooves of one polarity. Subsequent to this step, a second grooving operation would then define grooves of the opposite polarity, and simultaneously provide for series interconnection of the cells. Several metallization processes are being investigated, although the same electroless plating process as used in the bulk buried contact cells also looks very promising for this structure.

The thickness of each layer within the multilayer is chosen to be of a diffusion length or less. By connecting such multilayers of the same dopant type in parallel using the buried contact approach, the collection volume of the cell can be made as large as desired. This allows high efficiency devices to be fabricated on low quality material. Calculations show that efficiencies above 17% are feasible on material of very poor quality, such as might be characterized by a defect determined lifetime of a few tens of nanoseconds.

Other attractive features of this multijunction approach include an improved tolerance to horizontal grain boundaries than possible with normal single junction structures as well as an ability for current sharing between successive layers of the same polarity by carrier injection through the intervening layer. This means, for example, that the uppermost layer does not have to carry the entire current generated in it by the light absorbed in that layer. This is an important consideration for silicon, considering over half the photogenerated current comes from the first few microns of the silicon material. With the multijunction structure, this current can be injected to lower lying layers for lateral transport to the contacted areas. Due to the use of the buried contact approach, this lateral transport distance can be quite small (less than 1 mm) without appreciable obscuration of the front surface of the cell by metal shading.

Conclusion

Laser grooving has enormous potential for high performance commercial silicon solar cells. It provides a clean, small linewidth processing capability which allows recent laboratory design features to be incorporated into rugged production sequences.

The laser-grooved buried-contact approach appears destined to make an important impact in the bulk silicon area over the remainder of the present decade, offering substantial advantages over the screen printing sequence which formed the basis of past commercial practice. Towards the end of this period, a thin film silicon approach, also relying on laser-grooved buried contacts, may also be approaching the stage of large-scale commercialization. This offers the potential for high efficiency photovoltaic cells where the major cost is that of the supporting glass superstrate onto which the thin film silicon cells are deposited.

Acknowledgements

The authors would like to thank other members of the Centre for Photovoltaic Devices and Systems who have contributed to this work, particularly Abasafreke Ebong, Zhengrong Shi, Ying Hui Tang, Michael Taouk, Stuart Wenham and Fei Yun. This work was supported by the Australian Research Council, the Energy Research and Development Corporation, the New South Wales Office of Energy and Sandia National Laboratories. The Centre for Photovoltaic Devices and Systems is supported by the Australian Research Council under the Special Research Centres Scheme and by Pacific Power.

References

[1]   M.A. Green, S.R. Wenham, A. Wang, J. Zhao, G.-F. Zheng, W. Zhang, Z. Shi and C.B. Honsberg, "23.5% Efficiency and Other Recent Improvements in Silicon Solar Cell and Module Performance", paper to be presented at the 12th E.C. Solar Energy Conference, Amsterdam, April, 1994.

[2]   N.B. Mason and D. Jordan, "A High Efficiency Silicon Solar Cell Production Technology", 10th European Photovoltaic Solar Energy Conference, Lisbon, April, 1991, pp. 280-284.

[3]   T.M. Bruton, N.B. Mason and J.C. Summers, "Towards Production of High efficiency Terrestrial Solar Cells", Conf. Proceedings, 6th International Photovoltaic Science and Engineering Conference, New Delhi, India, February, 1992.

[4]   H.-W. Boller and W. Ebner, Proc. 9th E.C. Photovoltaic Solar Energy Conference, Freiburg, 1989, p. 411.

[5]   J. Wohlgemuth and S. Narayanan, "Buried Contact Concentrator Solar Cells", Conf. Record, 22nd IEEE Photovoltaic Specialists Conference, Las Vegas, pp. 273-277, 1991.

[6]   BP Solar data sheet, "Project Saturn", April, 1991.

[7]   S. Narayanan and M.A. Green, "High Efficiency Processing of Polycrystalline Silicon Substrates", Presented at 3rd International Photovoltaic Science and Engineering Conf., Tokyo, November, 1987.

[8]   J.C. Zolper, S. Narayanan, S.R. Wenham and M.A. Green, "16.7% Efficiency, Laser Textured, Buried Contact Polycrystalline Silicon Solar Cell", Appl. Phys. Lett., Vol. 55, pp. 2363-2365, 1989.

[9]   M.A. Green, S.R. Wenham, J. Zhao, A. Wang, F. Yun and P. Campbell, "High Efficiency Silicon Solar Cells", Proceedings of 6th Photovoltaic Science and Engineering Conference, New Delhi, February, 1992, pp. 863-868.

[10]   Patent applications PK9946, PCT/AU92,/00658 and subsequent applications.

Materials Science Forum Vols. 173-174 (1995) pp. 319-324
© 1995 Trans Tech Publications, Switzerland

A LASER SYSTEM FOR SILICON SOLAR CELL PROCESSING: DESIGN, SET-UP AND OPERATION

L. Pirozzi, U.B. Vetrella, M. Falconieri and E. Salza

ENEA, CRE Casaccia, via Anguillarese 301, I-00060 Roma, Italy

Keywords: High Efficiency Solar Cells, Texturing Doping

Abstract. In this paper we will show the results obtained by using a laser system in the fabrication process of mono and polycrystalline silicon solar cells. A laser system for silicon solar cell processing has been designed and assembled in our laboratories; this system consists of a Nd-YAG, 100 W multimode, 14 W single mode (TEM00), Q-switched laser, and of an X-Y 2 μm resolution motorized table. The most important variables affecting the characteristics of the scribing process have been studied, concerning the mechanical and optical arrangement of the system. Several batches of samples have been processed and tested, in order to optimize these variables. Possible damages arising from the laser treatments have been fully investigated by optical and electrical measurements. Texturization of the cell surface by a cross-hatched pattern of parallel and perpendicular grooves was obtained by realizing a pyramidal structure on the silicon surface. We achieved reflectance values of about 10% and even lower numbers when a suitable single anti-reflecting coating (ARC) was superimposed to such a structure. A further step was the grooving of solar cells silicon surfaces to produce metal gridline patterns, in the laser grooved, buried contact, high efficiency solar cell structure. In this way we have determined the best conditions to obtain grooves with the desired depth and width, and with the cleanest and sharpest shape. Eventually, we have set up the laser assisted heavy doping of the grooves for achieving the selective emitter structure by using the second harmonic (532 nm) of the Nd-YAG laser.

INTRODUCTION.

In photovoltaic applications, laser scribing is widely used in large area amorphous silicon integrated panels fabrication /1/, and seems to be a feasible way in the preparation scribed contacts of Copper Indium Diselenide solar cells /2/. As far the most important photovoltaic material, that is mono and poly crystalline silicon, laser systems are used by some industries to carry out important processes, like silicon ribbon trimming or gridline patterning. When applied at laboratory level, silicon solar cells laser processing reached incredible results /3/.

We have set up in our laboratory a Nd-YAG laser system, with the aim of carrying out a systematic study of its possible applications in photovoltaics, and in particular in mono and poly crystalline silicon solar cells processing. We used this system to make gridline patterns for buried metal contacts, to get silicon surface texturization with a low effective reflectance, and to study laser-induced doping in a suitable reactive controlled atmosphere. Thorough investigations about any possible performance loss induced by the laser process was moreover carried out.

SYSTEM DESIGN.

A laser system for solar cell processing consists mainly of three parts:
1) a laser source;
2) the focusing optics;
3) a sample or beam motion apparatus electronically controlled.

The choice of every part must be done taking into account both the general frame of electronic devices laser processing and the particular applications envisaged.

1). As far the laser source is concerned, the most widely used system for silicon electronic devices processing is a Q-switched Nd-YAG source working in TEM00, due to the sufficient absorption coefficient of silicon at the emission wavelength 1064 nm (α=11.1 cm^{-1}), the short pulse duration, which leads to low thermal damage, and the high repetition rate, that allows high processing speed /4/. In particular, the specifications of our laser source are:

Laser	ND-YAG Q-switched
Wavelength	1064 nm
CW Power Multimode	40 Watt
CW Power TEM00	14 Watt
Q-switched Power (1 KHz TEM00)	4 Watt
Pulse Width	100 nsec
Peak Power	40 KWatt
Beam Divergence	2.5 mrad

2). The choice of the focusing optics is related to the selection of the motion: beam motion requires a dynamic focusing system, while a fixed focus arrangement is sufficient if sample movements are selected. The latter option is more convenient for a laboratory level set up, taking into consideration that high precision, low cost, and simplicity are required. For these reasons, we choose a sample motion arrangement, so that our fixed focusing optics consists simply of a combined system made of interchangeable beam expanders and focusing objectives; the expansion ratio and the objective f/n being selected according to the particular application.

3). A computer controlled X-Y, DC operating table, with a total excursion of 140 mm and a resolution of 2 μm was used to move the sample under the beam; maximum speed is 50 mm/s. Table movements and laser switches were computer controlled, with the aid of suitable CAD software. The motorized table is positioned horizontally, placed at right angle with the laser, and a properly coated mirror is used to reflect the beam onto the sample; with this arrangement, we can avoid the use of any additional wafer holder, and we can follow real-time the process through a video camera placed vertically on the sample.

Further to this set up, the system was improved by adding a frequency doubler and a reaction chamber.

The green (532 nm) light has a penetration depth into the silicon of about 2 μm (α=8300 cm^{-1}), and permits a better focusing (by which we can get smoother grooves) and surface processes, such as laser induced doping. In our case the second harmonic of the fundamental 1064 nm Nd-YAG line is obtained by using an intra-cavity LBO crystal, cut at Brewster angle to minimize reflectance losses; a spatial selection of the doubled component only was made thanks to a system of prisms and mirrors properly coated and arranged. Maximum output power was 600 mW.

The reaction chamber has been realized to perform laser processes in a controlled atmosphere, such as laser induce doping. The chamber is made of high quality vacuum stainless steel in order to minimize desorptions from the walls and to be able to work with reactive gases, it has an O-ring seal and a quartz window on top; wafers up to 10x10 cm^2 can be held at a maximum temperature of 400 °C during the process. Pumping system has been designed to achieve a high vacuum value better than 10^{-5} mbar, and to be able to handle reactive gases used as dopants. The chamber is placed onto the motorized table, and every part of it is held at a suitable temperature, so to avoid any possible moist formation.

SYSTEM CHARACTERIZATION.

In order to determine the best working conditions, preliminary tests were performed on several samples of silicon with the aim to find out the influence of the laser parameters on the quality of the scribed patterns, and the presence of possible laser induced damages. This latter point is particularly important, as a laser assisted process does not seem to be free from drawbacks due to the basic mechanism of this process, based on the vaporization of the material.

Grooving set-up. Wafers used to characterize the system were commercially available monocrystalline and polycrystalline silicon; prior to the process, wafers were cleaned in a suitable chemical bath (HF, or HF+RCA). All the meaningful tests reported were performed at 1064 nm, spot size was 1.2 mm, lasing power was varied between 1.2 and 4.1 W.

To start with, we first determined the overlap value. Because Q-switching results in a series of discrete, high energy and short pulses, it is necessary to overlap pulses in order to create a continuous scribe line; after several trials, an overlap value of 75% was found as the best compromise between the speed and the quality of the lines; overlap values down to 50% were found to give satisfactory results anyway. We experimented also the effect of making the grooves by multiple passing: this option was rejected because of the too long time to draw a grid structure.

To keep the overlap to its optimum value, we used a repetition rate of 2000 Hz, and a slit translation speed of 10 mm/s; with these arrangement, the best shape and 'cleanness' of the grooves were obtained with two set-up: lens focus 27.2 mm, beam expander 6x; lens focus 46.4 mm, beam expander 10x.

After every test, groove characteristics were investigated by optical (Reichert-Young Polyvar Met) and scanning electron (SEM) microscopy. Results of these inspections are summarized in Table I.

By using the parameters reported above, grooves with a minimum widths in the range 25 to 35 micron and with a depth of about 60 micron were readily achievable.

From these measurements, we learned that 1064 nm line produces many silicon residues, deposited on the surface of the wafer and inside the grooves. To reduce the amount of this unwanted material, a nozzle was placed on the X-Y table, blowing air or nitrogen, and a suitable NaOH based etch was used to definitely clean the grooves after the process.

Differently, the use of the 532 nm green beam allowed grooves with a width of 20 μm and a depth of less than 25 μm; they appeared also very clean, with a limited presence of melted residues, so that a light chemical treatment was necessary.

We also processed wafers on which it was deposited an ARC layer, slight adjustments of the parameters around their optimal value were needed.

Laser induced damages. The effects of rapid thermal treatments and of thermal waves driven into the sample, in principle, could give origin to cell performances loss mainly through three mechanisms: creation of dislocations, creation of amorphous phases, diffusion of impurities due to chemical treatments.

To simulate dopant diffusion, after scribing, some wafers underwent a thermal treatment, 920 C for 30 minutes.

Laser induced damages were investigated by using Surface Photo Voltage (SPV) and Laser Beam Induced Current (LBIC) techniques. By using these measurements together with microscopy, we have been able to know not only the 'optical' width of grooves and the occurrence of any residue, but also their 'electrical' width and activity due to micro structural changes or impurity contamination, both induced by chemical or thermal treatments.

To carry out SPV measurements, thick polycrystalline wafers (t=1.2 mm) were measured with a large spot SPV equipment illuminating about 80% of the total wafer area, in order to determine the diffusion length /5/. Good quality material was chosen, with a measured starting diffusion length between 120 and 150 μm. After laser scribing, wafers were measured again, and no degradation was detected. The same measurements were repeated subsequently the high temperature step and HF removal of the surface oxide: no appreciable degradation of the diffusion length was observed this time either.

Our LBIC apparatus was equipped with three light beams with different wavelengths, namely 532 nm (doubled Nd-YAG laser), 632.8 nm (He-Ne laser), and 1064 nm (Nd-YAG laser), so to be able to probe the material at different depths; the minimum step and reproducibility of the x-y motorized system was 1 μm; dimensions of the beam spot could be made as small as 4 μm for He-Ne laser.

Results of LBIC scans gave no indication of detrimental effects induced by the laser process, by the chemical etch, or by the thermal treatment.

To summarize our set-up and results, in Table 1 we report the sample preparation procedures, the main laser parameters, and the results for several laser test patterns; values in the table show quite well our system characterization procedure, and clearly show that the optimization of the laser grooving process allowed us to obtain well-fashioned patterns, avoiding, at the same time, any detectable induced damages to the cell.

Table 1.   Sample preparation procedure, main process parameters, and results for a
           set of laser grooved test patterns at 1064 nm.

Chemical treatment	Thermal treatment	Laser power (W)	Translation speed (mm/s)	Electrical width (μm)	Optical width (μm)
HF+RCA	NO	1.25	10	50	38
		1.7	10	60	46
		4.1	10	120	75
		1.7	20	45	35
NaOH 12% 15', 80°C	NO	1.25	10	60	60
		1.7	10	75	75
		4.1	10	165	160
HF+RCA		1.7	20	60	60
NaOH 12% 15', 80°C	YES	1.25	10	65	60
		1.7	10	70	75
		4.1	10	160	160
HF+RCA		1.7	20	60	60

APPLICATIONS.

Our research is devoted to evaluate the feasibility of the laser assisted processes when applied to mono and polycrystalline high efficiency solar cells. In particular, when dealing with polycrystalline silicon, laser applications seem to be a solution to the difficulties connected with this material.

High efficiency devices, texturing, and laser induced doping are the subjects of our studies. The results obtained so far on the use of our system are shown in the next subsections.

High efficiency solar cells. Laser grooved solar cell technology, originally developed elsewhere /3,6/, is one the ways to obtain high efficiency devices (buried contact solar cells). According to this technology, on the surface of a lightly diffused wafer, grooves are opened through a silicon nitride layer by using a laser scriber. After a second heavy diffusion into the grooves, metal contact is formed into the scribes by electroless deposition.

We have used our Nd-YAG laser in the fabrication process of selective emitter, buried contact solar cells on mono and polycrystalline silicon.

After a POCl3 open tube diffusion, we deposited a Si3N4 layer by a plasma assisted CVD; then, by using the laser, we scribed grooves through the film, according to the procedure reported above. Subsequently the BSF formation, a Nickel electroless deposition was performed to get the front contact. In conclusion we have shown the feasibility of a silicon solar cell laser based fabrication process, without any detectable performance losses.

Laser texturization. It is well known that, due to the grain disorientation, chemical texturization of polycrystalline material results only in a partial texturing effect: laser texturing of the cell surface is a possible way of overcoming this difficulty. Both a V-grooved or a pyramidal surface structure, in fact, can be obtained simply by closely grooving or by forming a cross-hatched pattern by the laser.

In order to find the best working conditions, we realized several test patterns on polycrystalline substrates, formed by two sets of orthogonal, parallel grooves.

All the adjustable parameters, such as groove pitch (distance between groove axis), lasing power, substrate-to-lens distance and sample translation rate have been varied.

After the laser process, the sample was chemically cleaned to remove melted residues using the following recipe:

NaOH 15% (15°C 30 min) + HCl (5 min)

After the final etch, the height of pyramids was of about 25 μm.

To study the effects of our process, total reflectance measurements have been then performed on the samples using a Perkin Elmer 330 spectrophotometer, the obtained spectrum was then weighted with the solar spectrum to get the effective reflectance of the wafer, according to the formula:

$$\text{Reff} = \frac{\int_{\lambda 1}^{\lambda 2} \text{Rmeas}(\lambda)N(\lambda)d\lambda}{\int_{\lambda 1}^{\lambda 2} N(\lambda)d\lambda}$$

On top of same sample we deposited also an ARC film, to check the combined effect of such a layer together with the texturing. Typical curves of a good quality laser textured sample with and without an ARC are shown in Fig. 1, the sample being with about 950 Å of SiO on top.

Table 2.   Most significant numbers of the laser texturization set up and results with and. without an ARC.

Pattern	Groove Pitch (mm)	R_{eff} %	ARC (Å)	R_{eff} %
V - groove	40	10.0	950, SiO	4.7
	50	13.6	950, SiO	5.8
Pyramid	40	10.5	950, SiO	4.1
	40	10.5	1400, TiO$_2$	4.3
	40	11.3	800, TiO$_2$	3.6
	40	11.3	1600, TiO$_2$	5.2
	50	8.7	950, SiO	3.9
	50	8.7	800, TiO$_2$	4.4

In Table 2 we report some of the most interesting values obtained, with reflectance spectra reported for samples with a suitable ARC layer deposited on top, as well as some process variables. As it can be seen, laser texturization process lowers the effective reflectance to less than 10%, while the addition of an ARC pushes down this value to about 4%. It is worth noting that TiO$_2$ layers have been deposited by using a production process facility.

Laser parameters turned out to be slightly different from those reported above. In particular, we varied the slit speed up to 40 mm/s, the repetition rate was 5000 Hz; lens focus was kept at 23 cm and laser power below 2.5 Watts. Values of the pitch, have been continuously varied from 40 to 70 μm.

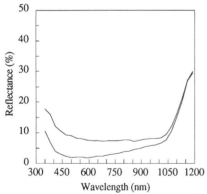

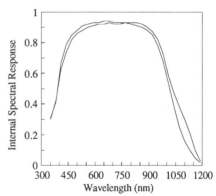

Fig. 1.   Reflectance curves for a textured sample   Fig. 2. Internal spectral responses of textured
with and without (lower curve) an ARC.   and untextured solar cells.

The use of the 532 nm green light allowed us to get patterns with melted material deposited into the grooves lower than what obtained with the 1064 nm beam; besides, this process gave also smaller pyramids height.

Smaller pyramids made therefore the effective reflectance lowering less efficient; on the other hand, this fact allowed us to test the effectiveness of our process in a screen printed production line, where too high patterns make metal contacts formation impossible.

Still keeping in mind the possible occurrence of laser induced defects, we tested their presence by preparing , on the same wafer, 2x2 sqcm textured and untextured solar cells. On these structures, we then measured the internal spectral response, which is a direct measurements of the changes in the bulk material.

Results are in Fig. 2: there is almost no difference between the two curves, that is, no damage or appreciable change was induced in the material by our laser texturing process.

Laser doping. In an innovative photovoltaic process, doping induced by laser can be applied when needs exist to dope a particular pattern in large area devices.

As an example, no further thermal treatment is necessary in a laser grooved, buried contact, selective emitter cell, as it could be replaced by a localized dopant diffusion induced by the laser. Another example of the applications of laser doping could be the formation of a selective BSF on the rear of the cell.

The basic idea for this process is that, under high fluence pulsed laser irradiation, the surface of the irradiated sample melts, and some material incorporation coming from the surrounding environment takes place /7/.

In our particular experiment, the 532 nm line is used to confine the whole process as close as possible to the surface, so to successfully promote the dissociation of molecules. Moreover, grooving and melting in the same step are achievable. Laser beam contemporary melts the silicon surface and creates dopants by pyrolysis from an adsorbed layer. The incorporation of desired atoms takes place during the melting time, dopants are driven in the liquid phase by fast diffusion. We use PCl_3 as dopant-bearing gas, the wafer is placed into the reaction chamber, and the process starts only after a sufficiently long period of high vacuum cleaning of the environment; the process is performed at a gas pressure of 70 mbar.

Experiments on this subject are still at a preliminary stage, and started from finding a range of laser parameters in such a way to be able to vary from a simply melted line to the usual grooved patterns.

Our goal is to study the properties of the junctions obtained with this method, and to apply it at a high efficiency process.

CONCLUSIONS.

In this paper we have reported the design and the set-up, of a Nd-YAG laser system to be used in solar cells processing. This system has been applied to make buried contacts in high efficiency solar cells, to texture silicon surface, and to obtain laser doped patterns. Results obtained so far are satisfactory, and show the feasibility of this process technique, mainly on polycrystalline silicon.

This work is partially supported by the E.C. in the JOULE II, "Multichess 2" Program.

REFERENCES.

/1/   H. A. Haulich: 10th E. C. Photovoltaic Solar Energy Conference, p. 1124 (Kluwer Academic Publishers, Lisbon 1991)
/2/   L. Quercia, S. Avagliano, A. Parretta, E. Salza and P. Menna: this conference.
/3/   M. A. Green: *High Efficiency Silicon Solar Cells* , p. 169 (Trans Tech Publications, Switzerland 1987).
/4/   D. L. King, B. R. Hansen and W. M. Lehrer: 21st IEEE Photovoltaic Specialists Conference, p. 278 (Orlando 1990).
/5/   A. M. Goodman, J. Appl. Phys. **32**, 2550 (1961).
/6/   M. A. Green: 10th E. C. Photovoltaic Solar Energy Conference, p. 250 (Kluwer Academic Publishers, Lisbon 1991).
/7/   G. G. Bentini, M. Bianconi, L. Correra, R. Nipoti, C. Summonte, C. Cohen and J. Siejka: E-MRS Meeting **15**, 251 (1987).

Materials Science Forum Vols. 173-174 (1995) pp. 325-330
© 1995 Trans Tech Publications, Switzerland

LIGHT AND ELECTRON BEAMS SCANNING OF SOLAR CELLS

S. Mil'shtein, B. Bakker, S. Iatrou, D. Kharas and R.O. Bell

EE Department, University of Massachusetts, Lowell, MA, USA

and

Mobil Solar Co., Billerica, MA, USA

Keywords: SEM, Dark Voltage Contrast, Solar Cells, p-n Junction, Cells Efficiency, Potential Distribution, Field Profile

Abstract. Simultaneous Light and Electron Beam Scanning (LEBEAMS) method was developed to investigate, on the microscale, the operation of various optoelectronic devices. Simultaneous light and electron beam scanning faces certain limitations due to the fact that the detectors of both the electron beam and optical system sense the reflected light and secondary electrons. We present a design where mutual interaction of light and electron operating subsystems is minimized. The LEBEAMS method is combined with Dark Voltage Contrast (DVC), a technique to study on a microscale the factors which impede the efficiency of the polycrystalline solar cells. This technique provides information on the distribution of the electrical field in a p-n junction of a solar cell under illumination and in the dark. Using a solar simulator as a light source we examined the Edge defined Film fed Grown (EFG) polycrystalline solar cell by SEM-DVC. Computer processing of the acquired SEM image and proper calibration results in detailed information of a field distribution in a p-n junction. With the light beam on a solar cell power is produced and the electric field distribution changes. We established a good correlation between the abruptness of the field and the efficiency of the solar cell. Low efficiency was found in samples where the original potential step was not steep, but rather graded.

Introduction

Recently Scanning Electron Microscopy (SEM) Dark Voltage Contrast (DVC) [1-2] was used to examine electric field distribution of solar cells. The quantitative DVC method consists of computer processing of the difference between an SEM image of a biased solar cell and an image of the same solar cell unbiased. Calibration of the resultant image depends on the reverse bias voltage and current flowing through the cell's p-n junction. Detailed study [2-3] revealed correlation between strength of internal electric field measured by DVC on a micro-scale and power produced by the solar cell. The motivation to test solar cells by scanning Light and Electron Beams (LEBEAMS) comes from the fact that solar cells do not work as power generators during DVC measurements, but rather as conventional reverse biased p-n junctions. Clearly operational conditions [4] of solar cells, namely illumination by light and somewhat elevated temperatures were missing in the DVC measurements. The current paper demonstrates for the first time how benefits of DVC measurements could be combined with light to emulate the operational conditions of the solar cell. The experimental setup is flexible enough to allow for the exchange of light sources, for example from solar simulator to laser. Use of a laser of given emission line would allow spectral response characterization.

Experiment

The LEBEAMS method was originally developed with simultaneous scanning of both light and electron beams [5]. However this carries limited meaning in our experiment, since scanning of the sample by a light beam of a few microns in diameter would produce immeasurable power. Therefore the area scanned by the electron beam was uniformly illuminated by light of a solar simulator, delivered through a multimode optical fiber.

Simultaneous use of optical and electron beam subsystems causes mutual interference which was described before [5]. We took special steps to prevent blinding of SEM detectors by light beams reflected from the solar cell surface. Shadowing of the SEM detectors was done to prevent photons from blinding the secondary electron detector. The intensity of the light of the solar simulator was controlled by a variable power supply. The effective intensity of light at the output of the fiber was calibrated to one sun. We conducted two sets of experiments. In the first set we measured the changing strength of electric field, i.e. gradient of the contrast distribution, in a cells p-n junction with increasing light intensity. In a second set we compared the DVC profiles in illuminated efficient cells and ones which performed poorly. Our samples were cut from the EFG polysilicon solar cell with the same characteristic diffusion length.

Fig. 1 presents micrographs of a solar cell in the dark (1a) and under illumination near one sun condition (1b). One can see in fig.1a a demarcation line which presents n-type (darker) and p-type (brighter) parts of a p-n junction. The contrast is due to a built-in potential in the p-n junction. As intensity of light on a semiconductor surface increases the contrast decreases and finally disappears (fig. 1b). It could be explained in the following manner: The number of hole-electron pairs increase with increasing light intensity. Electrons are injected into the n-side and holes into the p-side, wiping out the built-in potential and at the same time producing photovoltage at the terminals of the solar cell.

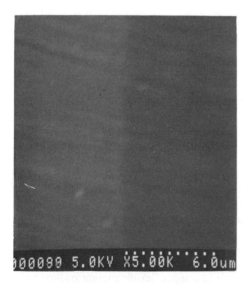 

Fig. 1a: SEM micrograph of solar cell (#6) in the dark.

Fig. 1b: SEM micrograph of solar cell (#6) under illumination of 1.05 sun.

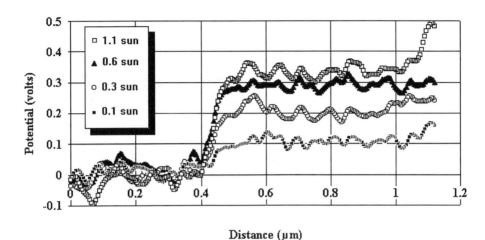

Distance (μm)

Fig. 2: Change of the potential distribution in a solar cell p-n junction under various intensities of illumination (measured relative to one sun conditions).

To make quantitative measurements of the change in contrast we developed the following calibration procedure. Specially created software allows us to subtract SEM images taken in the dark from images obtained under illumination. This effectively subtracts out surface features, scratches, contamination spots etc, resulting in an image that only carries information on the change in a p-n junction potential and its distribution. Obviously more intense light would produce a greater change in a potential step, which in turn is equal to a bigger voltage produced at the terminals of the cell. Thus the voltage produced per unit of illuminated area at the cells terminals is used in calibration of the subtracted image. The open circuit voltage V_{OC} and short circuit current I_{SC} of an illuminated cell were measured using a curve tracer in the DC mode.

Fig. 2 presents the plot of a potential distribution across the same p-n junction (cell #6) for different intensities of illumination. One can see that the step of the potential (i.e. produced voltage) is proportional to the intensity of light from the sun simulator. Fig. 3 compares the plot of change of a potential across a good cell #6 and less efficient cell #7 under illumination. Although the illuminated area for both cells is the same, cell #7 is about 15% less efficient than cell #6. This result correlates with weaker electric field in the p-n junction of cell #7, compared to the field in an efficient cell. One can readily identify a steep change of built-in potential in cell #6 against the very slow rise of the potential in a p-n junction of cell #7.

Conclusion

In our previous DVC-SEM studies [1-3] the solar cell was operated as a conventional p-n junction under reverse bias conditions. In the current study the solar cell was operated as a converter of solar energy, namely as an optoelectronic generator, with no external bias, to make operational conditions closer to reality. The solar cells were illuminated under one sun and were not artificially cooled down. Video-tape of a decreasing contrast (decreasing built-in potential) of a p-n junction with increased intensity of light demonstrated for a first time on a microscale the physical phenomena of solar energy conversion.

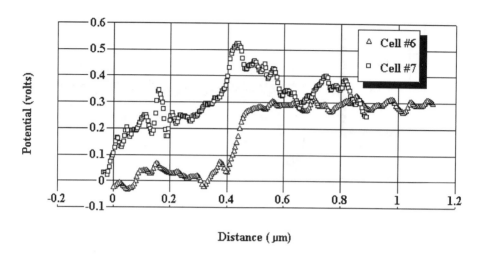

Fig. 3:   Distribution of potential in an efficient cell (#6) and less efficient cell (#7) under illumination of .56 sun.

The quantitative SEM-DVC measurement combined with measurements of power produced by solar cell (fig 2) confirmed the visualization recorded by SEM pictures (fig 1a, 1b). That as light intensity increases the built in field is wiped out and the change in the potential step across the junction increases proportionally. Efficiency of solar energy conversion was correlated with the strength of the electric field in a cell's p-n junction. The more efficient sample (cell #6) had electric field $E_{p-n}= 3.43*10^4$V/cm compared with less efficient cell #7 with $E_{p-n}= 1.88*10^4$ V/cm. It is necessary to note that both groups of cells were cut from material with the same diffusion length. Therefore we believe the difference in the performance of two groups of examined cells can be attributed to the strength of the electric field in a p-n junction.

In a previous study [3] we observed traps which were very active in a reverse biased junction. In a strong E-field the traps were observed to build-up a negative charge. Similar traps were not observed in the present study. As in the previous study [3] the forward biasing did not produce good conditions for trapping and consequently charge build-up. Cell #7 was intentionally implanted with Titanium which appears to reduce the efficiency of solar cells [6] and act as a strong trap [6]. Although we do not see strong localized traping as in [3], the shift of the potential outside of the p-n junction (see fig. 3, cell #7) is caused most likely by a random distribution of Titanium defects and their enhanced trapping under illumination.

References

1   S. Mil'shtein and S. Iatrou, First Semicond. Workshop, Hong-Kong, Aug (1992).

2   S. Mil'shtein, S. Iatrou, D. Kharas, R.O.Bell and D. Sandstrom, MRS **283**,921-925 (1993).

3   S. Mil'shtein, S. Iatrou, D. Kharas, and R.O.Bell, Proc. Int'l Conf. on Polycrystal. Silicon, Aug (1993).

4   S. Mil'shtein, D. Tripp and A. Karakashian, MRS **209**, 547-553 (1990).

5   P. Muzumdar, W. Rapose, H. Nayar and S. Mil'shtein, Bull. Amer. Phys. Soc. **35**, 1550 (1990).

6   J. Borenstein, J. Hanoka, B. Bathey, J. Kaleys and S. Mil'shtein., App. Phys. Lett. **62**, (14), 1615 (1993).

Materials Science Forum Vols. 173-174 (1995) pp. 331-336
© *1995 Trans Tech Publications, Switzerland*

CHARACTERIZATION OF TEXTURED TRANSPARENT CONDUCTIVE OXIDES FOR THIN FILM SOLAR CELL APPLICATIONS

J. Wallinga, J. Daey Ouwens, R.E.I. Schropp and W.F. Van der Weg

Department of Atomic and Interface Physics, Debye Institute, Utrecht University, P.O.Box 80.000, NL-3508 TA Utrecht, Netherlands

Keywords: Transparent Conductive Oxide, Optical Trapping, Amorphous Silicon, Solar Cells

Abstract. We studied the influence of textured Transparent Conductive Oxide (TCO) window layers on the performance of amorphous silicon solar cells. Incoming light is partly scattered by *textured* TCO-layers. By texturing the optical path of the light through the cell is considerably increased, especially for the long wavelength region (λ >500 nm). Various layers of fluorine-doped tin oxide (SnO_2:F) with a thickness of about 1 μm and tin-doped indium oxide (In_2O_3:Sn or ITO) of 100 nm thickness, deposited on a glass substrate have been examined. To determine the scattering characteristics due to texture, the TCO-layers are placed in a HeNe laser-beam (λ=632.8 nm). The laser light is partially scattered by the TCO. A silicon photodiode is used as a detector to measure the angle-dependent light intensity $I(\theta)$ of the transmitted light. We found the intensity distribution for several samples with a transparency of about 80% in the visible region to be Lambertian, $I = I_0 \cos\theta$. This means the scattering is random. Amorphous silicon solar cells have been deposited by rf plasma deposition on different textured TCO-layers. The short circuit current of the devices increases with the experimentally determined scattering parameter of the TCO. The open circuit voltage decreases for highly textured substrates. The laser scattering measurements provide a useful quality assessment tool for the scattering of TCO-layers.

Introduction

Amorphous silicon solar cells require a transparent conductive window layer as a front contact. Metal oxides like fluorine-doped tin oxide and tin-doped indium oxide are frequently used for this purpose [1]. They combine good transmission of light from 350 to 800 nm, typically over 80%, with a low sheet resistance, typically lower than 10 $\Omega/\square$.

The metal oxide is deposited on a glass substrate, for instance by evaporation or CVD [1]. Usually APCVD is used to deposit textured TCO. The size and shape of the grains depend on the deposition parameters. The amorphous silicon (a-Si:H) p-i-n structure is deposited on the front contact by PECVD. A metal back contact is evaporated to complete the solar cell. A schematic drawing of an a-Si:H solar cell is shown in Fig. 1.

Electron-hole pairs are created by absorption of light in the i-layer. They are separated by the electric field, caused by thin p- and n-type regions on both sides of this active layer. The localized states in the band gap of intrinsic a-Si:H regulate transport properties and recombination kinetics. This gives rise to a relatively low effective carrier collection length and implies that the i-layer should be very thin (<1 μm).

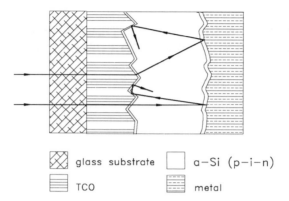

Figure 1: Solar cell on textured substrate

Light-induced defects [2], however, lead to a reduction of the carrier collection length. In order to obtain amorphous silicon cells with a high *stabilized* efficiency we have to develop structures in which the light is effectively trapped. In this way the i-layer thickness can be reduced without a loss in the total absorption.

For short wavelength light the absorption coefficient in a-Si:H is high. Therefore this light will be absorbed in one pass through the i-layer.

For long wavelength light ($\lambda >500$ nm) the absorption coefficient of a-Si:H is lower. The photon flux in this part of the solar spectrum is large. Scattering the light into the active structure by using a textured TCO is crucial to increase the effective path length for this part of the spectrum.

Scattering is usually quantified by the haze ratio. The haze ratio is defined as the amount of diffuse transmission divided by the total (diffuse plus specular) transmission of white light. Since only the scattering of light with long wavelengths is important in a-Si:H solar cells, we measured scattering with a HeNe-laser ($\lambda=632.8$ nm).

Theory

The amount of scattering depends on the feature size associated with the texture of the TCO and on the wavelength of the light. Optimum textures have dimensions somewhat larger than the wavelength of the light in the semiconductor [3]. Scattering is then described by Mie theory [4, 5].

Non-uniformly structured scattering interfaces cause a Lambertian intensity distribution [6]:

$$I(\theta) = I_0 \cos \theta. \tag{1}$$

The effective absorption A_{eff} in the i-layer is obtained by integrating the angle-dependent absorption $A(\theta)$ over the associated angle-dependent relative intensity $I(\theta)$ with respect to the intensity of the incident beam I_i:

$$A_{\text{eff}} = \int_{\theta=0}^{\pi/2} A(\theta) \frac{dI(\theta)}{I_i}. \tag{2}$$

The angle dependent absorption is given by

$$A(\theta) = 1 - e^{-\alpha t/\cos\theta}, \tag{3}$$

where α is the absorption coefficient and t is the thickness of the layer. For a Lambertian distribution

$$\frac{dI(\theta)}{I_i} = 2\cos\theta\sin\theta\,d\theta \tag{4}$$

is the fraction of the intensity per solid angle.

The fraction of the light that is scattered increases when the light is reflected at the semiconductor/metal back interface. This interface roughly replicates the front surface. The fact that the light is scattered enhances optical trapping in the structure due to total internal reflection at the front and at the back interfaces of the solar cell.

Theoretically the maximum total effective absorption is [7]:

$$A_{\text{eff}} = 4\alpha t\left(\frac{n_{\text{a-Si:H}}}{n_{\text{air}}}\right)^2. \tag{5}$$

This is an increase of 50 times with respect to the absorption of non-scattered light which passes through the i-layer perpendicular to the surface. In practical applications this number is about 4.

Measurements

As previously stated, for a-Si:H solar cells not the scattering of blue light, but only of red light is important. Therefore in this study we measure the scattering of red light from a 5 mW HeNe-laser (λ=632.8 nm).

To reduce the background we place the glass substrate with the TCO in a black box. The laser beam is incident from the glass side, perpendicular to the glass surface. The light is partially scattered and the angle-dependent intensity of scattered light is detected with a silicon p-i-n photo diode (BPW 34). The set-up can also be used for reflection measurements, and the angle of incidence can be varied.

From these measurements we find an angle-dependent intensity of scattered light, $I(\theta)$, which can be converted to the intensity per solid angle by integration over this angle. Fig. 2 shows the result of the scattering measurements of a flat ITO-sample and a textured tin oxide sample. The intensities we find in this way from textured samples follow the Lambertian distribution, Eqn. 1, for angles larger then 5°. The total intensity of the light scattered under small angles is small, see Fig. 3. Therefore we are able to define a scattering parameter u_0 (in a.u.) and to compare the textured samples to each other by means of:

$$\frac{dI(\theta)}{I_i} = u_0 f(\theta). \tag{6}$$

A higher value for u_0 means more effective scattering and $f(\theta) \sim \cos\theta\sin\theta\,d\theta$.

Initially the samples under investigation scatter only 1 to 20% of the incoming light. This means that the intensity of the specular transmitted beam is very high, compared to the scattered light. This intensity is not measured.

Fig. 3 is a plot of the scattered fluxes of the samples on which solar cells were deposited. This concerns one flat ITO-sample and three textured tin oxide-layers from different manufacturers.

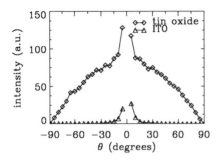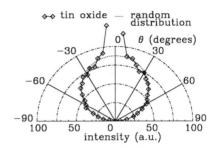

Figure 2: Left: scattered intensity of flat and textured TCO. Right: scattered intensity of textured TCO compared to a Lambertian distribution.

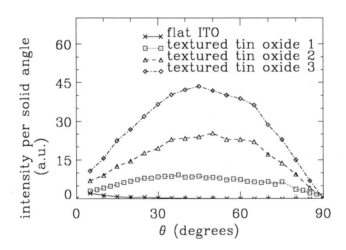

Figure 3: Scattered flux for flat ITO-sample and three tin oxide samples with different degrees of texture

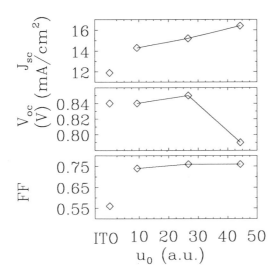

Figure 4: J_{sc}, V_{oc} and Fill Factor vs. scattering parameter u_0 for single junction a-Si:H solar cells with i-layers of 500 nm

From the measurements on the three textured samples we found values of u_0 of 9, 26 and 44 for sample numbers 1, 2 and 3, respectively. The short circuit current density J_{sc}, open circuit voltage V_{oc} and Fill Factor FF of solar cells deposited on these substrates are presented in Fig. 4.

From Fig. 4 we observe:

- an increase in J_{sc} with the amount of scattering u_0. This is a direct result of the increased absorption of long wavelength light.

- a decrease in V_{oc} for the sample with the highest u_0. This can be attributed to three facts.

    - if the transparent conductive oxide has a high degree of texture, the p-layer will not be deposited uniformly on the substrate [8]. This lowers the V_{oc}.

    - furthermore extra defects are incorporated in an i-layer which is deposited on the textured TCO [9]. The recombination increases and also by this mechanism the V_{oc} is lowered.

    - a third possible cause for the low V_{oc} of cells on highly textured substrates is the increase of the dark current.

- higher Fill Factors for solar cells on tin oxide, compared to indium tin oxide. The low Fill Factor for the cell on ITO compared to cells on tin oxide is thought to be caused by indium diffusion into the p- and i-layer during the deposition of the a-Si:H p-i-n structure [10].

Conclusions

The angle-dependent scattering as measured with the HeNe-laser at a wavelength of 633 nm, can be a better parameter for scattering than the haze for white light, because absorption enhancement is only necessary for long wavelengths. This makes the described method for scattering measurements a useful quality assessment tool for the scattering of transparent conductive oxides.

The measured intensity distribution of scattered light from tin oxide layers was proportional to $\cos\theta$, where θ is the angle with respect to specular transmission. Only small deviations from this shape were found. This means that in general, the scattering can be considered random.

From the presented solar cell characteristics we conclude that the short circuit current is proportional to the scattering parameter u_0 defined in this paper. For very high values of u_0, however, the enhancement in J_{sc} is compensated by a decrease in V_{oc}. Apparently at high u_0 the surface morphology also becomes an important quality in the analysis of TCO properties.

References

[1] K. L. Chopra, S. Major, D. K. Pandya, Thin Solid Films **102**, 1(1983).

[2] D. L. Staebler, C. R. Wronski, Appl. Phys. Lett. **31**, 292(1977).

[3] H. W. Deckman, C. R. Wronski, H. Witzke and E. Yablonovitch, Appl. Phys. Lett. **42**, 968(1983).

[4] G. Mie, Ann. Phys. **25**, 377(1908).

[5] H. Schade, Z. E. Smith, Appl. Opt. **24**, 3221(1985).

[6] H. Schade and Z. E. Smith, J. Appl. Phys. **57**, 568(1985).

[7] E. Yablonovitch, G. D. Cody, IEEE Trans. Electron Devices **29**, 300(1982).

[8] S. Tsuge, Y. Hishikawa, N. Nakamura, S. Tsuda, S. Nakano, Y. Kishi, Y. Kuwano, *Technical Digest of the 5th International PVSEC*, ed. by Y. Hayashi (International PVSEC-5, 1990), Kyoto, p.261.

[9] H. Sakai, T. Yoshida, T. Hama, Y. Ichikawa, Jpn. J. Appl. Phys. **29**, 630(1990).

[10] M. B. von der Linden, R. E. I. Schropp, W. G. J. H. M. van Sark, M. Zeman, G. Tao, J. W. Metselaar, *Proceedings of the 11th E.C. Photovoltaic Solar Energy Conference*, ed. by L. Guimarães, W. Palz, C. de Reyff, H. Kiess, P. Helm (Harwood Academic Publishers, 1992), p.647.

Materials Science Forum Vols. 173-174 (1995) pp. 337-342

APPLICATION OF THE STEADY-STATE PHOTOCARRIER GRATING TECHNIQUE FOR DETERMINATION OF AMBIPOLAR DIFFUSION LENGHTS IN Cu(In,Ga)(S,Se)$_2$-THIN FILMS FOR SOLAR CELLS

S. Zweigart[1], R. Menner[2], R. Klenk[1] and H.W. Schock[1]

[1] Institut für Physikalische Elekronik, Universität Stuttgart,
Pfaffenwaldring 47, D-70569 Stuttgart

[2] Zentrum für Sonnenenergie- und Wasserstoff-Forschung,
Hessbrühlstr. 61, D-70569 Stuttgart

Keywords: Steady-State Photocarrier Grating Technique, Ambipolar Diffusion Length Chalcopyrite

Abstract. Measurements of the ambipolar diffusion lengths with the steady-state photocarrier grating technique (SSPG) in polycrystalline chalcopyrite thin films will be presented in this contribution. The diffusion lengths will be compared to the collection of photogenerated carriers in solar cells as determined from spectral response measurements and to SEM micrographs of the film cross sections showing the typical crystallinity. Limitations of the SSPG-method for polycrystalline chalcopyrite thin films will be discussed.

Introduction

Solar cells based on Cu(In,Ga)(S,Se)$_2$ are among the most promising options for low cost large area solar module fabrication. Efficiencies exceeding 16 % have been obtained by several groups [1]. Heterojunctions made from absorbers with compositions close to pure CuInSe$_2$ usually exhibit very good photocurrent collection properties. Photocurrent losses, however, become more important if larger amounts of CuGaSe$_2$ are incorporated. This is done in order to increase the bandgap for a better match to the solar spectrum. The quantum efficiency measurements suggest that small minority carrier diffusion lengths cause these losses.

It has been demonstrated by Ritter, Weiser and Zeldov [2] that the steady-state photocarrier grating (SSPG) method, as developed especially for a-Si films, is suitable to determine ambipolar diffusion lengths on the order of 100 to 1000 nm, thus covering the expected range for Cu(In,Ga)(S,Se)$_2$. Although some measurements of Cu-poor CuGaSe$_2$ films were already published [3], there is a lack of measurements on nearly stoichiometric and Cu-rich films as used in CuGaSe$_2$ solar cell heterojunctions and for CuIn(S,Se)$_2$ absorbers.

Under Cu-rich growth conditions secondary copper selenide phases segregate at grain surfaces, which is detrimental to the cell performance [4]. They have to be removed e. g. by etching in KCN solution [5]. The etched films exhibit very good crystalline and electronic properties: large crystallites, nearly stoichiometric composition, good photocurrent collection efficiency and high p-type doping.

Bilayer recipes have been used to avoid the secondary copper selenide phases instead of using additional etching procedures. In this approach a Cu-poor layer is evaporated on top of a previously deposited slightly Cu-rich absorber layer, thereby consuming the excess copper selenides on the grain surfaces. The performance of solar cells made from these layers is often found to be limited by poor photocurrent collection efficiency, which is probably caused by short minority carrier diffusion lengths. Because the photoconductivity ratio in these absorbers is much higher than in the Cu-rich case the SSPG method promises to be useful in measuring these diffusion lengths. Investigation of minority carrier diffusion lengths in bilayer structures is also important with respect to high efficiency Cu(In,Ga)Se$_2$ devices where normally the Cu/In flux is varied during the absorber deposition in a manner similar to the bilayer recipe.

In this paper we will discuss the influence of composition and growth conditions on the transport properties of CuGaSe$_2$ bilayer thin films. The analyses were performed on layers grown the same way as absorbers for solar cells, but without back contact metallisation so that a direct comparison could be made between solar cell performance (quantum efficiency) and absorber layer properties (diffusion length).

Fabrication of Semiconductor Films and Sample Preparation

The CuGaSe$_2$ and CuIn(Se,S)$_2$ absorber layers were deposited by three (four) source evaporation onto heated 7059 Corning glass. For solar cells of the "substrate" and "superstrate" configuration Mo- and ZnO-coated glass was used, respectively. The film thickness varied between 1.9 and 2.1 μm. The spatial separation of the evaporation sources causes compositional gradients across the substrates, enabling us to compare the influence of the Cu/Ga-ratio under otherwise identical preparation conditions. The composition has been determined using EDX (energy dispersive X-ray spectroscopy). An important property of the CuGaSe$_2$ layers is the typical grain size, which is estimated from SEM (secondary electron microscopy) micrographs. The grain sizes increase with increasing Cu/Ga-ratio and with substrate temperature. At 450°C the Cu-poor and Cu-rich compositions give grains with range from $\leq$200 nm to $\approx$300 nm. The respective sizes at 650°C are from $\approx$300 nm to $\geq$1000 nm.

We evaporated molybdenum contacts to the Cu(In, Ga)(S,Se)$_2$ thin films for the SSPG-measurements. In order to minimize dark currents through not illuminated regions, the width of the semiconductor film was reduced mechanically to 2 mm parallel to the composition gradient. We used contacts with a width of 0.45 mm separated by 0.2 mm. In addition, contacts of the same design were directly evaporated to the glass substrate.

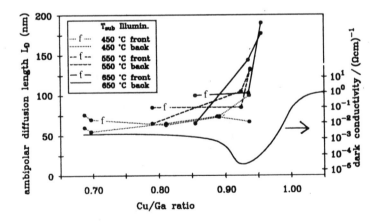

Fig. 1: *Diffusion length as determined by SSPG vs. Cu/Ga-ratio for bilayers grown at 450°C, 550°C and 650°C substrate temperature. The lines marked with f represent values obtained by illumination of the free surface, the other line the back illumination through the substrate. The solid line represents the corresponding dark conductivity for the 650°C sample.*

Experimental

The conductivity of the films has been measured in the dark as well as under an illumination of about 100 mW/cm^2 (AM 1,5). The dark conductivity exhibits the typical [6] increase over three to four orders of magnitude from Cu-poor to Cu-rich composition with a minimum near stoichiometry. A reasonable light to dark conductivity ratio could be found only in high resistivity Cu-poor samples. In contrast to the intrinsic a-Si films, for which the SSPG method was originally used, the chalcopyrite films are clearly p-type. That is the reason why the ratio between the photoconductivity and the dark conductivity never exceeded ten. This needs to be accounted with a correction factor γ_d in the β-formula [2]. It is not possible to measure the diffusion lengths in the samples with high conductivity because γ_d is quite low.

A SSPG setup as described in [2] has been employed to determine the ambipolar diffusion length using a HeNe-laser at 635 nm. The corresponding absorption coefficient in CuGaSe$_2$ is on the order of $3*10^4$ cm^{-1} [7]. The absorption of the laser light in the upper 500 to 1000 nm wavelength range causes the photoconductivity. The rest of the layer acts as a shunt and further reduces the factor γ_d. A reduction of the layer thickness was not desired because the aim was to measure diffusion lengths in absorbers relevant for solar cells. The measurements were evaluated by the method presented by Ritter, Zeldov and Weiser [2], as well as using the Balberg [3] method. There were no major differences in the results. The measured diffusion lengths were found to be independent of the light intensity and beam splitter ratio. They were also found to be independent of the applied electrical field which clearly indicates that we measured a diffusion and not a drift length.

Results and Discussion

Three different layers of CuGaSe$_2$, deposited at 450, 550 and 650°C substrate temperature, were analyzed at different compositions with illumination both from the free surface and through the

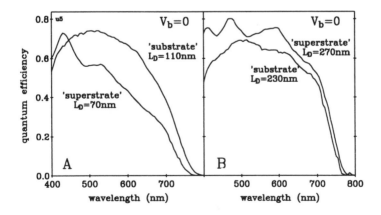

Fig. 2:   *Spectral quantum efficiencies of Cu-poor (A) and nearly stoichiometric (B) CuGaSe$_2$ solar cells in 'superstrate' (Au/CuGaSe$_2$/ZnO/Glass) and 'substrate' configuration (ZnO/thin CdS/CuGaSe$_2$/Mo/Glass) prepared in the same run. The diffusion lengths L_D refer to SSPG measurements of either the free surface or through the glass, whichever was closer to the heterojunction.*

transparent glass substrate. The results as depicted in Fig.1 clearly indicate that the typically low diffusion length in Cu-poor layers increases towards the stoichiometric point (Cu/Ga=1). Therefore we were able to measure films with higher Cu/Ga-ratios than ever reported before. The films with Cu/Ga-ratios of about one are relevant for thin film solar cells. As also described in Fig.1, the dark conductivity increases for compositions greater than Cu/Ga =0.95 because of the existence of secondary phases. This means that it is not possible to extend the measurements into Cu-rich regions where much higher diffusion lengths are to be expected.

The results also show a distinct increase of the diffusion length with the growth temperature. Differences in the measured diffusion lengths at identical positions were found for illumination from the free surface and the substrate side. The diffusion lengths in Cu-poor regions were shorter when measured from the substrate side. In the near stoichiometric films, on the other hand, they were longer. This is indicative for the special bilayer structure. The finding for the Cu-poor region is not unexpected because grain size and hence film quality may improve with increasing film thickness. In the nearly stoichiometric region, the bilayers were grown first with a Cu-rich and then a Cu-poor composition which explains the improved crystal quality of the bottom layer compared to the surface layer. The SEM cross sections also indicate this behavior. Especially in the Cu-rich regions, the crystallinity size is much larger than the measured diffusion length.

If the grain size affects the effective diffusion length, a transversal measurement like the SSPG would yield lower values than in the vertical direction. In order to test this, a comparison was made between SSPG measured diffusion lengths and the long wavelength part of the spectral quantum efficiency (which is indicative for the minority carrier diffusion) of solar cells based on the same absorber layers. Fig. 2 depicts the quantum efficiencies of both Cu-poor and nearly stoichiometric solar cells in the substrate and superstrate configuration. Both types of solar cells were prepared with the same $CuGaSe_2$ bilayer, so that in the superstrate cell the more Cu-rich layer forms the main absorption region adjacent to the ZnO. In the substrate configuration the transport properties of the Cu-poor layer are more relevant. It is quite clear from the curve shapes that the cell with the shorter SSPG diffusion lengths also show greater long wavelength response losses. A value for the minority carrier diffusion length can be estimated by fitting the long wavelength portion to a simple analytical approximation [8]. However, the accuracy of this method is limited due to uncertainties in the absorption edge data and because the approximation does not account for recombination in the space charge region. In our case the values are slightly larger than the vertical diffusion lengths as measured by SSPG.

Tab. 1: Diffusion lengths measured by SSPG for several chalcopyrite films.

Sample	Diffusion length measured by SSPG
KCN etched $CuGaSe_2$ films (Cu/Ga=1,05)	300 nm
CuIn(S,Se)$_2$, ~10% S, free surface illumination substrate illumination	100 nm 150 nm
CuInSe$_2$	> 500 nm

Tab. 1 shows the results from some SSPG measurements of the diffusion lengths for other chalcopyrite films. For KCN-etched $CuGaSe_2$ films there was no change of the measured diffusion lengths detected before and after etching. For Cu-rich samples diffusion lengths on order of 300 nm were found. In In-rich CuInSe$_2$ samples diffusion lengths longer than 500 nm were indicated. The conductivity is quite high in the CuInSe$_2$ films, due to the smaller bandgap, so measurements were not possible in Cu-rich samples. Initial measurements at liquid nitrogen temperature were tried on CuGaSe$_2$ and CuIn(S,Se)$_2$. They were identical to the measurements at room temperature which indicates that the limiting factor for the diffusion length, as measured by SSPG, is the grain size.

Conclusions

It is possible to measure diffusion lengths in chalcopyrite films which are relevant for thin film solar cells. The measured values in CuGaSe$_2$ films are 50 to 200 nm depending on the composition of the film and the direction of the illumination. A correlation can be established between the by SSPG measured diffusion lengths and the quantum response in the long wavelength region. The indicated diffusion lengths are always smaller than the grain sizes as determined from SEM micrographs. The limiting factor for the SSPG method in chalcopyrite films is the high dark conductivity and the low dark- to light-ratio.

Acknowledgments

This work has been supported by the German Ministry for Science and Technology (BMFT), contract #0328059D/E.

References

[1]   J. Hedström, H. Ohlsén, M. Bodegård, A. Kylner, L. Stolt, D. Hariskos, M. Ruckh and H. W. Schock, in *Proc. 23rd IEEE Photov. Spec. Conf.*, (IEEE, New York, 1993), p.364

[2]   D. Ritter, K. Weiser, and E. Zeldov, J. Appl. Phys. **62**, 4563 (1987)

[3]   I. Balberg, D. Albin, and R. Noufi, Appl. Phys. Lett. **58**, 140 (1991)

[4]   R. Klenk, T. Walter, H. W. Schock, and D. Cahen, Adv. Mater. **5**, 114 (1993)

[5]   R. Klenk, R. Mauch, R. Schäffler, D. Schmid, and H. W. Schock, in *Proc. 22nd IEEE Photov. Spec. Conf.*, (IEEE, New York, 1991), p. 1071

[6]   R. Klenk, R. Menner, D. Cahen and H.W. Schock, in *Proc. 21st IEEE Photov. Spec. Conf.*, (IEEE, New York, 1990), p. 481

[7]   D.Albin, J. Tuttle, J. Goral, A. Mason, R. Noufi and S.H. Risbud, in *Proc. 20th IEEE Photov. Spec. Conf.*, (IEEE, New York, 1988), p. 1495

[8]   W.W. Gärtner, Phys. Rev. B **116**, 84 (1959)

Materials Science Forum Vols. 173-174 (1995) pp. 343-348
© 1995 Trans Tech Publications, Switzerland

LASER RAMAN SCATTERING CHARACTERIZATION OF A^3B^5 COMPOUNDS AND PHOTOVOLTAIC STRUCTURES

V.M. Andreev, M.V. Cherotchenko, L.B. Karlina, A.M. Mintairov, V.P. Khvostikov and S.V. Sorokina

A.F. Ioffe Physical-Technical Institute RAS, Politechnicheskaya 26,
194021 St. Petersburg, Russia

Keywords: Raman Scattering, Phonons, Phonon-Plasmon Modes, Free-Carrier Distribution, Photovoltaic Structures, p-n Junction

Abstract. The Raman scattering spectra of different layers in liquid phase epitaxy (LPE) grown AlGaAs-heterostructure and Zn-diffused GaSb solar cells have been investigated. It is shown that Raman line-shape function of Si- or Ge-doped p-type GaAs and Zn-doped p-type GaSb ($p=10^{17}$-10^{19} cm^{-3}) can be fitted using a theory which takes into account only heavy-hole contribution in dielectric function of the crystal and that the measurements of Raman intensity at the longitudinal (LO) phonon and transversal (TO) phonon frequency can be used for the contactless determination of the free-hole density in p-type GaAs and GaSb. The method of simultaneous measurements of the free-carrier density distribution over the thickness of AlGaAs and GaSb solar cell heterostructures (including the position of p-n junction) and AlAs composition (for AlGaAs structures) is proposed.

Introduction

Raman scattering spectroscopy technique has highly promising capabilities for the characterization of semiconductor heterostructures of different devices - semiconductor lasers, high electron mobility transistors, short-wavelength photodetectors etc. [1-3]. In [2] we had shown that the combination of Raman spectra from optical phonons with layer-by-layer anodic oxidation makes it possible to measure with high accuracy the AlAs distribution over the thickness of multi-layer AlGaAs heterostructures with ultrathin (>5 nm) layers. In [4] Raman spectra from phonon-plasmon modes were used for the determination of the profiles of free-electron density across the thickness of epitaxial n-type $Al_xGa_{1-x}As$ layers.

In this paper we propose the Raman scattering technique developed for the determination of the free-hole density in p-type GaAs and GaSb. The results of simultaneous measurements of the free-carrier density distribution over the thickness of AlGaAs and GaSb solar cell heterostructures (including the position of p-n junction) and AlAs composition (for AlGaAs structures) are presented.

Experimental method

The p-type GaAs layers and $Al_xGa_{1-x}As$ solar-cell heterostructures were grown by LPE method [5]. The Si- or Ge-doped GaAs layers with different hole densities ($p=5*10^{17}$-$4*10^{19}$cm^{-3}) were grown on semi- insulating [100] GaAs substrates in temperatures range below 500^{0}C. The free-hole density and mobility values of GaAs samples obtained from the Hall measurements are indicated in Table 1. The solar cell heterostructures were grown on a n-type GaAs substrates with [100] orientation and consisted of a buffer n-type GaAs layer, n-type $Al_{0.2}Ga_{0.8}As$ back-side barrier, n-type and p-type GaAs photoactive layers, p-type $AL_{0.9}Ga_{0.1}As$ wide-gap "window" and p^+-type $Al_{0.2}Ga_{0.8}As$ contact layer. As a dopants Te for n-type and Ge and Mg for p-type layers were used. Ge was added in melt for growing of GaAs and of solid solutions with low AlAs concentrations ($x<0.3$), whereas Mg was added for growing of solid solutions with higher AlAs concentrations. The crystallization temperature range of the solar cell structures was chosen in interval - 720 and 620^0C. Therefore it was possible to prepare ultrathin wide-gap windows together with simultaneous formation of Mg-diffused p-n junction. The GaSb solar cells were prepared by Zn diffusion from Ga+Zn melt to Te-doped n-type GaSb ($n=4*10^{17}sm^{-3}$) substrates in the temperature range $450-600^0C$ in quasi-closed boats.

Table 1. Parameters of p-GaAs samples

Sample No	Hole density $[10^{18}sm^{-3}]$		Hole mobility $[sm^2/V*s]$	
	Hall	Raman	Hall	Raman
P1461	0.55	0.66	210	150
P1920	0.72	0.9	240	100
P1458	2.2	3	143	45
P1924	4.3	5	41	34
P1915	14	10	40	24
P1922	40	30	18	19

The position of p-n junction was checked by the electron microscope JSM-50A.

The Raman spectra were excited with an 2.41 eV argon laser line and measured on a DFS-52 double monochromator in a backscattering geometry in $z(xy)z$ polarization, where x,y,z correspond to the crystallographic directions [100], [010], [001]. The anodic oxidation was performed in a water solution of citric acid, ammonia and ethylene glycol. The details of Raman scattering measurements and anodic oxidation technique were described in [3,5].

In Raman scattering study of the p-type GaAs and GaSb samples the modeling of experimental lineshape by set of Lorentzian contours was used for separation of phonon-plasmon contribution in experimental spectra.

The expression for the differential Raman cross section from [4] and parameters of GaAs and GaSb given in [6,7] were used in the calculations of the phonon-plasmon mode (PPM) Raman lineshape. The contribution of only heavy holes to low frequency dielectric function of a crystal was accounted. This contribution was calculated in hydrodinamical approach [8], which is strictly valid for the hole plasma for commonly used free-hole densities in epitaxially grown p-type GaAs and GaSb. The heavy-hole density - p and damping - γ were varied in the calculations to obtain the best fit to PPM Raman lineshape deduced from experimental spectra. From obtained γ values the mobility's of holes were calculated using the expression $m=e/(m^*_{hh}*\gamma)$, where e is electron charge and m^*_{hh} is effective mass of heavy holes.

Free-hole density dependence of Raman spectra of GaAs and GaSb

The experimental spectra and comparison of PPM Raman lineshapes, obtained from Lorentzian contour modeling with calculated ones for some p-type GaAs samples have shown on Fig.1,a-d. It is seen from Fig.1,a-d , that the interaction of LO-phonons with free holes results in a strong dependence of the intensity and half-width of PPM band on hole density. For $p<10^{19}cm^{-3}$ the shift of PPM from LO-phonon frequency does not exceed several reciprocal centimeters. At higher doping level PPM shifts down to TO-phonon frequency. The single band of PPM in our Raman spectra of LPE grown GaAs doped with Ge or Si was observed. This fact is in a good agreement with the Raman investigations of MBE grown GaAs doped with Be [6,9]. The single-mode behavior of p-type GaAs is a consequence of strong plasmon damping of a hole gas [6]. However, Lorentzian contour modeling have shown the existence of the week second PPM

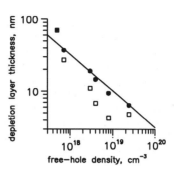

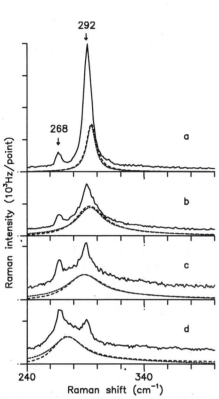

Fig.1 Experimental Raman spectra (solid curves) and phonon-plasmon Raman lineshapes, obtained from Lorentzian contour modeling (dotted curves) and theoretical calculations (dashed curves) for p-type GaAs layers with free-hole densities p,cm^{-3}:a - 7.2×10^{17},b - 4.3×10^{18},c - 1.4×10^{19},d - 4.0×10^{19}. The upper inset presents the comparison of calculated dependence of surface depletion layer thickness on hole concentration and measured one from Raman spectra of p-type GaAs layers: circles - d_o^t,squares - d_o^L .

in the spectra on Fig.1,c and d with frequency ~320 cm^{-1} and half-width ~100 cm^{-1}, which describe the high-frequency tail of the PPM profile. The obtained PPM Raman lineshape agree well with theoretical calculations as well as obtained values of hole densities agree well with Hall values (tabl.1).

It is interesting to compare our results with that of more detailed calculations of the hole susceptibility of GaAs, which account intra- and inter-valence-band transitions within the heavy- and light-hole bands in random phase approximation (RPA) using Linhard-Mermin expression [6]. It has been shown that the intra-light-hole and inter-heavy- to light-hole transitions strongly influence on Raman lineshape, in spite of the small density of the light holes and that the total hole densities deduced from Raman spectra have lower values than Hall ones. Our calculations have shown, that within our approach we can obtain the same Raman lineshape as in [6] but with greater values of the total hole density, which have very close values to Hall ones. So, we can conclude, that RPA overestimates the intra-band light-hole and inter-band contributions in the dielectric function of the hole gas in GaAs.

It is assumed that RPA can be used in the case $r_s \ll 1$ [10] where r_s is the dimensionless parameter representing the average distance between free carriers in a crystal, expressed in units of effective Bohr radius of electrons. For light and heavy holes in GaAs $r_s < 1$ is for p>10^{20}cm^{-3}. So for investigated range of p RPA are not strictly valid, what may be the reason for overestimation of contributions to dielectric function connected with light hole obtained in [6].

We used the intensity values of LO-phonon band, obtained from Lorentzian contour modeling, and the values of total intensity of Raman spectra at LO-phonon frequency 292 cm^{-1} in calculating the depth of surface depletion layer d_o^L and d_o^t for different hole densities. On the inset of Fig.1 we compare this values with the curve calculated from expression d_o[nm]=$31.8 \times 10^8 \times$ (p[cm^{-3}])$^{-0.5}$, obtained for the case of a constant charge density of the depletion layer and the surface potential equal a half of GaAs bang gap [11]. It is seen from this comparison that d_o^t values are in a very good agreement with d_o-curve. The lower d_o^1 values with respect to the d_o-curve reflect, to our opinion, deviation of hole density distribution in the depletion layer from uniform one. A good agreement of d_o^t values with d_o-curve had been also observed for MBE grown p-GaAs doped with Be [12]. This shows that the measurements of intensity of Raman spectra of p-type GaAs at the frequency of LO-phonon give the simple method for the determination of free-hole density.

The similar results have been obtained from measurements and calculations of Raman spectra and·depletion layer thickness of p-type GaSb. The experimental Raman spectra of differently doped p-type GaSb were obtained during anodic oxidation of Zn-diffused structures.

Measurements of free-carrier profiles

As an example, we used the measurement procedure mentioned above in determination of free-hole and free-electron density profiling of the AlGaAs and GaSb solar cell structures. For investigations of solar cell heterostructures we calculated the curves relating the intensity ratio $I_o^{LO}/I_{p(n)}^{LO}$ for excitation line 2.41 eV, where I_o^{LO} is intensity of LO-phonon band maximum of semi-insulating substrate and $I_{p(n)}^{LO}$ is intensity of Raman spectrum of p- or n-type GaAs and GaSb at frequency of LO-phonon. For GaSb the intensity I_o^{LO} were measured during anodic oxidation process at the stage when oxide front reaches p-n junction.

The Raman spectra of the most typical stages of the AlGaAs solar cell heterostructure oxidation are presented on Fig.2,a-e. Prior to the oxidation (Fig.2,a) the lines corresponding to a heavily doped surface contact layer of p^+-type $Al_{0.2}Ga_{0.8}As$ are observed in the spectrum [8]. When the oxide front reaches the wide-gap window (Fig.2,b) the only LO-phonon lines of

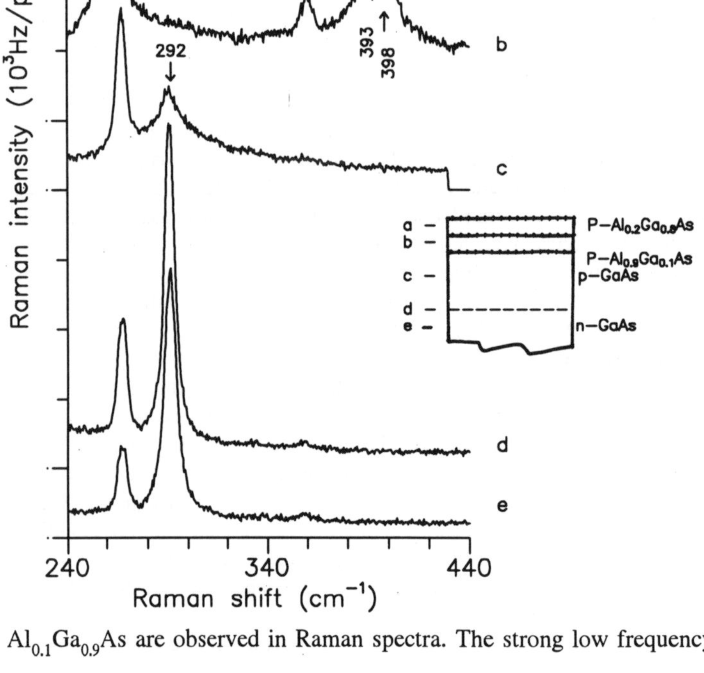

Fig.2 Raman spectra obtained with different penetration depths of the anodic oxide D (mkm): a - 0, b - 0.22, c - 0.66, d - 1.2, e - 1.74. The top inset shows a diagram of the arrangement of the layers in the solar cell structure studied.

$Al_{0.1}Ga_{0.9}As$ are observed in Raman spectra. The strong low frequency wing of its LO-phonon of

AlAs-type at 393 cm^{-1} corresponds to scattering of light within the intermediate layer with x=0.75.

The spectra on Fig.2,c-e demonstrate how Raman spectroscopy can be used in determination of the position of *p-n* junction in GaAs. The spectra on Fig.2,c and e corresponds to the *p*- and *n*-type regions of structure and the intensities of its LO-phonon lines correspond to $p=7*10^{18}$ and $n=3*10^{16}$ cm^{-3} . The spectrum on Fig.2,d corresponds to the case when the oxide front reaches the *p-n* junction - the region of the structure with the lowest free-carrier density. In this case the LO-phonon line has the highest intensity.

The resulting profiles, deduced from the Raman spectra measured at all stages of the oxidation are plotted on Fig.3. The vertical arrow indicates the position of *p-n* junction measured by electron microscope which is in a good agreement with our Raman measurements. Our Raman data profile shows a significant increasing of the free-electrons density in the *n*-type GaAs region adjoined the heteroface with back-side barrier *n*-type Al$_{0.2}$Ga$_{0.8}$As layer. The reason of this gradient is high supersaturation of a melt at the initial stage of LPE crystallization of *n*-type GaAs layer and, consequently, a better situation for incorporation of *n*-dopant atoms in solid phase [4]. The existence of such a gradient improves the conditions for movement of photogenerated carriers to *p-n* junction and leads to higher long-wavelength photosensitivity.

The results of measurements of free-carrier distributions in several GaSb solar cells are represented in Fig.4. This measurements shows that varying of the temperature and time of Zn diffusion in the ranges 550-600^{0}C and 7-10 minutes gives the possibilities of varying of *p-n* junction depth from 0.1 to 0.9 mkm.

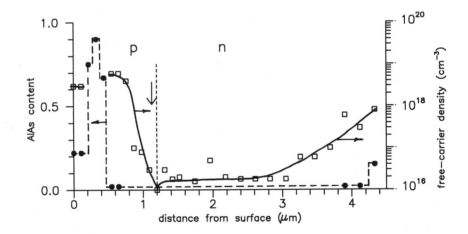

Fig.3 Experimentally measured distribution of AlAs and free-carrier density over the thickness of the solar cell structure studied. The solid and dashed lines are drawn for clearness. The vertical arrow indicates the position of *p-n* junction measured by electron microscope.

Summary

In conclusion we showed that Raman scattering spectroscopy technique can provide the simultaneous determination of both composition and free-carrier density distributions over the

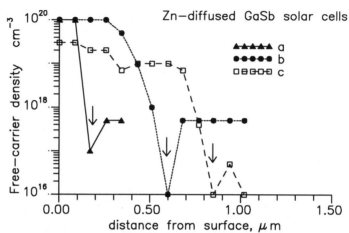

Fig.4. Experimentally measured distribution of free-carrier density over the thickness of the GaSb solar cell structures, obtained at different temperature $(T, {}^{0}C)$ and time (t,min) of Zn diffusion: a-550, 7; b-600, 7; c-600, 10. The vertical arrow indicates the position of *p-n* junction.

thickness together with location of *p-n* junction in GaAs/AlGaAs and GaSb heterostructures. Of course this method can be applied to other A^3B^5 based heterostructures of optoelectronic devices.

This work was partly supported by the Department of Defense, USA.

References

[1] G.Abstreiter, E.Bauser, A.Fischer, K.Ploog, Appl. Phys. 16, 345 (1978).
[2] ˙ V.M.Andreev, V.R.Larionov, A.M.Mintairov, T.A.Prutskikh, V.D.Rumyantsev, K.E.Smekalin, V.P.Khvostikov., Sov. Tech. Phys. Lett., 16, 325 (1990).
[3] A.V.Bobyl', P.S.Kop'ev, N.N.Ledentsov, A.M.Mintairov, V.M.Ustinov., Sov. Tech. Phys. Lett. 16, 803 (1990).
[4] A.M.Mintairov, K.E.Smekalin, V.M.Ustinov, V.P.Khvostikov. Sov.Phys.Semicond., 26, 347 (1992).
[5] V.M.Andreev, V.Yu.Aksenov, A.B.Kazantsev, T.A.Prutskikh, V.D.Runyantsev, E.M.Tanklevskaya, V.P.Khvostikov, Sov.Phys. Semicond. 24, 1096, (1990)
[6] Kam Wam and J.F.Young, Phys.Rev. B. 41, 10772, (1990).
[7] O.Madelung: Semiconductors. Group 1y Elements and 111-y Compounds (Springer-Verlag, Berlin, 1991).
[8] U.Nowak , W.Richter, G.Sachs, Phys.Stat.Sol.(b) 108, 131 (1981).
[9] T.Yuasa, M.Ishii, Phys.Rev.B. 35, 3962, (1987).
[10] D.Pines, P.Noziers, The Theory of Quantum Liquids, 1, Normal Fermy Liquids (Benjamin, New York 1966).
[11] V.M.Andreev, V.S.Kalinovskii, M.M.Milanova, A.M.Mintairov, V.D.Rumyantsev, K.E.Smekalin, E.O.Strugova, Semiconductors 27, 82, (1993).
[12] R.Fukasawa, M.Wakaki, K.Ohta, H.Okumura, Jpn.J.Appl.Phys., Part 1, 25, 652, (1986).

Materials Science Forum Vols. 173-174 (1995) pp. 349-354
© *1995 Trans Tech Publications, Switzerland*

RESONANT RAMAN AND PHOTOLUMINESCENCE OF CdTe FILMS FOR PV USING DIODE LASERS

A. Fischer[1], A. Compaan[1], A. Dane[2] and A. Aydinli[2]

[1] Department of Physics and Astronomy, University of Toledo, Toledo, OH 43606, USA

[2] Department of Physics, Bilkent University, Bilkent, Ankara 06533, Turkey

Keywords: CdTe, Diode Laser, Raman, Photoluminescence, Photovoltaics, Exciton

Abstract. Cadmium telluride, a leading candidate for polycrystalline thin-film solar cells, has a band gap which matches readily available diode lasers. We have constructed a simple grating-tuned, temperature regulated, single-mode diode laser operating in the range from 775 - 790 nm and used it as the source for resonant Raman and photoluminescence excitation spectra. The CdTe samples included high quality, single crystal CdTe as well as polycrystalline CdTe thin films deposited on glass by pulsed laser physical vapor deposition (LPVD) and by rf sputtering. We have used similar films to fabricate solar cells with efficiencies above 10% (air mass 1.5). Results show that tunable diode laser spectroscopy can be a powerful tool for analysis of defect states in semiconductor thin films and in completed photovoltaic devices.

Introduction

Traditionally Raman scattering and photoluminescence studies in semiconductors have been performed with argon or krypton ion lasers (at least since these lasers became widely available). Often for Raman scattering in thin films it is helpful to perform these studies with laser wavelengths near resonance with the fundamental band gap or higher band-to-band transitions to take advantage of resonance-enhancement of phonon Raman intensities and to limit the penetration depth of the laser. For photoluminescence (PL) it is particularly interesting to perform PL excitation studies near the fundamental direct gap using a tunable light source, usually a laser, in order to probe for states which feed into particular PL features. The tunable lasers have typically used organic dyes and more recently titanium-sapphire crystals pumped by argon or krypton lasers [1]. Such systems are complicated and expensive and not well suited for routine operation in an industrial environment such as might be desired for film characterization in a photovoltaics fabrication facility. However, the optimum band gap energies for PV use lie in a range for which diode lasers are now available. In fact recent developments have made such diode lasers available in stable, single mode configurations with plenty of output power for PL and even Raman measurements.

In this paper we describe Raman and PL results we have obtained from CdTe crystals and thin films using a single-mode diode laser which can be tuned by temperature variation and also by diffraction grating feedback [2]. A single diode could be tuned over about 15 nm in the region of the n=1 exciton of CdTe at liquid helium temperatures. For comparison we have obtained results with a HeNe laser and also a Kr ion laser.

Instrumentation

The Raman scattering and photoluminescence studies were performed using a Kr ion laser, a HeNe laser, and the diode laser described below. The samples were cooled with closed cycle helium refrigerators. Standard Raman spectrometers were used. At Bilkent University, we used a JY U1000 double spectrometer with a GaAs photocathode photomultiplier and photon counting electronics. At the University of Toledo we used an ISA S3000 triple spectrograph with a Princeton Instruments 298x1152 CCD detector.

The unusual feature of this work was the use of the laser diode as the excitation source. We used successfully a variety of diodes ranging from a 30 mW Sharp LTO24MD to a 3 mW Mitsubishi ML4402. Most of the data shown here were obtained with the ML4402 which had a nominal (room temperature) output wavelength of 783 nm. By cooling and warming the diode we easily operated it over the range from 775.6 nm to 786.9 nm. This range spans the location of the n=1 exciton in CdTe at ~20K which lies at 777.5 nm (1.595 eV) as well as various defect-bound exciton states.

The diode laser system used in this work was similar to one which we have used in an undergraduate laboratory to perform a saturated absorption experiment in rubidium vapor. Such an experiment is described in detail in a recent article by MacAdam, Steinbach and Wieman [3] who provide all the details for construction of a grating-feedback-controlled diode laser system with a gradient index lens for collimation, and a Peltier cooler for temperature control and stabilization. This laser, incorporating a piezodisc for wavelength scanning by fine angular control of the diffraction grating, can be scanned continuously through the Doppler broadened profile of Rb vapor. With a counter-propagating beam configuration it can be used for Doppler-free spectroscopy of the Rb $^2S_{1/2} \rightarrow {}^2P_{3/2}$ transition at 780.0 nm. MacAdam, et al [3] show that the various ^{85}Rb and ^{87}Rb hyperfine components are easily resolved with individual widths of ~50 Mhz (1.7×10^{-3} cm^{-1}). Clearly the laser linewidth and stability are easily sufficient for high resolution Raman and PL excitation spectroscopy! In order to extend our tuning range somewhat, we used a bottom plate for the Peltier cooler which had several loops of copper tubing circulating water from the building closed-loop water system. The diode was operated over a temperature range from about -5C to +40C. For most of the data presented here, we measured a laser power of 0.5 mW to 1 mW incident on the CdTe crystals or films from the 3 mW diode.

Although the laser operated in a single longitudinal mode, there is residual amplified spontaneous emission similar to that arising from dye lasers except that the internal facets of the diode give a residual mode structure with peak separations of 3 to 5 Å. We filtered out this emission by using a diffraction grating at the output of the laser and a ~1 mm aperture near the cryostat.

Results and Discussion

To provide some examples of the flexibility of the laser diode as an excitation source, we discuss three types of measurements: i) a comparison of near-band-edge PL excited by the HeNe laser and by the diode nearly resonant with the n=1 exciton in single crystal CdTe, ii) a series of PL spectra from single crystal CdTe taken while tuning the diode from 7766 to 7789 Å, and iii) a comparison of HeNe-excited and diode-laser-excited spectra from a polycrystalline thin film of CdTe.

At liquid helium temperatures the band edge of CdTe is 1.606 eV. The photon energy of the 632.8 nm line of the HeNe laser is 1.959 eV. Thus the photoabsorption process leaves 0.353 eV of excess kinetic energy in the free electron and hole system. An additional 10 meV of energy is released when the exciton is formed. The diode laser can be tuned to pump directly into the free exciton state and avoid this excess energy release. Figures 1 and 2 provide a series of spectra for four different nominal cryostat tip temperatures, 8, 17, 27, & 37 K. (From other evidence we have been able to conclude that the actual sample temperature was about 15 K higher than the tip thermometer indicated.)

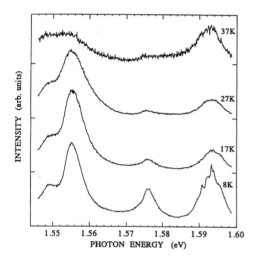

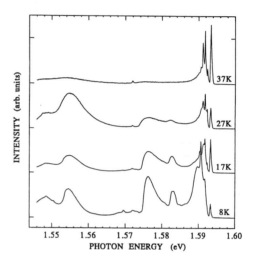

Fig. 1. PL spectra from single crystal CdTe (Eagle Picher) at nominal temperatures indicated. Excitation λ = 632.8 nm, 2.6 mW.

Fig. 2. PL spectra from single crystal CdTe excited by the diode laser (0.75 mW @ λ = 778.4 nm).

The collection of peaks from 1.590 to 1.596 eV result from several acceptor-bound excitons, the peak near 1.576 arises mainly from an optic-phonon-assisted free exciton decay, and the strong, broad peak near 1.556 eV has been attributed to a donor-acceptor pair recombination [4,5,6]. Comparison of the spectra of Fig. 1 and 2 show that the diode-laser-excited spectra (Fig. 2) provide noticeably greater detail in this region. We believe that this may arise from two factors--deeper penetration of the diode laser beam at λ = 7784 Å (1.593 eV), thus avoiding possible surface defect states, and the resonant excitation which avoids beam heating effects.

In addition to the better-resolved structure in the bound-exciton region, there is clearer definition of the phonon-assisted exciton band, X-LO (*vide infra*), small first and second-order optic phonon Raman peaks at 1.572 and 1.551 eV respectively, and an additional peak at 1.583 eV which is not seen in the HeNe data.

The phonon-assisted PL band near 1.576 can readily be used to obtain the exciton temperature. The shape of the band arising from the optic-phonon assisted decay from the n=1 free exciton is given simply by the density of states of the exciton band multiplied by the occupation probability for the excitons. This occupation probability is just the Bose-Einstein factor which for low densities and moderate temperatures is equivalent to the Boltzmann distribution. Thus the intensity of this feature is given by

$$I(v-v_{ph}) = I_0 \, \rho(v-v_0) \, \exp[-h(v-v_0)/k_B T_{ex}),$$ Eq. 1

where $h(v-v_0)$ is the exciton kinetic energy, v_{ph} is the phonon energy, and $\rho(v-v_0) \sim (v-v_0)^{1/2}$ is the density of states for the three-dimensional parabolic exciton band. T_{ex} is the inferred exciton temperature.

Figure 3 shows the fit of this PL band to the functional form of Eq. 1 for nominal tip temperatures of 8, 17, and 27 K. The temperatures indicated in the figure are the best-fit exciton temperatures, T_{ex}. We believe that the excitons are thermalized with the lattice and this is one evidence of the true sample temperature. Additional indication of a sample temperature approximately 15 K higher than the nominal tip temperature was found by measuring Stokes/anti-Stokes phonon ratios for tip temperatures of ~30K and above and with excitation wavelength of 752.5 nm to reduce resonance effects.

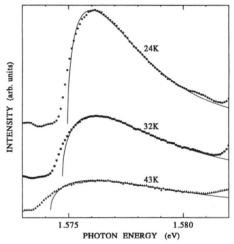

Fig. 3. Phonon-assisted free exciton decay peak with fit to expected functional form. Fitted exciton temperatures are indicated.

A major advantage associated with use of the diode laser is the ability to tune the excitation wavelength. An example of this is shown in Fig. 4 where the laser was tuned to five different wavelengths from 7766.4 Å to 7789.1 Å (1.5965 eV to 1.59187 eV). This tuning covers a region in which a strong resonance occurs for a peak at 1.583 eV. At least one author has identified this as arising from an exciton bound to an unknown defect [4]. The fact that the excitation of this feature is strong only over a narrow region from 1.594 to 1.592 eV will help to identify its origin. For example, it appears to be related to another bound exciton feature but not to the free exciton. Studies with other types of samples will be useful to identify its origin.

Rf sputtered CdTe thin films--One of the objectives of this work is to use the diode laser to facilitate studies of the properties of CdTe thin films used for photovoltaics applications. Fig. 5 shows results obtained from a CdTe film approximately 1.5 μm thick which was grown at 350 C by laser physical vapor deposition (LPVD) on alkali-free glass. The film received a post-growth treatment of a thin layer of CdCl$_2$ and a 400 C anneal in air for 15 minutes, as is typically done

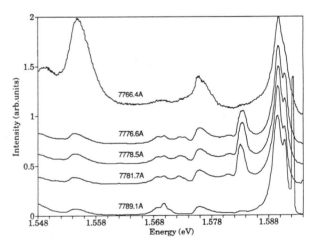

Fig. 4. PL spectra obtained with diode laser from single crystal CdTe at nominal temperature of 8 K. Power = 0.75 mW. Successive traces shifted vertically for clarity.

for the fabrication of complete solar cell structures [7]. The sample was mounted next to the single crystal sample on the tip of the closed cycle He refrigerator. Note that there exists a strong PL band in the region of the free exciton and bound exciton features of the single crystal. Such a band was not reported in epitaxial layers of CdTe on KBr substrates [8]. Typical grain sizes in this film were approximately 0.5 μm and grain boundaries probably lead to the broad PL bands observed. However, it is interesting to note that first and second order Raman peaks are clearly observable from this polycrystalline CdTe film. Because this film has about 1 mm of glass insulating it from the copper sample holder, it is quite likely that the film temperature is as much as 10 to 20 K higher than we found for the single crystal sample. Efforts are underway to improve the mounting geometry to achieve better cooling of the sample.

Conclusions

These early studies with a tunable diode laser indicate that it can readily be used as a source for both resonant Raman and PL excitation spectroscopy in semiconductors with band gaps in the region of 1.6 eV and below. Our results here on CdTe show that these lasers have considerable potential for use in materials characterization of semiconductors used for photovoltaics.

Acknowledgments

The support of the U.S. National Renewable Energy Laboratory is gratefully acknowledged. The efforts at Bilkent were facilitated by a Distinguished Senior Visitor Award from TUBITAK.

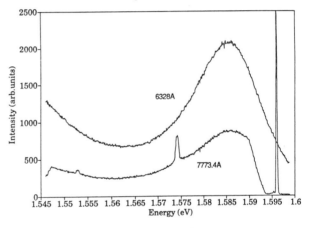

Fig. 5. Comparison of PL spectra (diode laser and HeNe laser) from an LPVD CdTe film on glass at a nominal temperature of 8K.

References

[1]   A. Compaan, Applied Spectroscopy Reviews **13**, 295 (1977).

[2]   A. Compaan, A. Fischer, A. Dane, and A. Aydinli, Bull. Am. Phys. Soc. **39**, 767 (1994).

[3]   K.B. MacAdams, A. Steinbach, and C. Weiman, Am. J. Physics **60**, 1098 (1992).

[4]   Z.C. Feng, A. Mascarenhas, and W.J. Choyke, J. Luminescence **35**, 329 (1986).

[5]   J. Lee, N.C. Giles, D. Rajavel, and C.J. Summers, Phys. Rev. B **49**, 1668 (1994).

[6]   J.S. Gold, T.H. Myers, N.C. Giles, K.A. Harris, L.M. Mohnkern, and R.W. Yanka, J. Appl. Phys. **74**, 6866 (1993).

[7]   A. Compaan and A. Bhat, Int. J. Solar Energy, **12**, 155 (1992); A.D. Compaan, C.N. Tabory, Y.Li, Z. Feng, and A. Fischer, *Proc. 23rd IEEE Photovoltaics Specialists Conference* (IEEE, New York, 1993) p. 394.

[7]   S.R. Das, J.G. Cook, N.L. Rowell, and M.S. Aouadi, J. Appl. Phys. **68**, 5796 (1990).

Materials Science Forum Vols. 173-174 (1995) pp. 355-360
© 1995 Trans Tech Publications, Switzerland

LASER-AIDED MEASUREMENTS OF ELECTRIC FIELDS ON III-V SEMICONDUCTOR STRUCTURES USING MODULATION SPECTROSCOPY: SOLAR CELL P-N JUNCTIONS AND [111] STRAINED LAYER SUPERLATTICES

R.G. Rodrigues, K. Yang, L.J. Schowalter and J.M. Borrego

Rensselaer Polytechnic Institute, Center for Integrated Electronics,
Troy, New York 12180-3590, USA

Keywords: Modulation Spectroscopy, Three-Beam Photoreflectance, Built-in Electric Fields, Solar Cells, Franz-Keldysh Oscillations, Quantum Wells, Strain-Generated Electric Fields, Optical Nonlinearities

Abstract. We report the results of a photoreflectance (PR) study of different III-V semiconductor samples featuring surface electric fields. In spectra obtained from p-n junction solar cells we identify Franz-Keldysh oscillations. Using a laser beam attenuated to several values of intensity (three-beam PR) to vary the surface electric field, we are able to obtain the impurity concentration of the base material. In the spectra obtained from both [100]- and [111]-oriented p-i-n samples with strained-layer quantum structures grown in their intrinsic region we observe features due to transitions in the confined layers. Once more using three-beam PR, we observe energy shifts in these quantum transitions and are able to confirm a non-linear optical behavior that had been predicted theoretically for the [111]-oriented samples.

Introduction

A laser light can often be used to slightly modulate or even to substantially screen the built-in surface electric fields that occur in semiconductor samples. This screening is due to the presence of photo-generated free carriers. Many semiconductor device structures feature built-in electric fields whose presence is crucial for their performance. It is therefore desirable to have a tool to directly detect and measure these fields. The most common electric fields originate in the alignment of the Fermi level between dissimilar materials forming an interface. At present there is also a strong interest in the study of the fields resulting from polarization effects in semiconductor layers subject to mechanical efforts [1,2,3].

The modulation spectroscopy techniques of electroreflectance (ER) and photoreflectance (PR) are well established characterization tools of semiconductor bulk materials, requiring no sample preparation, and capable of accurately determining the energy location of electronic transitions [4]. Furthermore, the presence of a strong electric field is revealed in these modulation spectra by the occurrence of Franz-Keldysh oscillations (FKOs) whose period allows a measurement of the intensity of the field. Due to its non-destructive character and high sensitivity, modulation spectroscopy has lately been applied to the study of semiconductor devices, and especially to microstructures, where its use in the detection of quantum transitions has quickly become a standard [5,6].

The usual configuration of ER requires electrical contacts to introduce a perturbation in the reflectivity of the sample. We find the collector contact grid of finished solar cells to be suitable for this purpose. However, we extract similar information from our samples using PR, a technique in which the perturbation is introduced by a modulated laser beam. Our ER and PR spectra measured at room-temperature from a number of solar cells of GaAs and InP consistently display the bandgap of each material and exhibit FKOs, from which electric fields in the order of 10^5V/cm are estimated. Using the FKOs we looked for alterations in the intensity of the surface electric field introduced either by a DC voltage, or by a CW laser optical pump (three-beam PR) [7].

Using PR, we are also able to identify quantum transitions in the spectra of [111] InGaAs/GaAs strained layer superlattices on which the presence of strain generated electric fields is expected. These superlattices are of potential practical interest, as it has been predicted that their bandgap will change largely in the presence of free carriers, originating non-linear optical effects.

Experimental Conditions

We tested five n^+-p-p^+ InP solar cells made at RPI (described in [8]), three InP solar cells with linearly graded junction made at NREL, and three p^+-n-n^+ GaAs solar cells made at RTI [9], in all of which the active layers were grown using MOCVD. The basic structure of the InP solar cells made at RPI is shown in Fig. 1. For the purpose of comparison, some of the cells are not AR coated. The GaAs solar cells feature a 0.07µm AlGaAs window, a 0.25 µm emitter, a 3µm base, and 0.2µm BSF AlGaAs layer on a 0.5µm buffer layer. The quality of the solar cells was preliminarily evaluated by a measurement of their quantum efficiency. The high values of quantum efficiency   that we measured reassure us of the good quality of the material and of the p-n junction.

    We also tested a total of 11 samples containing quantum wells, grown on [111]B and on [100] substrates. The optimized MBE procedure used to grow on [111]B GaAs surfaces has been described elsewhere [10]. The main rules are to grow on GaAs substrates tilted 3° towards the <211> direction, to grow at the low-temperature end of the √19x√19 surface reconstruction region (about 540°C), and under an As/Ga flux ratio of 10. The structures of the relevant samples for this study are presented in Table 1.

Table1 MBE quantum structures in the InGaAs/GaAs system.

Ref.	Orientation	# QWs	QW width (Å)	QW spacing (Å)	In conce- ntration (%)	Type
G342	[111]B	10	~ 60	~ 140	7.5	p+-i-n+
G409-b	[100]	10	~ 70	~ 140	~ 6 ? (PR)	p+-i-n+

    Our photoreflectance setup (see Fig. 2) employs a Schoeffel GH252 high intensity grating monochromator with a focal length of 0.25m, illuminated by a 350W quartz-tungsten-halogen lamp. Its output beam is the probe beam for PR and is focused onto a spot of approximately 6mmx1mm. The whole area of this spot is modulated by the chopped light of a 1mW HeNe laser. The power of the modulation beam is, in some cases, enough to screen the surface fields on the samples to some extent (which we evaluate to be under 2% in any case). Prohibitive noise levels prevent its further attenuation. The detector is a silicon photodiode, preceded by a glass filter that rejects the scattered laser light of 632.8nm. The signal from the detector is input to a PAR preamplifier and the output of this is fed both to a DC voltmeter and to a PAR model 124 lock-in amplifier. These two instruments feed their output to the A/D board of a PC which also reads the wavelength information from the monochromator. A second HeNe laser, this one not chopped , is available to shine additional CW light onto the probe beam spot. Together with a collection of neutral density filters, this laser can be used to screen the surface field of the samples and in this way we can obtain PR spectra under different electric field intensities without need for electrical contacts (three-beam PR). Table 2 lists the power densities of the beams used in ER and in PR experiments. In our standard ER experiments the modulation beam and the screening beam are replaced by a signal generator with DC bias capability.

Table 2   Power densities of the beams used in ER and in PR experiments.

Test	Probe Beam (at central wavelength) [mW/cm²]	Modulation Beam [mW/cm²]	Screening Beam Maximum [mW/cm²]
Three-beam PR of Solar Cells	~ 0.150	~ 5	~ 140
All Other	~ 0.015	~ 0.06	~ 35

Solar Cells

In this section we only display spectra obtained from InP samples. The spectra obtained from GaAs samples are analogous to these. Fig. 3 represents a PR spectrum obtained from a plain low-doped epitaxial layer. In it we identify the well-known low-field-case sharp feature which can be curve fitted to extract the bandgap energy and broadening parameter as detailed in [4]. A rule of thumb given in this reference tells us that the bandgap energy is located between the energies of the two most intense peaks of opposite sign and closer to the most intense of the two. The electric field present in this case is only the field corresponding to the pinning of the Fermi level at the surface (modulated by the laser-light generated carriers) which is not intense enough to produce FKOs under our measurement conditions. Fig. 4 represents a PR spectrum obtained from a finished InP solar cell. In this spectrum we identify the fundamental bandgap critical point and, in a range of energies of the probe beam starting immediately above this point, we identify FKOs. These oscillations are due to the intense electric field of the surface p-n junction. The presence of both of these features is an evidence of the high quality of the semiconductor material.
    The presence of metal fingers at the surface does not affect the energy location of features in our spectra. At the energy range studied, no effects of the 2eV bandgap AlGaAs window could be found in

analogous spectra of GaAs solar cells. We do not find any extra difficulty in measuring the intensity of the probe beam from AR coated surfaces although the intensity range of the reflected probe beam is reduced to 25 - 30% of the range measured from uncoated surfaces. Despite the fact that the base layers of the solar cells have a thickness a few times larger than half the wavelength of the beam used in the measurements, we find no oscillations attributable to Fabri-Perot interference. From the FKOs we can extract an electric field intensity. However, as detailed in [11], this will not be the equilibrium electric field of the junction unless the pump and probe beams are negligibly weak. This beams will in general produce a photovoltage which acts as an applied direct bias to decrease the field intensity.

In reference [12] a method is described which allowed the determination of the doping concentration of InP layers on which a Schottky barrier was fabricated by deposition of transparent InSnO. The method consists of applying several values of reverse bias to the Schottky barrier and of, for each value of bias, determining the junction electric field from the FKOs. We applied the same procedure to our finished homojunction solar cells and compared the doping concentrations obtained with data measured using other techniques. Fig. 5 represents ER spectra of an InP solar cell, obtained for different values of reverse bias. As the reverse bias increases, so does the electric field at the junction, and so does the distance between consecutive peaks of the FKOs. As detailed in [12], the amplitude of these oscillations decreases as the ratio V_{ac}/V_{dc} decreases (the amplitude of the modulating voltage, V_{ac}, was kept constant while the bias voltage, V_{dc}, was varied). The squared magnitude of the maximum electric field at a p-n junction is linear on the applied bias, the slope of the relationship yielding the net doping concentration. Fig. 6 represents our experimental data for the InP solar cell, obtained from the spectra of Fig. 5, together with the fit to this linear relationship. The same measurements were carried out for a GaAs solar cell and a good linear fit was also found for this cell.

In a solar cell production line it is desirable to have access to tools that allow the characterization of the devices in every stage of the fabrication process, including those before the deposition of electrical contacts. The electroreflectance technique just described, utilized to study the variation of the electric field at the p-n junction of solar cells, makes use of the contacts found in finished cells, and thus can not be used to evaluate a junction right after its fabrication, a capability that would be most worthy. An alternative way to vary the electric field at the junction is to shine a powerful light beam onto the sample and use the photovoltaic effect. Roughly, the square of the maximum electric field at the p-n junction of a solar cell illuminated by above-bandgap light of variable intensity, I, varies linearly with - log(I). We used a constant intensity laser beam and dimmed it using a collection of neutral density filters with density D, given by $D = -\log_{10}T$, where T is the filter transmissivity. In this way, the square of the maximum electric field has an approximately linear variation with D, the slope of the relationship once more yielding the net doping concentration. Due to the relatively small range of electric fields that can be scanned in this forward bias mode, a very clean PR signal that fits well to the model for a number of oscillations is necessary if one is to obtain meaningful and reproducible measurements. Fig. 7 represents the square of electric field intensity versus the density of the filter used to dim the screening beam of three-beam PR for an InP solar cell. Using a denser filter one permits a more intense field to subsist. Fitting FKOs from biased ER spectra and three-beam PR spectra to a simplified model [13], we obtain the data we present in Table 3 for the net impurity concentrations in the base of solar cells. Unlike the C-V method, the ER method is not dependent on the total area of the solar cell, and the doping concentration calculated respects only to the area under the spot of the probe beam.

Table 3 Impurity concentration on the base of solar cells, measured using the period of FKOs and compared to other measurements.

Solar Cell	ER/FKO Impurity Concentration (cm^{-3})	Three-Beam PR/FKO Impurity Concentration (cm^{-3})	Other Method Impurity Concentration (cm^{-3})
InP • abrupt junction • linear junction	$2.7*10^{16}$ $1.2*10^{16}$	$2.0*10^{17}$ not measured	$4.5*10^{16}$ (C-V) $1.0*10^{17}$ (manufacturer)
GaAs	$2.8*10^{16}$	$1.4*10^{17}$	$2.8*10^{17}$ (C-V)

Quantum structures

The quantum Stark effect is responsible for important energy shifts of transitions associated with the ground state of quantum wells when these are subject to electric fields [14]. In [100]-oriented samples, where the superposition of lattice mismatched layers can not result in any piezoelectric fields, a shift is expected towards higher energy when the field of the diode containing the quantum wells is screened by a laser beam. In [111]-oriented samples, the lattice mismatch between juxtaposed thin layers results in strain-generated electric fields (SGEF) across those layers [1]. Our samples are designed such that, in the confined regions, the SGEF opposes the diode field in which the quantum wells are imbedded. Consequently, a different

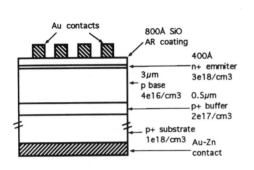

Fig. 1 The structure of the InP
solar cells tested

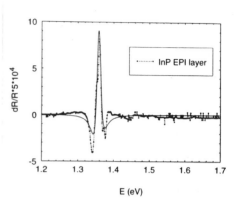

Fig. 3 PR spectrum of a lightly
doped InP epitaxial layer

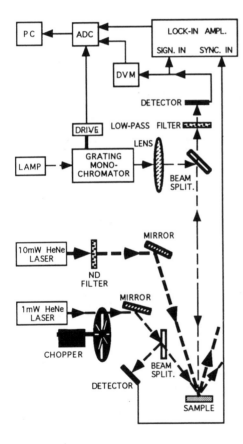

Fig. 2 Schematics of the setup
used for PR measurements

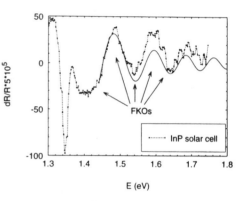

Fig. 4 PR spectrum of an InP
solar cell together with a fit
to a simple FKO model

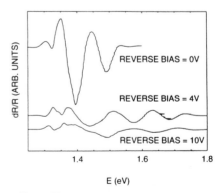

Fig. 5 ER spectra of an InP solar cell for
different values of applied reverse bias

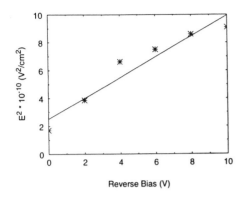

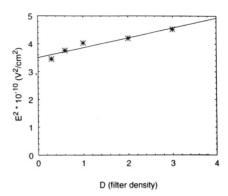

Fig. 6 Square of electric field intensity versus reverse bias for an InP solar cell.

Fig. 7 Square of electric field intensity versus density of the filter used to dim the screening beam of three-beam PR for an InP solar cell.

behavior can be expected from the [111] samples when the diode field is screened by a laser beam. Fig. 8 represents three-beam PR spectra of a [111]B-grown p+-i-n+ InGaAs/GaAs strained-layer superlattice structure (G342) under different intensities of CW laser illumination. Fig. 9 represents analogous spectra for sample G409-b, grown on a [100] substrate, but otherwise similar to G342. We attribute the features in Figs. 8 and 9 located between the bandgap energies of GaAs and InGaAs to transitions between confined levels. These levels are expected to be localized due to the strong intensity of the diode field, forming what is known as a Stark Ladder [14]. The screening laser beam makes it possible for one to observe shifts in quantum transitions in these samples. In particular, looking at feature B in Fig. 8, which we assign to the ground transition in the quantum wells, we can notice an initial red shift with increasing power of the screening beam, followed by a blue shift. In Fig. 9 the corresponding feature always shifts to the blue with increasing power of the screening beam as expected from theory [15]. We interpret the non-linearity of the [111] sample using the band diagram of Fig. 10. Previous work [3] has demonstrated that reverse biasing the p+-i-n+ diode it is possible to reach a situation in which the field of the diode perfectly cancels the SGEF and the quantum well reaches a flat-band situation. At this point the exciton energy position reaches a local maximum as a function of reverse bias intensity. Our experimental results now show that a local minimum of the exciton energy position as a function of forward bias exists in our sample. This suggests that another way to reach the flat-band situation for the quantum well is to engineer the sample in such a way that under forward bias the free carriers present screen the SGEF faster than they do the diode field. At this flat-band point the energy location of the exciton will reach another local maximum. Between the forward bias local maximum and the reverse bias local maximum a local minimum such as the one we detected must exist.

Conclusions

The electric-field-related features of the lineshapes found in PR spectra of solar cells and quantum structures were interpreted in terms of established theory. The standard technique of PR is found to be a valuable probe for buried electric fields of different nature, especially when aided by an extra CW laser beam, which can be used to substantially screen the electric fields present at the surface (three-beam PR). Both FKOs in bulk material and transitions associated with confined levels in quantum structures are found to be affected by the screening beam.

Acknowledgements

ONR Grant No. N00014-92-J-1277 provided partial support for this work. L. Schowalter is thankful to D. Smith, I. Campbell, and B. Laurich at Los Alamos National Laboratories for useful conversations. R. Rodrigues acknowledges support from Programa CIENCIA, JNICT, Portugal, and is thankful to Prof. F. H. Pollak at CUNY-Brooklyn and to Dr. E. Mendez at IBM for very enlightening conversations. The solar cells used in this work were kindly supplied by Prof. I. Bhat at RPI (InP cells), by Dr. M. Wanlass at NREL (InP cells), and by Dr. R. Venkatasubramanian at RTI (GaAs cells).

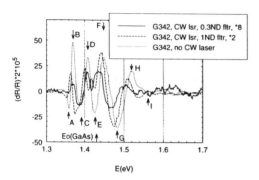

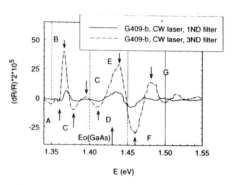

Fig. 8 PR spectra of a [111]B-grown p+-i-n+ InGaAs/GaAs strained-layer superlattice structure under different intensities of CW laser illumination.

Fig. 9 PR spectra of a [100]-grown p+-i-n+ InGaAs/GaAs strained-layer superlattice structure under different intensities of CW laser illumination.

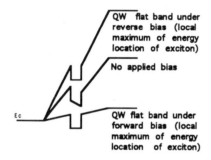

Fig. 10 Proposed band diagram of a [111]B-grown p-i-n InGaAs/GaAs strained layer QW structure under bias.

References

[1] D. L. Smith, Solid State Comm., **57**, 919 (1986).

[2] C. Mailhiot and D. L. Smith, Phys. Rev. B, **35**, 1242 (1987).

[3] E. A. Caridi, T. Y. Chang, K. W. Goossen, and L. F. Eastman, Appl. Phys. Lett. **56**, 659 (1990).

[4] D. E. Aspnes, in *Handbook on Semiconductors*, edited by T.S. Moss (North-Holland, New York, 1990), Vol. 2, p. 109.

[5] O. J. Glembocki and B. V. Shanabrook, in *Semiconductors and Semimetals*, edited by R. K. Willardson, A. C. Beer and E. R. Weber, Vol. 36, edited by D. G. Seiler and C. L. Littler (Academic Press, Boston, 1992), p. 221.

[6] F. H. Pollak and H. Shen, Mat. Sci. Eng. R **10**, 275 (1993).

[7] H. Shen, M. Dutta, L. Fotiadis, P. Newman, R. Moekirk, W. Chang, and R. Sacks, Appl. Phys. Lett., **57**, 2118 (1990).

[8] S. Tyagi, K. Singh, H. Bhimnathwala, S. K. Ghandhi, and J. M. Borrego, Proc. 21st IEEE PVSC (1990).

[9] R. Venkatasubramanian, Research Triangle Institute, Research Triangle Park, North Carolina 27709-2194.

[10] K. Yang and L. J. Schowalter, J. Vac. Sci. Technol. B **11**, 779 (1993).

[11] V. Airaksinen and H. Lipsanen, Appl. Phys. Lett., **60**, 2110 (1992).

[12] R. Bhattacharya, H. Shen, P. Parayanthal, F. Pollak, T. Coutts, and H. Aharoni, Phys. Rev. B, **37**, p.4044 (1988).

[13] D. E. Aspnes, Phys. Rev. B,**10**,4228 (1974).

[14] C. Weisbuch and B. Winter, *Quantum Semiconductor Structures*, p. 87, (Academic Press, Boston, 1991).

[15] G. Bastard, E. Mendez, L. Chang, and L. Esaki, Phys. Rev. B, **28**, 3241 (1983).

AUTHOR INDEX

KEYWORD INDEX